UNDERSTANDING MEDICINAL PLANTS

By

Dr. Pooja

Deptt. of Botany
R.C.C. College
Ghaziabad (U.P.)
(India)

DISCOVERY PUBLISHING HOUSE PVT. LTD.
NEW DELHI-110 002

First Published - 2011

Reprinted - 2016

ISBN: 978-81-8356-861-6

Understanding Medicinal Plants

Published by:
DISCOVERY PUBLISHING HOUSE PVT. LTD.
4383/4B, Ansari Road, Darya Ganj
New Delhi-110 002 (India)
Phone: +91-11-23279245, 43596064-65
Fax: +91-11-23253475
E-mail: discoverypublishinghouse@gmail.com
sales@discoverypublishinggroup.com
web: www.discoverypublishinggroup.com

Printed at:
Infinity Imaging Systems
Delhi

PREFACE

The present title *"Understanding Medicinal Plants"* provides a structured approach to learning by covering all the important topics in a uniform, systematic format. The book has been comprehensively designed incorporating recent advances in this fast moving field. It is written to provide accessible information on medicinal plants in compact form for undergraduate students in biology and related life sciences. It will be useful for both beginning students and those who are more advanced. In addition, busy lecturers who require a quick reference compendium will find it useful, particularly for tutional planning. Simple, yet hopefully clear figures and tables are provided throughout the book.

The over-riding goal of this book, and indeed of the whole *Understanding series,* is to present the essential information concering microbiology in a compact, readily accessible form which leads itself to student learning and revision. The convergence of various approaches has generated a rich panorama of detail, the significance of which we are still attempting to unraval. The present text has been written as an introduction to this rapidly growing field.

To make the work more comprehensive and informative, the author has consulted many authoritative books, research journals, abstracts, monographs etc., so there can be no claim to originality except in the manner of treatment.

The author expresses his thanks to his friends and colleagues whose continue inspirations have initiated him to bring out this book.

The author expresses his gratitude to Mr. Wasan and staff of M/s Discovery Publishing House Pvt. Ltd. for their whole hearted co-operation in the publication of this book.

In the mean time, the author will remain sincerely responsible for any shortcomings of the book and be grateful to the readers for their suggestions and constructive criticism for the continuous betterment of the book. He takes this opportunity to appeal to the readers to send their suggestions straightaway to his Publisher.

Author

CONTENTS

1

Nicotiana tabacum

Nicotiana tabacum is a stout annual of the Tomentosae family, approx 1–3 m high. The stem is erect with few branches. Leaves are ovate, elliptic, or lanceolate, up to 100 cm or more in length, usually sessile or sometimes petiolate with frilled wing or auricle. The inflorescence is a panicle with distinct rachis and several compounds branches. Flowers are light red, light pink, or white. The fruit is a capsule approx 15–20 mm long, narrowly elliptic ovoid or orbicular. Seeds spherical or broadly elliptic, 0.5 mm long, brown, with fluted ridges.

Origin and Distribution

Nicotiana tabacum originated from the borders of Argentina and Bolivia. It has been cultivated in pre-Columbian times in the West Indies, Mexico, Central America, and the northern region of South America. It is now found worldwide as a cultivated crop.

Traditional Uses

Argentina. Leaves are smoked by adults in ayahuasca mixture during healing rituals.

Brazil. Dried leaf is used as an insecticide. Leaves are heated and the juice is squeezed out, mixed with ash from bark of *Theobroma subircanum* or other *Theobroma* species to make an intoxicating snuff. Snuff is not a hallucinogenic, but it does produce inebriation. The Tukanoan peoples of the Vaupes rub a decoction of the leaf briskly over sprains and bruises. The leaf juice is taken orally to induce vomiting and narcosis.

Colombia. The Witotos and Boras used the fresh leaf as poultice over boils and infected wounds. The Tikuna men mix the crushed

Fig. 1.1. Species of tobacco, Nicotiana tabacum.

leaves with oil from palms and used as a hair treatment to prevent baldness. The juice is taken orally by the Tukanos to induce vomiting and narcosis.

Cuba. Extract of the leaf is taken orally to treat dysmenorrhea.

East Africa. Dried leaves of *Nicotiana tabacum* and *Securinega virosa* are mixed in a paste and used externally to destroy worms in sores.

Ecuador. The Jivaros use the leaf juice for indisposition, chills, and snake bites and to treat pulmonary ailments. The Aguarunas apply leaf juice by clyster, alone or mixed with ayahuasca, to induce vomiting before tobacco-ayahuasca enema. The Kulina customarily smoke all night when taking ayahuasca.

Fiji. Fresh root is taken orally for asthma and indigestion. Fresh root juice is applied ophthalmically as drops for bloodshot eyes and other problems. Seed is taken orally for rheumatism and to treat hoarseness.

Guatemala. Leaves are applied externally by adults for myasis, headache, and wounds. A mixture of the leaf with menthol VapoRub is applied externally for children for cough. Hot water extract of the dried leaf is applied externally for ring-worms, fungal diseases of the skin, wounds, ulcers, bruises, sores, mouth lesions, stomatitis, and mucosa. The leaf is taken orally for kidney diseases.

Haiti. Decoction of the dried leaf is taken orally for bronchitis and pneumonia.

India. Juice of *Securinega leucopyrus* is mixed with the dried leaf and applied externally for parasites. Fresh leaf is mixed with corn-cob or *Amorphophallus paeonifolium* to treat asthma.

Iran. Infusion of the dried leaf is applied externally as an insect repellent. Ointments made from crushed leaves are used for baldness, dermatitis, and infectious ulcerations and as a pediculicide. Juice is

Fig. 1.2. Species of tobacco, N. rusticum.

Fig. 1.3. Indian smoking a tabac (pipe)..

applied externally as an insect repellent. Leaf is added to the betel quid and used as a mild stimulant.

Kenya. Water extract of the dried leaf is applied ophthalmically for corneal opacities and conjunctivitis.

Malaysia. Infusion of the dried leaf is taken orally as a sedative.

Mexico. Extract of the plant, massaged on the abdomen with saliva, is used to facilitate expulsion of placenta. Exudate from the leaf and stem is used as a dentifrice for gum inflammation.

Nepal. Leaf juice is applied externally to treat scabies[NT463].

Nicaragua. Leaves are chewed for toothache and applied externally for aches, pains, bites, stings, and skin rashes.

Nigeria. Hot water extract of the fresh leaf is taken orally as a sedative.

Papua New Guinea. Dried plant, mixed with the bark of *Galbuli ma belgraveana* and *Zingiber officinale* is taken orally for head lice. Young leaf tip is chewed to relieve stomachache. Decoction of the young leaf is taken orally to treat gonorrhea.

Paraguay. Extract of the plant is administered orally to cows as an insecticide and insect repellent. Smoked or chewed dried leaf will spoil the milk of nursing mothers. Dried resin accumulated in the stem of a smoking pipe is applied externally against botfly larvae and severe pediculosis.

Peru. Decoction of the leaf with ayahuasca beverage (*Banisteriopsis caapi* and *Psychotria viridis*) is taken orally for hallucinating effect during shamanic training. A diet of cooked plantain and smoked fish follows each use. Hot water extract of the dried flower and leaf is used externally for snake and spider bites. The Witotos and Boras used the fresh leaves as poultice over boils and infected wounds. The Tikuna men mix the crushed leaves with oil from palms as a hair dressing to prevent baldness. The Jivaros use the tobacco juice for indisposition, cold, chills, and snake bites and to treat pulmonary ailments.

Sierra Leone. Leaf is chewed and rubbed on area to dress umbilical stump.

Tanzania. Leaves are placed in the vagina to stimulate labor.

Turkey. Powdered leaf is applied externally for wounds.

United States. Extract of the plant is taken orally to treat tiredness, ward off diseases, and quiet fear.

Fig. 1.4. Mayan priests blew smoke to the four winds during religious ceremonies.

Medicinal Use

8-Oxodeoxyguanosine Level Increase

Nitrosamine 4-(methylnitrosamino)-1-(3-pyridyl)-1-butanone, administered intragastrically to A/J mice at doses 0.25 or 0.5 mg/mouse, three times a week for 3 weeks, produced a significant elevation of 8-oxodeoxyguanosine in lung DNA 2 hours after the last administration. A single-dose treatment (4 mg/mouse) produced an insignificant increase of lesion in the lung DNA. In the liver, the increase was only significant by multiple doses of the higher dose. At 4 and 24 hours after treatment, the 8-oxodeoxyguanosine levels declined to the basal levels in both liver and lung. Administration of a single dose (20 mg/animal) to F344 rats, produced a significant increase of 8-oxodeoxyguanosine in rat lung DNA and an insignificant increase in liver DNA. The 8-oxodeoxyguanosine level in rat kidney, a nontarget tissue, was inert to nitrosamine treatment.

Acetylhydrolase Activity

Smoke extract, administered to rats at a dose of 0.5 cigarettes/kg for 5 days, did not alter plasma platelet-activating factor acetylhydrolase activity during the treatment. Combination of smoke extract and 17-α-ethynylestradiol decreased plasma platelet-activating factor acetylhydrolase activity by 90%. The effect of medroxyprogesterone on plasma platelet-activating factor acetylhydrolase activity was not influenced by smoke extract. Administration of an increasing concentrations of extract to confluent cultures of collagen-producing mouse fibroblasts or primary osteoblasts from chick embryo calvarias, produced no effect on glycolysis and DNA synthesis. There was no influence on cell proliferation in fibroblasts, but it was inhibited at the highest extract concentration in osteoblasts. Adenosine triphosphatase (ATPase) activity was not detectable in fibroblast medium but was decreased in osteoblast medium. At the highest concentration, [^{3}H]hydroxyproline and [^{3}H]proline contents in the cell layers were decreased to the following respective values: fibroblasts 56% and 45% and osteoblasts 50% and 29%, respectively. When incubation with smokeless tobacco (STE) was discontinued for 1 day, recovery did not occur.

Age-related Alterations

Smoke, administered to old CBA/CA mice, produced spontaneous death more frequently than in the control animals. Prevalence of hepatocellular carcinoma was higher and body weight was lower in the smoke-treated mice than in controls. There were differences in organ indices. The reactivity against sheep erythrocytes antigen was lower in mice kept in smoke than in controls. The ratio of normal reactivity (against sheep erythrocytes) and autoreactivity (against mouse erythrocytes) showed a decrease in the smoke-treated mice.

Airspace Permeability Increase

Cigarette smoke condensate in rat lungs produced a 59.7% increase in epithelial permeability over control values, peaking 6 hours after instillation and returning to control values by 24 hours. Instillation of human recombinant tumor necrosis factor (TNF)-α produced an increase in epithelial permeability in the rat lung. Only a vestigial amount of TNF-α was detected in bronchoalveolar lavage fluid in vivo or in culture medium from bronchoalveolar lavage leucocytes from smoke-treated animals. Anti-TNF antibody did not abolish the increased epithelial permeability produced by cigarette smoke condensate. One hour after instillation of cigarette smoke condensate there was a marked decrease in the reduced glutathione content in the lung in association with increased oxidized glutathione levels. Reduced glutathione levels in bronchoalveolar lavage fluid decreased following cigarette smoke condensate.

Aldosterone Synthesis Inhibition

Nicotine, anabasine, and cotinine, in freshly isolated rat adrenal cells at concentrations up to 100 m*M*, did not inhibit stimulated corticosterone production but inhibited aldosterone production in a dose-dependent manner. The relative inhibitory potency was: cotinine > anabasine > nicotine. Cotinine, anabasine, and nicotine, at a concentration of 100 m*M*, inhibited adrenocorticotropic hormone (ACTH)-stimulated aldosterone synthesis by 75, 44, and 21%, respectively. Angiotensin (ANG)-II-stimulated aldosterone synthesis was inhibited by 92, 78, and 62%, respectively. The plasma cotinine concentration range attained in tobacco smokers was between 1 and 10 m*M*. When tested with [^{3}H]corticosterone and [^{3}H]progesterone as exogenous substrates, 1–10 m*M* cotinine produced a significant dose-dependent inhibition of ACTH- and ANG-II-stimulated aldosterone synthesis.

Allergenic Activity

Cigarette smoke condensate, in cell culture at concentrations of 6.6–20 μg/mL, produced an inhibition of cell surface antigen-presenting major histocompatibility complex class I expression and immunoglobulin (Ig) synthesis. Intraperitoneal administration to C57BL/6 mice before challenge with ovalbumin antigen produced a decrease of antiovalbumin-specific antibody response. This inhibition affected Ig protein synthesis then membrane bound major histocompatibility complex class I expression. Supplementation with selenium significantly reduced the inhibitory effects. IgE antibodies against crude tobacco leaf were present in smokers, non-smokers, and ex-smokers, and the atopic individuals were far more likely to show such responses than nonatopic individuals. It has also been established that IgE antibodies can be detected against at least three specific tobacco leaf allergens, namely crossed immuno-electrophoresis (CIE) antigens 19, 23, and 30, but these IgE antibodies did not correlate with any type of clinical smoke sensitivity. There is no concrete evidence for the presence of IgE antibodies in man against smoke extract. There is preliminary evidence that smoke challenge under controlled conditions in an environmental chamber does not induce significant decreases in forced expiratory volume in 1 second (FEV1) or peak flow in smoke-sensitive subjects, even though they complain of symptoms. Tobacco leaf was immunogenic in rabbits, but it is not known if any tobacco incineration products *per se* are immunogenic in man. Tobacco glycoprotein from the cured leaf, administered intradermally by three injections to mice, produced a long-lasting IgE antibody response. No hemagglutinating antibodies were produced. Smoke from Kentucky Reference IRI cigarettes, administered in a Prototype Mark II Walton Horizontal Smoke Exposure Machine to Balb/c mice within 24 hours after birth for up to 10 weeks, produced no difference in the magnitude of the splenic plaque-forming cells response in smoke-exposed and untreated control animals immunized with sheep red blood cells on sequential days up to day 9 postpartum. On day 10, the plaque-forming cells' response of smoke-exposed mice was reduced by 33%, on day 14, there was a 60% reduction, whereas animals exposed to smoke from 4 to 10 weeks showed a 90% reduction of the splenic plaque-forming cells response. Tobacco smoke, administered to normal CFW mice, produced a significantly increased susceptibility to the lethal effects of histamine. The lethal dose $(LD)_{50}$ for mice subjected to smoke was 45 mg/kg of histamine, whereas in normal CFW mice the LD_{50} was 1.1 g/kg. The histamine susceptibility of smoked mice was markedly diminished by injecting the animals with isoproterenol. Normal CFW mice and sham control mice exhibited an epinephrine-induced hyperglycemia, whereas the blood glucose values for smoked mice given epinephrine were similar to those for sham mice given only saline. Results indicated that tobacco smoke can contain a component that causes an autonomic imbalance, hence rendering the mice more susceptible to histamine. Tobacco leaf extract, administered to CFW female mice with developed type 1 skin and mast cell sensitivities, produced IgG1- and IgE-tobacco-mast-cell-sensitizing antibodies detected by 2 and 48 hours after immunization.

Alveolar Macrophages Fluorescence

Cigarette smoke, administered to rats at a dose of two cigarettes for 1 or 5 days, produced an increased fluorescence after 1 day of exposure and enhanced after 5 consecutive days. Larger and more granular/complex alveolar macrophages were more fluorescent than smaller and less granular/complex cells. Smoke-exposed rats (5 days of exposure) lavaged immediately after the exposure had less cells in their bronchoalveolar lavage fluid than control animals. Rats lavaged 3 smoke-free days after the exposure, produced an increase in cell recovery, probably resulting from to less airway obstruction.

Analgesic Activity

Nicotine or tobacco smoke, administered to male Sprague- Dawley rats, produced analgesia measured by tail-flick latencies. A second treatment, 24 hours after the first, failed to produce analgesia, thereby demonstrating the rapid development of tolerance. The restraint, which was a necessary part of the tobacco smoke exposure, also produced analgesia, although of a more transient nature and lesser magnitude than that resulting from tobacco smoke exposure. Tolerance also developed to restraint stress-induced analgesia. The long-term (43 weeks) daily exposure of rats to tobacco smoke or restraint stress resulted in the development of cross-tolerance. Long-term tobacco smoke exposure resulted in increased tail-flick latency when the animals had been withdrawn from tobacco smoke for 24 hours.

Anesthetic Activity

Nicotine, administered to mice, increased the latent time of biting the clip in the tail press test ($p < 0.001$) and retarded tail withdrawal latency in the tail immersion test ($p < 0.01$), compared to controls.

Antibacterial Activity

Tincture of the dried leaf (10 g plant material in 100 mL ethanol), on agar plate at a concentration of 30.0 μL/disc, was inactive on *Escherichia coli*, *Pseudomonas aeruginosa*, and *Staphylococcus aureus*. Methanol extract of the dried leaf, on agar plate at a concentration of 25 mg/mL, was active on *Bacillus subtilis*, *Corynebacterium pyogenes*, *Pseudomonas aeruginosa*, *Serratia marcescens*, *Shigella dysenteriae*, and *Staphylococcus aureus* and inactive on *Escherichia coli*, *Klebsiella pneumoniae*, and *Proteus vulgaris*. Smoke condensate, in murine alveolar macrophage cell line (MH-S) cells, significantly enhanced the replication of *Legionella pneumophila* in macrophages and selectively down-regulated the production of interleukin (IL)-6 and TNF-α induced by bacterial infection.

Anti-convulsant Activity

Methanol (50%) extract of the dried leaf, administered to mice, was active vs leptazol-induced convulsions. Ethanol (70%) extract of the fresh leaf, administered intraperitoneally to mice of both sexes at variable doses, was active vs metrazole- and strychnine-induced convulsions.

Antifungal Activity

Hot water extract of the dried leaf, in broth culture, was inactive on *Epidermophyton floccosum*, *Microsporum canis*, *Trichophyton mentagrophytes* var. *algondonosa*, and *Trichophyton mentagrophytes* var. *granulare*. Undiluted methanol extract of the fresh leaf, on agar plate, was active on *Aspergillus fumigatus*. Saponin solution of the fresh seed, administered to infected plants, was active on *Puccinia recondita*.

Antiglaucomic Activity

Water extract of the callus tissue, administered intravenously to rabbits at a dose of 250 μg/animal, produced a 55% drop in intraocular pressure.

Antioxidant Activity

Smoke, administered by inhalation to mice for 10 weeks, produced an increase of catalase, glutathione peroxidase and glutathione reductase activity. The activity of superoxide dismutase was unaltered. Administration to DBA/2, C57BL/6J, and ICR mice at a dose of 3 cigarettes/day 5 days/week for 7 months, produced a decrease in the antioxidant defense of bronchoalveolar lavage fluids in the DBA/2 and C57BL/6J strains and an increase in ICR mice. Lung elastin content was significantly decreased in DBA/2 and C57BL/6J mice but not in ICR mice. Emphysema was present in DBA/2 and C57BL/6J but not in ICR mice. When administered to pallid mice with a severe serum α(1)-proteinase inhibitor (α[1]-PI) deficiency for 4 months, there was an acceleration of the development of the spontaneous emphysema assessed with morphometrical and biochemical (lung elastin content) methods. Rutin and chlorogenic acid from STE leaf extract, in murine mast cell culture, reduced reactive oxygen species levels and inhibited histamine release in antigen-IgE-activated cells. They augmented the inducible cytokine messages, i.e., IL-10, IL-13, interferon (IFN)-γ, IL-6, and TNF-α in IgE-sensitized mast cells after antigen challenge. Results indicated that tobacco polyphenolic antioxidants differentially affected two effector functions of antigen-IgE-activated mast cells[NT042]. Cigarette smoke, administered to Sprague–Dawley rats for 2 hours/day for 4 weeks, enhanced *N*-methyl-D-aspartate receptor (NMDAR) subunits 2A and 2B concentrations in the hippocampus. Lipid peroxidation and antioxidant enzyme activities did not show any change. The results indicated that cigarette smoke induces NMDAR 2A and 2B expression in the hippocampus not because of an increased lipid peroxidation but because cigarette smoke has no effect on lipid peroxidation and antioxidant enzyme activities in the hippocampus. Tobacco smoke, administered to rats for 2 days or 8 weeks (6 hours/day, 3 days/week), significantly increased the number of cells recovered by bronchoalveolar lavage (BAL). Manganese(III)meso-tetrakis(*N,N'*-diethyl-1,3-imidazolium-2-yl) porphyrin 10150 significantly decreased BAL cell number in tobacco smoke-treated rats. Squamous cell metaplasia, following 8 weeks of tobacco smoke exposure, was 12% of the total airway epithelial area in animals exposed to tobacco smoke without AEOL 10150, compared with 2% in animals exposed to tobacco smoke, but treated with AEOL 10150 ($p < 0.05$). Cigarette smoke, administered to albino rats for 30 minutes/day for 30 days, increased the lipid peroxide levels in liver, lung, and kidney of smoke-treated rats. No changes were found in the brain and heart. The activity of the antioxidant enzymes was also elevated in the livers, lungs, and kidneys of the test animals. Brain and heart did not show any change in the activities of all of these antioxidant enzymes, except an increase of glutathione-*S*-transferase in brain. The level of reduced glutathione was lowered in the livers, lungs, and kidneys of the test animals when compared to controls. There were no significant changes in brain and heart. Whole cigarette smoke, administered to rats daily for 1, 2, 7, or 14 days, increased expression of manganese superoxide dismutase, glutathione peroxidase, and metallothionein. Copper–zinc superoxide dismutase and catalase expression did not change from control levels. The distribution of manganese superoxide dismutase expression was similar in control and smoke-exposed animals. Catalase, copper–zinc superoxide dismutase, glutathione peroxidase, and metallothionein showed widespread expression in the lung by *in situ* hybridization. Copper-zinc superoxide dismutase, glutathione peroxidase, and metallothionein were highly expressed in bronchial epithelium. Catalase expression levels were similar in all cell types. Results indicated that most of these antioxidant enzymes and scavengers showed prominent bronchial expression but that manganese superoxide dismutase showed a unique pattern, with intense hot spots in the epithelium of the small airways. Water-soluble substances in cigarette smoke in nerve terminals prepared from the rat cerebral cortex, significantly reduced the spontaneous increase in thiobarbituric acid-reactive substances in synaptosomes in a dilution factor-dependent manner. The aqueous extract also inhibited the elevation of lipid peroxidation induced by 2,2'-azobis (2-amidinopropane) dihydrochloride, a peroxyl radical generator. Smoke substances scavenged superoxide radicals generated from stimulated human leukocytes and from the xanthine/xanthine oxidase system. These effects were not mimicked by nicotine. The antioxidant effects of smoke substances were preserved for several days at 5°C or –80°C. Cigarette smoke, administered to rats for 3 months, decreased activities of glutathione reductase, catalase, and brush-border enzyme γ-glutamyl transpeptidase and the levels of glutathione in the kidney. The activities of glutathione peroxidase and lipid peroxide levels were increased. Urinary excretion of γ-glutamyl transpeptidase, glutathione, and lipid peroxide were also higher.

Antistress Effect

Smoke of the dried leaf, administered to rats at variable doses, produced equivocal activity. Smoke, administered to C57B1/6 and Balb/c mice, and MNRA and MR rats in a chamber for 19 days, produced no emotional-stress reaction. A difference in the free preference of space containing tobacco smoke was observed among inbred animals with active and passive emotional-stress reaction phenotypes.

Antiviral Activity

Leaf, on agar plate at a concentration of 2%, was active on herpes simplex 1 virus and reduced plague formation in monkey kidney cells. Undiluted leaves produced strong activity on Coxsackie B5 virus, herpes simplex 1 virus, and measles virus. Viral reproduction was inhibited.

Anti-yeast Activity

Tincture of the dried leaf (10 g plant material in 100 mL ethanol), on agar plate at a concentration of 30 μL/disc, was inactive on *Candida albicans*.

Apoptosis

Smoke extract, in rat lung alveolar L2 cells, induced apoptotic cell death. A concentration of 0.25% resulted in a 50% increase of caspase-3 and matrix metalloproteinase activities. Chloroform extract of the cigarette smoke, in a human gastric epithelial cell line (AGS) for 5 hours, induced apoptosis in a dose- and time-dependent manner in AGS cells and a decrease of bcl-2 and an increase of caspase-3 activity. Pretreatment with Z-DEVD-FMK (specific inhibitor of caspase-3) dose-dependently blocked the DNA fragmentation induced by the chloroform extract. Chloroform extract time- and dose-dependently increased the level of cytochrome C in the cytoplasm, which might activate caspase-3.

The ethanol extract was not active. Mainstream cigarette smoke, administered to Sprague–Dawley rats for 18 or 100 days, produced a significant and time-dependent increase in the proportion of apoptotic cells in the bronchial and bronchiolar epithelium. Oral *N*-acetylcysteine did not affect the background frequency of apoptosis but significantly decreased smoke-induced apoptosis. A mixture of sidestream and mainstream smoke, administered to Sprague–Dawley rats for 28 days, produced a more than 10-fold increase in the frequency of pulmonary alveolar macrophages undergoing apoptosis. Exposure to smoke produced an acute increase of cells positive for proliferating cell nuclear antigen.

Cigarette smoke, administered to rats at concentrations of 2 and 4% for 1 hour daily for 1, 3, 6, and 9 days, produced a time- and concentration-dependent increase in apoptosis in the rat gastric mucosa. The effect was accompanied by an increase in xanthine oxidase activity. The increased apoptosis and xanthine oxidase activity could be detected after even a single exposure. In contrast, the p53 level was elevated only in the later stage of cigarette smoke exposure. The apoptotic effect could be blocked by pretreatment with xanthine oxidase inhibitor (allopurinol, 20 mg/kg intraperitoneally) or a hydroxyl free radical scavenger (dimethyl sulfoxide [DMSO], 0.2%, 1 mL/kg in-travenously). Neither of these treatments had any effect on the p53 level of the mucosa.

Aromatase Inhibition

Aqueous extract of cigarette smoke, administered to female adults at a concentration of 25 μL/plate, was active on granulosa cells[NT595]. Synthesis and testing of a series of acylated nornicotines and anabasines for their ability to inhibit aromatase showed an interesting correlation of activity with the length of the acyl carbon chain, with maximum activity at C-11. The acylated derivatives showed activity, which was significantly greater than that of nicotine and anabasine. In vivo studies in rats indicated that administration of this inhibitor delayed the onset of nitrosomethyurea (NMU)-induced breast carcinoma and altered the estrous cycle by suppression of the aromatase enzyme system. Toxicity studies indicated relatively low toxicity with LD_{50} for *N-N*-octanoylnornicotine at 367 mg/kg body weight.

Arrhythmogenic Effect

Water extract of the dried leaf, administered intravenously to cats at doses of 10–20 mg/kg, produced weak activity.

Arterial Endothelial Injury

Environmental smoke, administered to ovariectomized rats treated with subcutaneous placebo or 17-β-estradiol pellets for 6 weeks, produced a more than fourfold increase of carotid artery low-density lipoprotein (LDL) accumulation compared with filtered air exposure. The effect was largely mediated by increased permeability. No protective effect of estradiol was observed. Acute smoke exposure of a buffer solution containing LDL produced a more than sixfold increase in the highly reactive carbonyl glyoxal. Perfusion of this solution through carotid arteries produced 105% increase in permeability. Perfusion of glyoxal alone produced a 50% increase in carotid artery permeability.

Aryl Hydrocarbon Hydroxylase Induction

Smoke of cured leaf, administered by inhalation to mice and rats at an undiluted concentration, produced an induction in lungs and kidneys. There was no induction in bowels and liver.

Atherogenic Effect

Smoke, administered by inhalation to mice for 10 weeks, produced an increase of lipid peroxidation and a decrease of reduced glutathione level in the heart. Levels of total cholesterol, LDL cholesterol, and triglycerides were increased. The high-density lipoprotein (HDL) cholesterol levels decreased in serum.

Bacterial Colonization of Lower Respiratory Tract

Cigarette smoke, administered for 3 days before and after intratracheal instillation of bacterial suspension containing six bacterial species (*Staphylococcus aureus*, *Staphylococcus epidermidis*, *Streptococcus pneumonia*, *Proteus mirabilis*, *Haemophilus influenza*, *Peptostreptococcus* spp.) to male Wistar albino rats with or without vitamin E supplements (100 mg/kg/day), significantly increased the colony numbers of all isolated bacteria species in smoke-treated rats than in the control group and in the smoke- and vitamin E-supplemented rats ($p < 0.05$). Only *Staphylococcus aureus* and *Staphylococcus epidermidis* were isolated from vitamin E-supplemented rats.

Behavioral Changes

Nicotine, administered to rats for 7 days, significantly increased rat locomotion activity. This sensitization to nicotine was blocked by mecamylamine (1 mg/kg) and by the administration of 6 mg/kg of the following cembranoids: eunicin, eupalmerin acetate (EUAC), and (4R) -2,7,11 -cembratriene-4- 6-diol (4R). None of these compounds modified locomotor activity of nonsensitized rats. Nicotine, administered by continuous infusion to adolescent rats on postnatal days 30 to 47.5, using a dosage regimen that maintains plasma levels similar to those found in smokers or in users of the transdermal nicotine patch, produced a decrease of grooming in female rats on 44 day of administration, an effect not seen in males. This effect is opposite to the effects of nicotine in adult rats. Two weeks after cessation of nicotine administration, females showed deficits in locomotor activity and rearing. The males again were unaffected. The behavioral deficits appeared at the same age at which gender-selective brain cell damage emerges. Nicotine exposure enhanced passive avoidance, with the effect intensifying and persisting throughout the posttreatment period.

Smokeless tobacco extract (STD), administered by gavage to pregnant Sprague–Dawley rats on gestational days 6–20 at doses equivalent to 1.33 (STD-1), 4 (STD-2), and 6 mg nicotine/kg (STD-3) three times daily, reduced maternal weight gain at the 2 higher doses. During the preweaning period, significant pup weight reductions were noted in the STD-2 pups until postnatal day 6 and in the STD-3 group until postnatal day 15. In the STD-1 group, no statistically significant weight reduction was noted. The incidence of deaths was increased in a dose-related manner. No significant differences were noted for pinna detachment and incisor eruption. Smoke treatment significantly affected earlier eye opening and vaginal patency. No significant effects were seen on negative geotaxis, but for surface righting, a decreased success rate was noted. Open field activity increased from the preweaning to postweaning periods. During the preweaning period, the STD-3 offsprings were more active, and during postweaning, the STD-1 offsprings were more active. No difference was noted in vertical activity or in the number of stereotypical movements. No treatment-related difference was noted in the active avoidance shuttle box.

Benzo(a)pyrene Hydroxylase Induction

Masheri, a pyrolyzed tobacco product, administered orally to Swiss mice, Sprague- Dawley rats, and Syrian golden hamsters at a dose of 10% diet for 20 months, produced a significant induction of cytochrome P450 and benzo(a)pyrene hydroxylase in proximal and distal parts of the three species.

Benzopyrene Hydroxylase Induction

Methyl chloride extract of the leaf, administered intragastrically to rats at a dose of 3 mg/animal daily for 21 months in DMSO solvent, was active. The rats were divided into two groups. One was fed a vitamin A diet, and the other was fed a vitamin A-deficient diet. Treatment with the extract increased pulmonary and hepatic benzopyrene hydroxylase level over controls in both groups. The vitamin A-deficient group had significantly higher hepatic and lower pulmonary levels when compared to the vitamin-fed group.

Benzphetamine Demethylase Stimulation

Methyl chloride extract of the leaf, administered intragastrically to rats at a dose of 3 mg/animal daily for 21 months in DMSO solvent, was active. The rats were divided into two groups. One was fed a vitamin A diet, and the other was fed a vitamin A-deficient diet. In the vitamin-deficient group, treatment with the extract increased benzphetamine demethylase level vs control.

Biochemical Effect

Tobacco smoke was administered by inhalation in Hamburg II machine to senescence accelerated SAM-P/8 and SAM-R/1 mice 10 minutes/day, 5 days/ week for 5 weeks. The treatment increased lung weight and the ratio of albumin to total protein in the bronchoalveolar lavage fluid, a decrease elastase inhibitory capacity/trypsin inhibitory capacity in bronchoalveolar lavage fluid, and a decrease in the glutathione (GSH) content and the GSH/-sulfhydryl (SH) ratio of the lung, compared with those not exposed. There was focal infiltration of macrophages into alveoli with hyaline membrane and thickened alveolar wall in SAM-P/8 with tobacco exposure.

Birth-weight Effect

Cigarette smoke, administered to pregnant rats daily for a 2- hour period throughout gestation, significantly decreased the average birth-weight of pups compared with both pair-fed and *ad libitum* control groups. The body weights of the pups exposed to smoke were no longer significantly different from those in the pair-fed and *ad libitum* control groups at weeks 1 and 2 after birth, respectively. The study indicated that fetal growth retardation caused by exposure to cigarette smoke during pregnancy does not persist after birth.

Breathing Inhibition

Cigarette smoke from low-nicotine research cigarettes and gas phase smoke obtained by passing the smoke through a glass-fiber Cambridge filter was administered to anesthetized Sprague–Dawley rats at a dose of 6, 50%. Inhalation of gas phase smoke alone evoked a transient inhibitory effect on breathing, prolonging expiratory time (Te) to a peak of 159 ± 6% of the base line. The response was similar to that triggered by inhaling the unfiltered smoke (Te = 177 ± 12%). The bradypnea started within 1–4 breaths after the onset of smoke inhalation, lasted for three to five breaths and was completely abolished by vagotomy. This inhibitory effect of gas phase smoke on breathing was also largely prevented after a pretreatment with either intravenous infusion or aerosol inhalation of a hydroxyl radical scavenger, dimethylthiourea.

Bupivacaine Kinetics

Smoke, administered by Hamburg II smoking machine to mice for 4 or 8 days, produced no effect after 4 days and a significant increase of metabolism and elimination of bupivacaine in the treated group. The smoke acted at different levels, i.e., metabolism, elimination and binding of bupivacaine, increased the permeability of the cell membranes, and facilitates the penetration of bupivacaine and desbutylbupivacaine in erythrocytes.

Cadmium-induced Alterations

Mainstream smoke generated from University of Kentucky 2R1 reference cigarettes, administered to cadmium-treated young female Long–Evans rats at a dose of 10 puffs daily for 12 weeks, produced no significant changes in lung function or morphometry.

Carboxyhemoglobin Effect

Cigarette smoke was administered by inhalation with a modified Walton horizontal smoke exposure machine to mice at intermittent doses. During the first 30 seconds of each 1-minute cycle, the subjects were exposed to smoke diluted either 1:10 or 1:5 with air. This treatment produced carboxyhemoglobin values significantly higher than those in marijuana smoke-treated mice. Mice exposed to six or eight puffs of tobacco smoke had mean carboxyhemoglobin values of 24.6 and 28.5% saturation, respectively. No acute lethal effects were observed in mice receiving multiple daily episodes of eight puffs per episode of marijuana smoke, whereas mice exposed to a single eight-puff episode of tobacco smoke suffered approx 50% acute lethal effects.

Carcinogenic Activity

Methyl chloride extract of the leaf, administered intragastrically to rats at a dose of 3 mg/animal daily for 21 months in DMSO solvent, was active. The rats were divided into two groups. One was fed a vitamin A diet, and the other was fed a vitamin A-deficient diet. Cumulative tumor incidence in vitamin-deficient rats was significantly greater than that in vitamin-fed rats. Vitamin-deficient rats had a preponderance of pituitary adenomas, whereas vitamin-fed rats showed lung and forestomach tumors. Smoke and snuff of the dried leaf, administered to adults at variable doses, were active. Forty-one users of STE and 38 cigarette smokers, produced a decrease of total [4-(methylnitrosamino)-1-3 -pirydyl)-1-butanone and its glucuronide] in users of STE and smokers. The 1-hydroxypyrene level was unchanged in tobacco smokers and reduced in smokers who used medicinal nicotine (nicotine patch). The overall mean total [4-(methylnitrosamino)- 1 -3-pirydyl)-1-butanone and its glucuronide] levels among smokers who used the nicotine patch was significantly lower than among smokers who used the OMNI cigarette. A mixture of side-stream and mainstream smoke, administered to male A/J mice at concentrations of 99, 120, and 176 mg/m^3 of total suspended particulate material for 5 months, produced significantly more lung tumors by the highest smoke concentrations, although response to the high dose was slightly less than response from the medium dose. Lung tumor incidences were in all three groups significantly higher than in controls. Lung displacement volume and plasma cotinine level were increased in a dose-dependent manner. Tobacco specific nitrosamines were administered to a line of human papillomavirus-immortalized bronchial epithelial cells at concentrations of 100 or 400 μg/mL for 7 days. The transformed cells produced progressively growing subcutaneous tumors on inoculation into nude mice. Immunofluorescence staining for keratin expression confirmed the epithelial nature of the tumor cells.

There was an increased expression of p16, β-catenin and proliferating cell nuclear antigen (PCNA) in the established cell line. Smoke, administered to male strain A/J mice at a concentration of 140 mg/m^3, 6 hours a day, 5 days/ week for 4–5 months, increased plasma β-carotene level and lung tumor multiplicities and incidences but had no significant effect on lung β-carotene levels. β-carotene supplementation failed to modulate tumor development under all exposure conditions. Smoke was administered to male strain A/J mice on a diet of Bowman–Birk protease inhibitor concentrate (BBIC) at a concentration of 1% in AIN-93G diet either during smoke exposure, after exposure or for 9 months. The mice were exposed to 89% cigarette sidestream and 11% mainstream smoke 6 hours a day, 5 days/week for 5 months and then allowed to recover for another 4 months in air. As a positive control, the mice were injected with 3-methylcholanthrene and fed a diet containing 1% BBIC. These animals were sacrificed 5 months later. In the animals treated with 3-methylcholanthrene, BBIC decreased lung tumor multiplicities, whereas in the smoke-exposed mice, BBIC did not modulate lung tumor development. Aqueous extract of the STE, administered orally to female Sprague–Dawley rats at a dose of 25 mg/kg for 90 days, produced an accumulation of indistinct filamentous material in the perisinusoidal spaces, disintegration of lipids, and a significant increase in heat stress/shock protein 90 expression. Environmental tobacco smoke, administered by whole body exposure to male strain A/J mice at concentrations of 87 mg/m^3 of total suspended particulate matter, 16 mg/m^3 nicotine, and 246 ppm CO for 6 hours/ day, 5 days/week 5 months, produced lung tumors. More than 80% of all tumors were adenomas, and the rest were adenocarcinomas. Mainstream smoke, administered to male Balb/c mice for 4 months starting 10 or 30 days before the administration of ethyl carbamate, produced 7.6% decrease of lung adenoma multiplicity. The number of ethyl carbamate-induced lung tumors was not significantly affected by exposure to cigarette smoke when ethyl carbamate was injected intraperitoneally in single doses of 0.5 or 1 g/kg. Smoke condensate, administered dermally to Balb/c mice, increased the density and changed the morphology of Langerhans' cells. The number of Langerhans' cells in epidermal sheets of treated mice was significantly higher than in the controls and remained elevated for 35 weeks. Langerhans' cells became less dendritic, or even rounded in shape, and smaller in size. The function of the morphologically altered Langerhans' cells was impaired. There was skin tumor development in all treated mice. After stopping the treatment, Langerhans' cells number in skin tumors and around lesions remained increased. Tumor regression occurred in 23% of tumors. The remaining tumors showed a 50% reduction in size. Smoke, in pulmonary and liver microsomes of male NMRI mice, induced the enzymatic activities cata-lyzed by CYP1A1, CYP2B, CYP2C, CYP2D, CYP2E1, and CYP3A in liver microsomes. Immunoquantification of lung and liver CYP1A1, 2E1, and 3A demonstrated the following: CYP1A1 was induced in lung and liver; CYP3A subfamily was induced in liver and not detected in lung; and CYP2E1 was slightly induced in liver, whereas its pulmonary expression was more largely increased (6.8 fold) than CYP1A1 (twofold). This indicates that CYP2E1, which is known to be expressed in human lung, could actively participate in pulmonary carcinogenesis induced by cigarette smoke. Six tobacco-specific *N*-nitrosamines and two major volatile *N*-nitrosamines of cigarette smoke, administered to female A/J mice, were active. *N*-nitrosodimethylamine was the most potent inducer of lung adenoma in the A/J mouse model, followed in order of decreasing potencies by 4-(methylnitrosamino)-1-(3-pyridyl)-1-butanone (NNK), 4-(methylnitrosamino)-1-(3-pyridyl)-1-butanol (iso-NNAL), *N*-nitrosopyrrolidine (NPYR), *N*'-nitrosonornicotine (NNN), and *N*'-nitrosoanabasine (NAB). NNK and 4-(methylnitrosamino)-4-(3-pyridyl)butyric acid (iso-NNAC) were inactive. Iso-NNAC, administered to strain A mice, was inactive as a tumorigenic agent, and it did not induce DNA repair in primary rat hepatocytes. Acetone/methanol extracts of sidestream and mainstream smoke condensates of a filtered commercial brand of blond cigarettes, administered on the shaved skin of female NMR1 mice twice a week for 3 months, produced no significant difference between the life span of mainstream smoke-treated and untreated mice. The life spans of sidestream smoke- treated mice were significantly shorter than those of mainstream smoke-treated mice. The numbers of tumors or lesions in mainstream smoke-treated mice were not increased dose-dependently. The initiation of lesions in sidestream smoke-treated mice was dose-dependent. The sidestream smoke-treated mice developed two to six times more skin tumors than the mainstream smoke-treated mice. Comparing the treated groups with the negative controls, the over-all carcinogenic effect observed was statistically significant. Comparing both treated groups with each other, the overall carcinogenic effect of sidestream smoke was much higher than that of mainstream smoke. *N*-nitrosamines: 3- (methylnitrosamino) - propionic acid (NMPA), NNK, and iso-NNAC, evaluated in A/J mice, produced the following results in the lung (total dose in micromol mouse/lung tumors per mouse): NMPA (200/7.1 ± 2.9), NNK (2/15.7 ± 4.1), iso-NNAC (200/0.24 ± 0.43), and saline control (0.2 ± 0.4). Brown and black varieties of a pyrolyzed tobacco (masheri), was administered to Sprague–Dawley rats, Swiss mouse, and Syrian golden hamsters at a dose of 10% of diet.

In Sprague–Dawley rats, only brown masheri was used, whereas in Swiss mice and Syrian golden hamsters, both varieties were used. Forestomach papillomas were induced in 37% of the rats, 42–47% of the mice, and 25–43% of the hamsters. No malignant changes were observed in any of the groups except two of the 23 male hamsters that showed forestomach carcinoma in the black masheri diet group. NNN, NNK, NPYR, 5'-carboxy-*N*'-nitrosonornicotine (CNNN), *N*-nitrosoproline (NPRO), and 1 -(3-pyridyl)-2-buten-1-one (PBO), administered to A/J mouse, produced the following results (dose in micromol per mouse/lung tumors per mouse): NNN (100/1.8 ± 1.4), NPYR (100/3.9 ± 1.5), CNNN (200/0.3 ± 0.5), CNNN (100/0.5 ± 0.6), NPRO (100/0.6 ± 0.7), NNK (20/7.2 ± 3.4), PBO (20/0.7 ± 1), and saline control (0.5 ± 0.7). NNK and NPYR were more tumorigenic than NNN. CNNN was nontumorigenic. Smoke condensate, administered intragastrically to Swiss mice, did not produce any tumor. Bidi concentrate, at the same dose, induced liver hemangiomas, forestomach papilloma, and carcinomas of the esophagus and fore-stomach. Bidi concentrate had a higher benzo[a]pyrene level than cigarette smoke condensate. NNK, iso-NNAL, and NNN, administrated dermally to female SENCAR mice at an initiator dose of 28 μmol/mouse in 10 subdoses administered every second day. Promotion commenced 10 days after the last initiator dose and consisted of twice weekly application of 2. μg of tetradecanoylphorbol acetate for 20 weeks. NNK induced a 79% incidence of skin tumors with an average of 1.6 tumors/mouse and a 59% incidence of lung adenomas. Iso-NNAL and NNN were not active.

At a total initiator dose of 28 and 5.6 μmol/mouse, NNK induced a 59 and 24% incidence of skin tumors, respectively. In this dose response bioassay, NNK at a total initiator dose of 28 μmol induced a 63% incidence of lung adenomas. The numbers of lung adenomas induced at the lower doses employed were not significant. NNK, at a total initiation dose of 1.4 μmol, did not exhibit significant tumorigenic activity. Fresh, whole cigarette smoke of Kentucky reference 2R1 cigarettes was administered intranasally to 2053 (C57BL/Cum × C3H/AnfCum)F1 female mice daily 5 days/week, for 110 weeks, produced a weak carcinogenic activity in mouse lung tissue, and an increased incidence of pigmented alveolar macrophage accumulation, otitis media, head and neck fibrosarcoma, and deposition of smoke particulates approx 125–200 mg total particulate matter/lung/day. The only lung cancers observed in 19 of 978 smoke- exposed mice were diagnosed as alveolar adenocarcinomas. A significant increase in the incidence of lung cancer was observed in one subset, but this difference was not found in the population as a whole or as a result of any other analyses. Fresh mainstream smoke from one University of Kentucky reference cigarette (2R1), administered intranasally to male C57BL mice and F-344 rats daily under standardized conditions, produced in mice, significantly elevated levels of blood COHb and pulmonary aryl hydrocarbon hydroxylase activity, and a fivefold to sevenfold increase in the number of bronchoalveolar lavage cells. The proportion of neutrophils increased to 18 ± 3% in smoke-exposed mice as compared to less than 1% in controls.

Cessation of smoke treatments returned the proportion of neutrophils to those of controls within 5 weeks. Smoke exposure of rats for up to 32 weeks induced appreciable changes in the number and proportion of macrophages and neutrophils. Large brown macrophages were observed in smoke-exposed groups of both species. Bronchoalveolar lavage cells from smoke-treated mice but not rats released greater amounts of superoxides than controls under resting and phagocytically stimulated conditions. The activity of *N*-acetylglucosaminidase was increased in both species. The activity of 5'-nucleotidase was significantly reduced in macrophages from mice but not rats. The activity of leucine aminopeptidase remained unaltered in both species. Tobacco alcoholic extract, administered intragastrically or in the diet of male Swiss mice, increased the incidence of lung and liver tumors. An additive effect of tobacco extract and hexachlorocyclohexane on liver tumor induction was found. The weakly acidic fraction of cigarette smoke condensate subfractions II, administered dermally with 0.003% benzo[a]pyrene to noninbred Ha:ICR Swiss albino mice, produced significant cocarcinogenic activity (subfractions A-C and F-J); subfractions A, F, and H were the most active. Catechol was a major component of subfraction A and was also detected in subfractions B–D and F. Major components of the other subfractions included hydroquinone (B), coniferyl alcohol (C and H), hydroxyphenyl alcohols (D), alkyl-2-hydroxy-2-cyclopenten-1-ones (C, D, and F), hydroxyacetophenones (F), phenolic cyano compounds (F), and fatty acids (F). The results indicated the importance of catechol as a cocarcinogen in the weakly acidic fraction of cigarette smoke condensate. Methanol (80%) insoluble fraction of aqueous extract of tobacco combined with the methanol soluble fraction was administered to mice treated with 7,12-dimethylbenz[a]anthracene.

The subfraction (D) with a presumptive molecular weight greater than 13,000 produced a significantly higher tumor incidence and tumor yield, together with a significantly shorter latent period than the other subfractions. Subfraction D contained approx 12% of the total 80% methanol insoluble material. All of the other subfractions exhibited significant but less pronounced tumor copromoting activity. Undifferentiated carcinomas of the salivary glands were found in two of 44 strain A mice injected intraperitoneally with an *N*-nitrosonornicotine. The presence of intranuclear rodlets in the salivary carcinomas provided the first demonstration of such structures in a non- neuronal tumor in mice. Two types of rodlets were exhibited: one was composed of fibrillar filaments arranged in bundles, and the other was much thicker and branching in form. Results indicated that these intranuclear rodlets were closely associated with nuclear chromatin or nucleoli. NNK, NNN, and 4-(*N*-methyl-*N*-nitrosamino)-4-(3-pyridyl)butanal (NNA), administered to strain A mice, were active. NNK induced more lung adenomas per mouse than NNN, and NNA was less active than NNN. Two cases of undifferentiated carcinoma of the salivary glands occurred in the NNN experimental groups. Smoke components, administered dermally to female ICR/Ha Swiss mice three times a week with 5 μg benzo[a]pyrene per application, enhanced the carcinogenicity of benzo-[a]pyrene by catechol, pyrogallol, decane, undecane, pyrene, benzo[e]pyrene, and fluoranthene. The following compounds inhibited benzo[a]pyrene carcinogenicity completely: esculin, quercetin, squalene, and oleic acid. Phenol, eugenol, resorcinol, hydroquinone, hexadecane, and limonene partially inhibited benzo[a]pyrene carcinogenicity. No direct correlation existed between tumor-promoting activity and cocarcinogenic activity. The cocarcinogens pyrogallol and catechol did not show tumor-promoting activity. Decane, tetradecane, anthralin, and phorbol myristate acetate showed both types of activities. Acetone and alcohol extracts of the flue-cured tobacco, administered dermally to mice, produced weak activity.

Chloroform/water extract produced tumor in 38% of the animals and was about fivefold more active than smoke condensate derived from an equal weight of tobacco. NNN induced adenomas of the lung in mice. Bioassays of NNN in rats indicated carcinogenic activities on the esophagus and the nasal cavity. Weakly acidic fraction of cigarette smoke particulate matter were tested on initiated mouse skin by long-term application. Two of these subfractions (40% of weakly acidic fraction) were inactive, and the 18 and 35% of weakly acidic fractions showed tumor-promoting activity. Catechol, hydroquinone, 3-hydroxypyridine, 6-methyl-3-hydroxy-pyridine, linolenic acid, and linoleic acid were inactive as tumor promoters in the experimental animal. [5-(3)H]-(S)-NNN, in rat esophagus culture at a concentration of 1 μM, was predominantly metabolized to products of 2'-hydroxylation, 4-oxo-4-(3-pyridyl)butanoic acid (ketoacid), and 4-hydroxy-1-(3-pyridyl)-1-butanone. The major metabolite of (R)-NNN under these conditions was 4-hydroxy-4-(3-pyridyl)butanoic acid, a product of NNN 5'hydroxylation. The 2'-hydroxylation: 5'-hydroxylation metabolite ratio ranged from 6.22 to 8.06 at various time intervals in the incubations with (S)-NNN.

The corresponding ratios were 1.12-1.33 in the experiments with (R)-NNN. These differences were statistically significant ($p < 0.001$). Because 2'-hydroxylation is thought to be the major metabolic activation pathway of NNN in the rat esophagus, the results demonstrated that (S)-NNN is metabolically activated more extensively than (R)-NNN and, therefore, may be more carcinogenic. [5-(3)H]-(R)-NNN, [5-(3)H]-(S)-NNN, or racemic [5-(3)H]NNN, administered intragastrically to rats at dose of 0.3 mg/kg, was metabolized to hydroxylacid and ketoacid. Products of 2'-hydroxylation predominated in the urine of the rats treated with (S)-NNN, whereas products of 5'-hydroxylation were more prevalent in the rats treated with (R)-NNN. 2'-Hydroxylation:5'-

hydroxylation metabolite ratios ranged from 1.66 to 2.04 in the urine at various times after treatment with (S)-NNN, while the ratios were 0.398–0.450 for the rats treated with (R)-NNN ($p < 0.001$). The results indicated that the carcinogenicity of (S)-NNN, the predominant enantiomer in tobacco products, may be greater than that of (R)-NNN or racemic NNN.

Cardiovascular Effect

Water extract of the dried leaf, administered intravenously to cats and rats at doses of 10–20 mg/kg, induced variable electrocardiogram patterns, including increases in the height of the QRS complexes, S-T segment elevation, occasional extrasystoles, and arrhythmias.

Cataractogenic Effect

Cigarette smoke was administered to male Wistar rats for 90 days, with or without vitamin E supplement. The treatment significantly increased iron levels in the lenses of smoke-treated group. Smoke-treated rats and smoke-treated and vitamin-supplemented rats had significantly higher cadmium levels. Vitamin E treatment prevented iron accumulation in smoke-treated and vitamin-supplemented rats. Distinct histopathological changes observed in smoke-treated rats were not present in smoke-treated and vitamin- supplemented rats. Cigarette and fire-wood smokes condensates were evaluated on isolated capsulated rat lenses incubated for varying periods, with and without antioxidants, in the presence and absence of light. The smoke condensates permeated the lens capsule and impart color and opacify to the lens in a light- and dose-dependent manner. Antioxidants offer partial inhibition against the damages. Smoke-induced damage possibly occurs through systemic absorption and transport of toxic components to several tissues, specially into the lens. The turnover is slow, leading to chronic accumulation causing oxidative damage to the constituent molecules and consequently to lenticular opacity.

Cell Differentiation Induction

Water extract of the dried leaf, administered to CBA/N strain mice at a concentration of 0.5%, was active on lymphocytes B. A tobacco-specific carcinogen NNK, in cell culture, produced activation of the phosphatidylinositol 3'-kinase/Akt pathway (P13K/Akt). The pathway was evaluated in isogenic immortalized or tumorigenic human bronchial epithelial cells in vitro and in progressive murine lung lesions. Compared with immortalized cells, tumorigenic cells had greater activation of the P13K/Akt pathway, enhanced survival, and increased apoptosis in response to inhibition of the pathway. In vivo, increased activation of Akt and mammalian target of rapamycin were observed with increased phenotypic progression.

Cell Metabolism Induction

Methanol extract of STE, in collagen-producing cells, stimulated glycolysis by 80% in cartilage but was not affected in the other tissues. Medium alkaline phosphatase activity was unaffected. In the frontal bone and cartilage, [^{3}H]hydroxyproline and [^{3}H]proline contents were decreased. Neither was affected in the aorta.

Cell Proliferation

Snuff extract in combination with DMBA, in primary embryonal mouse tongue culture, inhibited the proliferation of cells and decreased ornithine decarboxylase and hydrocarbon hydroxylase activities, compared to control. NNN, in embryonic mouse tongue epithelial cells, increased [^{3}H]dT uptake, and ornithine decarboxylase and aryl hydrocarbon hydroxylase activities. NNK produced further increases of [^{3}H]dT uptake, cell count, and ornithine decarboxylase. The extract had an inhibitory effect on cell count, [^{3}H]dT uptake, ornithine decarboxylase, and aryl hydrocarbon hydroxylase activities when administered alone or in combination with NNN or NNK. Tobacco and tobacco smoke constituents, in ascites sarcoma BP8 cell culture, indicated that the most active constituents were unsaturated aldehydes and ketones, phenols, and indoles. Smoke, administered to Sprague–Dawley rats at a dose of 10 cigarettes (smoke only), or the smoke of 10 cigarettes after iv infusion of endothelin A antagonist BQ-610 (smoke and BQ-610), produced significant cell proliferation in the airway epithelium and wall, in the peribronchiolar arterial endothelial compartment, and in the endothelial and wall compartments of the perialveolar ductular arteries. Pretreatment with BQ-610 reduced the peribronchiolar arterial endothelial and the perialveolar ductular arterial wall proliferation to control levels and reduced, but did not totally abrogate, the smoke-induced proliferation of the airway epithelial, airway wall, and perialveolar ductular arterial endothelial compartments. Results indicated that cigarette smoke-induced cell proliferation of the airways and pulmonary arterial vessels is at least partially mediated through stimulation of the endothelin-A receptors.

Cell Signaling

Environmental tobacco smoke, administered to rats during gestation, the early neonatal period, or both, elicited induction of total adenylyl cyclase. In the brain, the specific coupling of β-adrenergic receptors to adenylyl cyclase was inhibited in the smoke-treated groups, despite a normal complement of β-receptor binding sites. In the heart, smoke evoked a decrease in M2-receptor expression. In both tissues, the effects of postnatal smoke, mimicking passive smoking, were equivalent to adenylyl cyclase level or greater than (M2-muscarinic cholinergic receptors) those seen with prenatal smoke mimicking active smoking. The effects of combined prenatal and postnatal exposure were equivalent to those seen with postnatal exposure alone. The smoke exposure evoked changes in cell signaling that recapitulate those caused by developmental nicotine treatment.

C-fos Expression

Mainstream smoke trapped in PBS solution, in quiescent Swiss 3T3 cells, produced dose-dependent expression of *c-fos* mRNA and protein. *C-fos* transcripts in cells exposed to 0.03 puffs (approx 1 cm^3) of smoke per medium, accumulated slowly but were still seen after 8 hours. The maximum expression rates were between 2 and 6 hours of exposure. An increase of *c-fos* message stability was observed in addition to slight transcriptional activation of the c-fos promoter.

Chemiluminescence

Cigarette smoke in determined quantity was streamed through physiologic saline solution, blood plasma, or ex vivo excised rat lung. The solutions and the supernatants from lung tissue homogenate produced a significant increase in chemiluminescence after *t*-butyl hydroperoxide induction.

Chromosome Aberrations Induction

Water extract of the dried leaf, in cell culture at a concentration of 15 mL, was active on Chinese hamster ovary cells. The number of aberrant metaphases increased in cultures with 15 mL tobacco extract per milliliter of growth media. Water extract of the dried leaf, administered to mice at a dose of 9.40 g/kg, 6 days a week for 10 months, was active on bone marrow. A combination of *Piper betle*, *Areca catechu*, and *Nicotiana tabacum* was used. Seed, administered orally to adults with oral cancer and oral submucosal fibrosis and to healthy chewers, was active. An average of 6 quids of tobacco leaf, *Areca* nuts, and lime were chewed daily.

Chronic Bronchitis

Cigarette smoke, administered to rats for 2 weeks, produced a significantly higher mean basal secretion of fucose in "bronchitic" rats than in the controls. In control and bronchitic animals, acute administrations of cigarette smoke, blown directly through the laryngeo-tracheal segment after equilibration, produced significant transient increases in the secretion of fucose, hexose, and protein but not albumin. Cigarette smoke, administered to the specific pathogen-free rats at a dose of 25 cigarettes daily for 14 days and concurrently given *N*-acetylcysteine (NAC) as 1% of their drinking water, increased the thickness of the epithelium by 37–72% at three of the airway levels studied. The number of secretory cells was increased at all airway levels distal to the upper trachea 102–421%. Secretory cells containing neutral glycoproteins were reduced in number, but this was more than offset by a large increase in the number of secretory cells containing acidic glycoproteins at all airway levels.

Chronic Obstructive Pulmonary Disease

Smoke-conditioned media, administered intranasally to Balb/c mice for 40 days, significantly increased bronchoalveolar lavage neutrophils, lymphocytes, chemokine, TNF-α, and mucin. There were changes in pulmonary reactivity to methacholine, inflammation, and cellular lung changes characteristic for human chronic obstructive pulmonary disease. A polyphenol reagent isolated from cigarette smoke condensate primed purified human neutrophils. A mouse monoclonal antiidiotypic antibody directed against the polyphenols-reactive determinants on a rabbit polyclonal antitobacco glycoprotein antibody was generated and also primed neutrophils.

After priming by the isolated polyphenol reagent or tobacco antiidiotypic antibody, there was a 2.5-fold to threefold increase in CD11b/18 expression and doubling of the number of formyl-methionyl-leucyl-phenylalanine receptors on the cells. The primed cells produced a twofold increase in production of superoxide and release of neutrophil elastase after stimulation with formyl-methionyl-leucyl-phenylalanine. The inflammatory process contributing to progression of chronic obstructive pulmonary disease in ex-smokers may be in part driven by tobacco antiidiotypic antibodies.

Chylomicron Metabolism

Smoke, administered to rats injected intravenously with ^{14}C- and ^{3}H-labeled chylomicrons, produced no difference in the initial plasma clearance time of labeled chylomicrons between smoke-treated and control animals. Hepatic uptake of chylomicron cholesterol was slower in smoke-treated animals than in controls. More labeled chylomicrons remained in the heart of smoke-treated rats than controls.

Cimetidine Kinetics

Low- or high-nicotine tar, administered orally and parenterally to rats for 10 minutes immediately after administration of cimetidine, produced lower plasma level after orally administered cimetidine in the absorption phase in the smoke inhaling groups than in the non-smoking control group. It was particularly marked in the high-nicotine tar cigarette smoke-inhaling group. No significant difference was found in cimetidine plasma level between the cigarette smoke inhaling group and the nonsmoking control group when administered intraperitoneally or intravenously. The cigarette smoke inhalation produced suppression or a delay in cimetidine absorption from the gastrointestinal tract, and the degree of influence was dependent on the content of nicotine tar in the cigarette smoke.

Clastogenic activity. Water extract of the dried leaf, administered to mice at a dose of 9.4 g/kg, 6 days/week for 10 months, was active on bone marrow. A combination of aqueous extracts of *Piper betle*, *Areca catechu*, and *Nicotiana tabacum* was used. Seed, administered orally to adults, produced an increase in micronuclei in healthy chewers and chewers with oral submucosal fibrosis. An average of 6 quids of tobacco leaf, *Areca* nuts, and lime were chewed daily. Mainstream smoke, administered by whole-body exposure to male albino Swiss mice, produced a significant increase in the number of micronucleated bone marrow polychromatic erythrocytes. A high correlation was observed among the number of micronucleated polychromatic erythrocytes and the content of tar and nicotine. Smoke, administered to pregnant BDF1 (C57B1 × DBA2) mice at a dose of 600 cm^3 of smoke, four times of 15 minutes each with 1 minute intervals on day 16/17 of gestation, produced a two- to three-fold increase in the number of micro-nucleated polychromatic erythrocytes in fetal liver and liver of newborn mice (1–5 hours after birth). Administration of smoke for 60 minutes/day repeatedly from day 11 of gestation produced slightly greater micronucleus response in fetuses. Smoke, administered transplacentally to newborn mice during the last trimester of pregnancy, produced greater clastogenic activity than in their 6-month-old mothers. Smoke, administered to BDF1 mice at a dose of 600 cm^3, two to six exposures of 30 minutes each, produced 3.5-fold increase of the number of micronucleated polychromatic erythrocytes in born marrow after 24 hours of exposure, and a two- to fivefold increase in the peripheral blood of mice treated twice daily for 30 minutes, starting after 48 hours of exposure.

Central Nervous System Depressant Activity

Ethanol (70%) extract of the fresh leaf, administered intraperitoneally to mice of both sexes at variable doses, produced strong activity. Initial excitation was followed by marked sedative effect.

Collagenase Activity

Smoke was administered to Guinea pigs at a dose of 20 cigarettes per day for 8 weeks. At 6 and 8 weeks of exposure, lungs exhibited interstitial and peribronchiolar inflammation and moderate emphysematous changes. There was patchy expression of collagenase mRNA mainly in macrophages but also in alveolar epithelial and interstitial cells. Immunoreactive protein was detected in alveolar macrophages, in alveolar walls, and in interstitium. Collagenolytic activity increased beginning in the fourth week of exposure. Collagen concentration decreased from 50.7 ± 8.5 mg/g dry weight in control lungs to 40.2 ± 5 and 42.9 ± 6 at 6 and 8 weeks of exposure, respectively.

Comutagenic Activity

Cigarette smoke condensate, in *Salmonella typhimurium* strains TA98 and TA98/1.8DNP6, specifically enhanced the mutagenicity of polyaromatic amines, such as 2-aminofluorene, 2-acetylaminofluorene, 4-acetylaminofluorene, and 2-aminoanthracene. Both black and blond tobacco proved to interact synergistically with 2-aminoanthracene mutagenicity. Administration of 2-aminoanthracene/smoke condensate mixtures, previously shown to be comutagenic in vitro, failed to demonstrate a synergistic effect in sister chromatid exchange induction in bone marrow cells of mice.

Connective Tissue Breakdown

Whole cigarette smoke, administered to C57-BL/6 mice, produced a dose-response increase in lavage neutrophils, desmosine, and hydroxyproline, but not lavage macrophages (MACs). The effect was evident after 6 hours of exposure to two cigarettes. Pretreatment with an antibody against polymorphonuclear leukocytes (PMNs) reduced lavage PMNs to undetectable levels after smoke exposure. It did not affect MAC numbers and prevented increases in lavage desmosine and hydroxyproline. Intraperitoneal injection of a commercial human α1-antitrypsin (α1AT) 24 hours before smoke exposure increased serum α1AT levels threefold and completely abolished smoke-induced connective tissue breakdown and the increase in lavage PMNs, without affecting MAC numbers.

CuZn-superoxide Dismutase Activity

Cigarette smoke was administered to the osteogenic disorder Shionogi (ODS) rats at doses of 4 mg/day, S4 or 40 mg/day, and S40 of ascorbic acid and exposed to smoke daily for 25 days. The treatment produced a significant decrease of CuZn-superoxide dismutase (SOD).

Cyclo-oxygenase Activity

Aqueous cigarette tar extracts, in rat pulmonary alveolar macrophages, increased cyclo-oxygenase activity threefold above the initial activity within 2 hours of incubation and gradually decreased below the initial activity after 8 hours of incubation. Accumulated levels of prostaglandin-2 increased dramatically after 12 hours of incubation.

Cytochrome C Oxidase Inhibition

Smoke extract, in the mouse brain mitochondria culture in the presence or absence of vitamin C for 60 minutes, inhibited mitochondrial Adenosine triphosphatase (ATPase) and cytochrome C oxidase activities in a dose-dependent manner. The effect of extract on mitochondria swelling response to calcium stimulation was dependent on calcium concentrations. The extract treatment induced mitochondrial inner membrane damage and vacuolization of the matrix, whereas the outer mitochondrial membrane was preserved. Nicotine produced no significant damage.

Cytochrome P450 Induction

STE, in murine system at doses of 50 or 100 mg/kg body weight/ day, elevated microsomal cytochrome b5, cytochrome P450 (CYP), and malondialdehyde levels. These results indicated the inhibitory potential of STE on garlic-induced hepatic glutathion-*S*-transferase (GST)/GSH system besides significant augmentation on garlic-, mace- or black mustard-induced microsomal cytochromes. Methyl chloride extract of the leaf, administered intragastrically to rats at a dose of 3 mg/animal daily for 21 months in DMSO solvent, was active. The rats were divided into two groups, one was fed a vitamin A diet, and the other a vitamin A-deficient diet. Treatment with the extract increased pulmonary CYP over controls in vitamin-deficient animals and among treated animals more so in the vitamin- deficient group. Crude cigarette smoke condensate, in human liver microsomes, inhibited P450 1A2 cytochrome. The tobacco-specific nitrosamines were activated by a number of P450 enzymes. P450 1A2, 2A6, and 2E1 activated nitrosamines to genotoxic products. Aged and diluted sidestream cigarette smoke, administered to timed pregnant rats and their pups four times a days from gestational day five to postnatal day 21, produced no alterations in mRNA in the fetal lung beginning at gestational day 17. Continued exposure significantly induced CYP1A1 but not other P450 genes as early as one day after birth. Results indicated that smoke-induced pulmonary CYP1A1 in the first day of life fetal cytochrome P450 genes were not induced by maternal exposure to smoke. In the fetal lung, CYP1A1 and 1B1 can be induced by β-naphthoflavone. Mainstream cigarette smoke, administered to F344 rats at a dose of 100 mg total particulate matter/m^3 for 2 or 8 weeks, induced CYP1A1 in respiratory and olfactory mucosae, liver, kidney, and lung. CYP1A2 levels increased slightly in the liver and olfactory mucosa. CYP2B1/2, which increased in the liver, decreased in the upper and lower respiratory tissues. Intense immunoreactivity was found in epithelia throughout the nasal cavity of smoke-exposed rats. Ethoxyresorufin *O*-demethylase activity (associated with CYP1A1/2) decreased approximately two-fold in olfactory mucosa but increased in nonnasal tissues. Methoxy- and pentoxyresorufin *O*-dealkylase activities (associated with CYP1A2 and CYP2B1/2, respectively) decreased in olfactory and respiratory mucosae and lung (CYP2B1/2) but increased in liver. Masheri, a pyrolyzed tobacco product, administered orally to Swiss mice, Sprague–Dawley rats, and Syrian golden hamsters at a dose of 10% diet for 20 months, produced a significant induction of cytochrome P450 in proximal and distal parts of the three species.

Cytotoxic Effect

Gas phase of mainstream cigarette smoke, in monolayer culture of mouse lung epithelial cells, produced an increase in cytotoxicity in a dose-dependent manner. Cell viability of cultures exposed to gas phase with only the nonorganic components was equivalent to controls. Removal of volatile organic constituents resulted in almost elimination of cytotoxicity of the smoke. Smoke condensate and tobacco extract, at high concentrations in Lewis lung adenocarcinoma cells and mice spleen lymphocytes, were cytotoxic. Smaller doses increased thymidine incorporation in both cell types. Lymphocytes were more susceptible to the toxic effect of tobacco products than lung cells. When smoke condensate and tobacco extract were mixed with Lewis lung adenocarcinoma cells and then inoculated into mice, they did not modify the size of the local Lewis lung adenocarcinoma-induced tumors or the number or appearance time of lung metastasis, although there was an increase in spleen weight. Mainstream and sidestream smoke from the same cigarette, in mono- layer cell culture of mouse fibroblast-like L-929 cells, produced a decrease of cytotoxicity with increasing smoke age (up to 8.7 seconds), smoke dilution, and the quantity of activated charcoal in filters. Acetate filters had little effect on cell mortality, and the age-of-smoke effect was not evident for mainstream smoke generated with a low puff volume and rapid dilution. The cytotoxicity of sidestream smoke also decreased rapidly with increasing smoke age and dilution. Smoke condensate from the mainstream smoke of TOB-HT, IR4F, and 1R5F cigarettes, in human bronchial/tracheal epithelial cells, coronary artery endothelial cells, coronary artery smooth muscle cells, foreskin keratinocytes, and WB-344 rat liver epithelial cell line exposed for 1 hour, produced no inhibition of gap junction intercellular communication by TOB-HT in any of the human cell types tested at concentrations where 1R4F and lR5F did inhibit ($p <$ 0.05). TOB-HT did not elevate lactate dehydrogenase release, when tested at concentrations where lR4F and lR5F did. Smoke condensate from TOB-HT cigarettes is less damaging to the structure or function of the cellular plasma membranes of a variety of human cell lines than from 1R4F and 1R5F tobacco burning reference cigarettes.

Dermal Tumor

Smoke condensates were administered externally to female SENCAR mice at doses of 10, 20, or 40 mg Eclipse or 1R4F cigarette smoke condensates three times a week for 29 weeks. The treatment was initiated with a single topical application of 7,12-dimethylbenzanthracene. The treatment did not alter body weight, survival, or other indicators of subchronic toxicity. In 7,1 2-dimethylbenzanthracene-initiated mice, there were significant increases in both the number of tumor-bearing animals and dermal tumors at all 1R4F doses and the high-dose Eclipse.

Desensitization of Nicotinic Receptor

S- (-)-Nicotine (10 and 100 n*M*) diminished [^{3}H]overflow from ^{3}H-dopamine-preloaded rat striatal slices after subsequent super-fusion with 10 μM *S*-(-)-nicotine (46 and 74%, respectively) or 10 μM *S*-(-)-nornicotine (59 and 81%, respectively). S-(-)-nornicotine (1 and 10 μM) diminished the response to subsequent superfusion with 10 μM *S*-(-)-nornicotine (85 and 97%, respectively) or 10 μM *S*-(-)-nicotine (82 and 88%, respectively). Thus, similar to *S*-(-)- nicotine, *S*-(-)-nornicotine-desensitized nicotinic receptors, but with approx 12-fold lower potency. Cross-desensitization suggested involvement of common nicotinic receptor subtypes.

Diltiazem Kinetics

Cigarette smoke, administered to rats for 10 minutes using a Hamburg II smoking machine, inmediately after oral administration of diltiazem (10 mg/kg), produced plasma diltiazem levels in the rats exposed to cigarette smoke reached the maximum (4.3 μg/kg) after 4 hours. In the nonsmoking nonrestrained rats, plasma diltiazem levels increased rapidly and reached the maximum (7.1 μg/kg) 2 hours after administration and decreased gradually thereafter. In the nonsmoking restrained rats, plasma diltiazem levels increased rapidly but showed almost constant levels between 1 hour and 8 hour after administration. The maximum level (5.4 μg/kg) was shown after 2 hours. Results indicated that absorption of orally administered diltiazem is inhibited and delayed by cigarette smoke.

DNA Adduct Formation

Mainstream Eclipse or 1R4F cigarettes smoke condensate was administered dermally to SENCAR mice at doses of 30, 60, or 120 mg/animal three times a week for 30 weeks. The treatment produced distinct time and dose-dependent diagonal radioactive zones in the DNA from lung, heart and skin tissues of 1R4F-treated mice. The relative adduct labeling values of lung, heart, and skin DNA from reference smoke condensate-treated animals were significantly greater than those of the solvent controls were. No diagonal radioactive zones were observed at any dose from the DNA of animals treated with smoke condensate from Eclipse. Smoke condensate from cigarette and a reference tobacco-burning cigarette (1R4F) were administered dermally to CD-1 mice three times a week for 4 weeks at mass up to 180.0 mg "tar" per week per animal. Distinct diagonal radioactive zones in the DNA from both skin and lung tissues of animals dosed with reference cigarette smoke was produced. No corresponding diagonal radioactive zones were observed from the DNA of animals dosed with the test cigarette smoke or acetone (solvent control). The relative adduct labeling values of skin and lung DNA from reference-treated mice were significantly greater than those of the test cigarette-treated mice. The relative adduct labeling values of the test cigarette-treated animals were no greater than those of solvent controls. Smoke condensate, administered topically to female ICR mice at four doses equivalent to three cigarettes daily, elicited aromatic adducts in most tissues, but not in white blood cells. Cigarette smoke administered by inhalation, cigarette smoke condensate administered intraperitoneally, or neutral fraction in genetically responsive C57BL/6 (B6) and nonresponsive DBA/

2 (D2) mice for 3–16 days, produced no detectable levels of benzo[a]pyrene-7 ,8-diol-9, 10-epoxide (BPDE)-DNA in lungs or liver. Aryl hydrocarbon hydroxylase (AHH) activity was induced in the lungs of B6 mice. Benzo[a]pyrene, administered intraperitoneally to mice at doses of 20–80 mg/kg, produced dose-dependent amount of BPDE-DNA-adducts in lung and liver. Administration of 4 mg/kg benzo[a]pyrene produced no effect. In B6 mice aryl hydrocarbon hydroxylase was induced in lungs and livers, but not in D2 mice, although the levels of BPDE-DNA-adducts were higher than in B6 mice. Sidestream cigarette smoke, administered by a whole-body exposure to female Sprague–Dawley rats for 6 hours/day for 4 weeks, produced one major and several minor smoke-related adducts in lung, trachea, heart, and bladder. Mainstream cigarette smoke, administered by whole-body exposure to female BD6 rats, 1 hour/day, 5 days/week for 8 months, produced no significant increase of DNA-protein cross-links in liver, lung, or heart. Cigarette smoke induced formation of DNA adducts in the lung and heart but not in the esophagus or liver. The combined ingestion of ethanol resulted in a significant formation of smoke-related DNA adducts in the esophagus and dramatic increase in the heart.

DNA Damages

Environmental smoke was administered by whole-body exposure to adult female Balb/c mice in a regimen consisting of sequences of a 30 minutes exposure followed by a 90 minutes nonexposure. This regimen was performed once for the single exposure and repeated three times for the triple exposure. The exposure increased 8-hydroxy-2'-deoxyguanoside levels in the heart, lung, and liver. In some instances, the increased level returned to normal by the end of the nonexposure period, whereas other tissues showed a further increase following nonexposure. NNK, administered to pregnant Swiss mice in a single or multiple doses, significantly increased levels of 8-oxo-2'-deoxyguanosine in maternal lungs by 23 and 32%, respectively. In maternal liver, a 38% increase was observed after multiple dose treatment. In the fetuses, a 45% increase in 8-oxo-2'-deoxyguanosine levels was observed in liver after multiple doses Aqueous extract of smoke condensate, in rat lung culture in the absence of microsomes, produced radioactively labeled bulky reaction products accumulating in a time- and dose-dependent manner. Pretreatment of extract with radical scavengers/reducing agents (ascorbic acid, GSH), diminished adduct formation in a concentration-dependent manner. Adduct fractions derived from in vitro and in vivo experiments showed similar chromatographic behavior. Smoke condensate, administered dermally to female ICR mice at a dose being equivalent of 4.5 cigarettes for 6 days, produced DNA damages in lung, heart, skin, and kidneys higher than in the liver. Spleen DNA was virtually adduct free. Preference for heart and lung was observed for mice treated for 1 and 3 days. Cigarette tar, administered dermally to ICR mice at doses being equivalent to 1.5, 3, 6, and 9 cigarettes for 4, 3, 5, and 7 days, respectively, produced 12 distinct ^{32}P-labeled DNA adduct spots, and diagonal radioactive zone. One derivative in particular (adduct 1) increased rapidly during the early treatment phase and persisted to 8 days after treatment. The prominent adduct 1 was observed in the same location on the fingerprints of DNA samples from human smokers. Co-chromatography experiments suggested identity of human and mouse DNA adduct 1. Several other human and mouse adducts (adducts 3, 5, 6, and 9) appeared identical, and the diagonal radioactive zone was also present on DNA adduct maps from smokers. Mainstream whole-smoke DMSO and phosphate buffer solutions induced DNA single-strand breakage in mice testicular cells. DMSO solution of cigarette smoke produced stronger cytotoxicity and genotoxicity than the phosphate buffer solution. Cigarette smoke, administered in combination with asbestos to rats for 1,2, and 14 days, increased in 2'- dioxyuridine-5'-triphosphate (dUTP)-biotin nick end labeling-positive, necrotic epithelial cells.

DNA Deletion

Filtered and unfiltered smoke and smoke condensate were administered by whole-body exposure for 4 hours to pregnant pink eye-unstable C57BL/6J mutant mice or 15 mg/kg of smoke condensate during 10th day of gestation. A significant increase in the number of DNA deletions in the embryo, as evidenced by the spotted offspring in both smoke-exposed groups, was observed.

DNA Synthesis Stimulation

Smoke of the dried leaf, administered to rats at variable doses, was active. The effect of a combination of cigarette and hashish smoke on stress response was measured in rats held in a wire cage inside of a larger cage with a cat. Brains were measured for protein and catecholamine levels. Whole Kentucky reference 2A1 cigarette smoke, administered to hybrid strain BC3F1/Cum (C57BL/Cum X C3H/AnfCum) mice, increased DNA replicative activity more than twofold within 1 week of beginning smoke exposure and remained elevated as long as smoke exposure was continued.

Treatment of lung tissues in vitro with either the lung carcinogen 4-nitroquinoline-1-oxide or methyl- methane sulfonate stimulated unscheduled DNA synthesis. Until the 10th to 12th week of smoke exposure, at which time the accumulated deposition of total particulate material in the lung was approx 40 mg, the level of unscheduled DNA synthesis (UDS) stimulated by the alkylating chemicals declined to approx 50% of that seen in lung tissue from sham-exposed control mice. If the mice were removed from smoke exposure, DNA replicative activity returned to normal levels within 1 week, but the UDS response to DNA damage remained depressed up to 5 months after ending smoke exposure.

Dominant-lethal Mutations

Whole tobacco smoke was administered to male Balb/c, DF1 and H mice at two doses: low 1-hour treatment/day and high 2-hour treatment/ day, 5 days/week for 8 weeks. Dominant-lethal mutations in both experimental groups of Balb/c, BDF1, and H mice were significantly induced ($p < 0.001$), but some strain differences existed. In Balb/c and BDF1 mice, the smoke-induced dominant-lethal mutations were found mainly in spermatocytes, spermatogonia, and gonial stem cells. In H mice, only spermatids and spermatocytes were affected. Exposure produced a twofold to threefold increase in the number of micronucleated polychromatic erythrocytes in the bone marrow of Balb/c and BDF1 mice in both dose groups. In H mice, this effect was observed only on days 19 and 38 of sampling. No cumulative or dose-dependent effects were detected.

Dopamine Protection

Cigarette smoke, in 1-methyl-4-phenyl- 1,2,3 ,6-tetrahydropyridine (MPTP) mouse model, partially protected against corpus striatal dopamine depletion by MPTP. This protection was associated with monoamine oxidase (MAO) inhibition in brain and liver and CYP induction. β-naphthoflavone pretreatment also partially protected against MPTP-induced depletion of striatal dopamine. The results indicated that both MAO inhibition and CYP induction may play a role in any biochemical protection afforded by cigarette smoke exposure against the development of Parkinson's disease. Nicotine, 4-phenylpyridine, and hydrazine, administered to mice, prevented the decrease in dopamine metabolite levels induced by MPTP, but there was no significant effect on dopamine levels. The compounds did not inhibit monoamine oxidase (MAO) activity in cerebral tissue in vivo. In vitro, an extract produced significant inhibition of MAO A and B activities in the brain.

Dopaminergic Activity

Smoke was administered intranasally to mice for 20 minutes twice daily for 3 days before methamphetamine treatment. The treatment significantly attenuated the neurotoxicity as judged by a lesser depletion of dopamine, dihydrophenylacetic acid, and homovanillic acid. The lesser effect of methamphetamine on the content of serotonin level was unaltered by prior inhalation of smoke. Tobacco glycoside, administered to mice, increased behavior via dopamine 2 neuronal activity but not dopamine 1 activity in a dose-dependent manner. The results indicated that smoking can affect the human brain function via not only the nicotinic cholinergic neuron but also the dopamine 2 neuron.

Duodenal Ulcer

Tobacco cigarette smoke, administered to mepirizole-treated rats, inhibited hyperemia at the ulcer margin after exposure.

E-cadherin Expression

Smoke extract, on pig airway epithelial cells and mouse trachea, produced a decrease of E-cadherin expression on membrane and an increase of cytoplasmic expression at 12 and 24 hours after exposure. The expression at 24 hours was higher than at 12 hours.

Electron Transport Inhibition

Acetonitrile extracts of cigarette tar inhibited stage three and four respiration of intact mitochondria. Exposure of respiring submitochondrial particles to the acetonitrile extracts of cigarette tar results in a dose-dependent inhibition of oxygen consumption and reduced nicotinamide adenine dinucleotide oxidation. Intact mitochondria were less sensitive to extracts of tar than submitochondrial particles. The nicotinamide adenine dinucleotide (NADH)-ubiquinone (Q) reductase complex was more sensitive to inhibition by tar extract than the succinate-Q reductase and cytochrome complexes. Nicotine or catechol did not inhibit respiration of intact mitochondria. Treatment of submitochondrial particles with cigarette tar resulted in the formation of hydroxyl radicals.

Embryotoxic Effect

Water extract of the dried leaf, administered by gastric intubation to pregnant rats, was active. Preimplantation and implantation periods were most sensitive.

Emphysematous Influence

Smoke was administered to mice at a dose of one cigarette daily for up to 6 months. Some animals received 20 mg of human A1AT

(Prolastin) every 48 hours. The treatments produced 63% protection against increased airspace size and abolished smoke-mediated increase in plasma TNF-α. Smoke, administered by whole-body exposure to B6C3F1 mice and Fischer-344 rats at a concentration of 250 mg total particulate matter/m^3 for 6 hours/day, 5 days/week for 7 or 13 months, produced an enlargement of parenchymal air spaces in both mice and rats. The alveolar air space was increased significantly only in mice. Tissue loss was decreased at both time points in mice, but not in rats. Morphometric differences in the mice at 13 months were greater than at 7 months. Inflammatory lesions within the lungs of mice contained significantly more neutrophils than those lesions in rats. Smoke, administered to macrophage elastase-deficient (MME-/-) mice, produced no increase in the numbers of macrophages in their lungs and did not develop emphysema. Cigarette smoke, administered to male weanling rats at a dose of 20 nonfiltered commercial cigarettes/day, 5 days/ week for 6 weeks, significantly decreased vitamin A levels in serum, lung, and liver. Histological examination revealed the presence of interstitial pneumonitis along with severe emphysema. There was a significant inverse relationship between vitamin A concentration in the lung and the severity of emphysema ($p < 0.03$). Detachment or hyperplasia (and metaplasia) of the tracheal epithelium and liver vacuole formation also were evident in the smoke-treated rats. Cigarette smoke was administered to rats immunized with rabbit antineutrophil antibody or antimonocyte/macrophage antibody and exposed to cigarette smoke 7 days/week for 2 months. Specific suppression of neutrophil accumulation and neutrophil-related elastinolytic burden in the lungs of the antineutrophil antibody-treated smoke-exposed rats was observed, in contrast to specific suppression of macrophage accumulation and macrophage-related elastinolytic burden in the lungs of the antimonocyte/ macrophage antibody-treated smoke-exposed rats. Cigarette smoke exposure-induced lung elastin breakdown and emphysema in the lungs were not prevented in the lungs of antineutrophil antibody-treated smoke-exposed rats but was clearly prevented in lungs of the antimonocyte/ macrophage antibody-treated smoke-exposed rats. Results indicate that macrophages rather than neutrophils were the critical pathogenic factor in cigarette smoke-induced emphysema. Tobacco smoke was administered to weanling Wistar rats on vitamin E-depleted or normal diet for 4 weeks. The smoke induced emphysematous changes with significant increases in the mean linear intercept and the destructive index. This was supported by an increase in elastase-like activity and a decrease in elastase inhibitory capacity in bronchoalveolar lavage fluid in the normal diet group. In addition to vitamin E depletion, elastase-like activity, elastase inhibitory capacity in bronchoalveolar lavage fluid, and destructive index were comparable to that of tobacco-exposed animals on a normal diet. Mean linear intercept was markedly decreased with thickened epithelium and shrunk alveolar space.

Endogenous Formation of Tobacco-specific Nitrosamines

(S)-Nicotine and $NaNO_2$ were administered intragastrically to rats at doses of 60 μmol/kg and 180 μmol/ kg, respectively, and (S)-nicotine administered at a dose of 12 nmol/kg twice daily for 4 days. The treatments produced no metabolites of NNK;, and its glucuronide in the urine of treated rats. This indicated that endogenous conversion of nicotine to NNK did not occur. The urine contained NNN, *N*'-nitrosoanabasine (NAB), and *N*'-nitrosoanatabine (NAT). (S)-Nicotine used in this experiment demonstrated that it contained trace amounts of nornicotine, anabasine, and anatabine. (S)-Nicotine and synthetic (R,S)-nicotine with $NaNO_2$ supplement, administered to rats, produced NNN-, NAB-, and NAT-detectable levels in the urine of the rats treated with the (S)-nicotine and $NaNO_2$. NNN, but not NAB or NAT, was present in the urine of the rats treated with synthetic (R,S)-nicotine and $NaNO_2$. NNN probably formed via nitrosation of metabolically formed nornicotine. These results demonstrated that endogenous formation of tobacco-specific nitrosamines occurred in rats treated with tobacco alkaloids and $NaNO_2$.

Endothelial Dysfunction

Cigarette smoke-treated Krebs buffer was evaluated on rat aortic rings. Agonist-stimulated endothelium-dependent vasorelaxation was measured. Relaxations to receptor-dependent agonists, acetylcholine, and adenosine 5'-diphosphate (ADP), as well as to a receptor-independent agonist, A23187 (Ca^{2+} ionophore) were significantly impaired by cigarette smoke. Cigarette smoke did not impair relaxations to sodium nitroprusside, indicating preserved guanylate cyclase activity. Cigarette smoke did not affect endothelial nitric oxide synthase (NOS) catalytic activity in homogenates from endothelial cells or aortas previously exposed to cigarette smoke-treated Krebs buffer. Treatment with superoxide dismutase or ifetroban and in a lesser degree by indomethacin prevented cigarette smoke-induced endothelial dysfunction.

Enzymatic Activity

STE, administered to lactating mice and suckling neonates at doses of 50 or 100 mg/kg body weight/ day, inhibited phytic acid-induced

GST/GSH system efficiency and significantly augmented phytic acid- or butylated hydro-xyanisole-induced microsomal phase I enzymes. STE, in murine system at doses of 50 or 100 mg/kg body weight/day, elevated microsomal cytochrome b5, CYP, and malondialdehyde levels. These results indicated the inhibitory potential of STE on garlic-induced hepatic GST/GSH system besides significant augmentation on garlic-, mace- or black mustard-induced microsomal cytochromes.

Epstein–Barr Virus Early Antigen Induction

Methanol extract of the dried leaf, in cell culture at a concentration of 1 μg/mL, was inactive. The assay was designed for tumor-promoting activity. Two diastereoisomers of 2,7,11-cembratriene-4,6-diol (α- and β-CBT) from the neutral fractions of cigarette smoke condensate, in Raji cells, produced potent inhibitory effects on the induction of Epstein–Barr virus (EBV)-EA by 12-*O*-tetradecanoylphorbol-13-acetate (TPA). The doses of α- and β-CBT required for 50% inhibition of EBV-EA induction by TPA were 7.7 and 6.7 mg/mL, respectively. Application of α- and β-CBT to mouse skin before treatment with TPA, inhibited TPA-induced ornithine decarboxylase activity in a dose-dependent manner. Application of 16.5 μM/mouse of α- and β-CBT resulted in a 50% and 40% reduction, respectively, of the maximum ornithine decarboxylase activity induced as a result of treatment with TPA. In initiation-promotion experiments, α-CBT markedly inhibited the promoting effect of TPA on skin tumor formation in mice that were initiated with 7,12-dimethylbenz[a]anthracene. The β-CBT was less effective. Application of 3.3 μM of α-CBT 40 minutes before treatment with TPA (1 μg) resulted in a 53% reduction in the number of papillomas per mouse.

Estrogenic Effect

Tobacco smoke, administered to rats, induced no changes in the rat uterus weight or in oestrus cycle. It decreased estradiol (E2) concentration in the uterus tissue and increased and later decreased the proliferation index and percentage of the cells in the S-phase. The results indicated a phasic character of changes in the reproductive system under the effect of tobacco smoke and corroborated the concept of the role of smoking in the shifting the type of hormonal carcinogenesis from promotional to genotoxic. Mainstream cigarette smoke, administered to female rats aged 2.5–3 and 6 months, 2 hours/day for 3 weeks or 3 months, did not induce any changes in uterine weight or estrous cycle. It decreased estradiol (E2) concentration in uterine tissue (especially in adult rats or in young rats after 3 months of experiment). No signs of aneuploidy were found in the uterus through proliferation index. The percentage of cells in S-phase were increased by 3 weeks and decreased by 3 months of experiment.

Foot-and-mouth Disease

Tobacco transgenized with foot-and mouth disease virus (FMDV) serotype O using a recombinant tobacco mosaic viruses (TMV)F11 and TMVF14 was administrated by parental injection to guinea pigs. The treatment produced a protection against FMDV challenge in case of using TMVF11, TMVF14, or the mixture TMVF11/TMVF14, but not wtTMV. The TMVF11/TMVF14 mixture protected all animals when challenged in 150 guinea pig 50% infection dosage. Oral administration of the mixture (3 mg total) protected 3/8 guinea pigs against the same FMDV challenge. Most of the suckling mice parentally injected with antiserum from guinea pigs immunized with the TMVF11/TMVF14 mixture, but not with wtTMV, were also protected against FMDV challenge with 10 suckling mouse LD_{50}.

Gastric Mucosal Hyperemia Inhibition

Tobacco cigarette smoke, administered to rats at doses of 3 and 18 mL/minutes, significantly attenuated hyperemia and aggravated hypertonic saline-induced lesion in a dose-dependent manner. Administration of 18 mL/min of tobacco cigarette smoke and the dose of iv nicotine blocked injury-induced hyperemia. The treatment also aggravated saline-induced gastric damage and gastric mucosal damage induced by acidified aspirin or acidified ethanol.

Gastric Mucus Synthesis Inhibition

Cigarette smoke, at concentrations of 2 or 4%, was administered to intact animals and animals with ulcers. The treatment significantly reduced the thickness of the mucous-secreting layer and gastric mucosal ornithine decarboxylase activity in animals with or without ulcers. The extract significantly reduced mucous synthesis and ornithine decarboxylase activity but not its mRNA expression in MKN-28 cells. Cigarette smoke and its extract, in human MKN-28 cells, markedly decreased mucus synthesis in vivo and in vitro and suppressed ornithine decarboxylase activity.

Gastric Secretory Stimulation

Smoke of the dried leaf was administered to smoking adults with duodenal ulcers at variable doses. The treatment reduced the anti-secretory effect of cimetidine from 86 to 38% and inhibited pepsin secretion by 14% during smoking compared with 80% by nonsmokers taking cimetidine. In a study with healthy adults, smoking had no effect on cimetidine-induced inhibition of pentagastrin-stimulated gastric acid secretion. The results were significant at $p < 0.05$.

Gastric Ulcerations

Cigarette smoke was administered to rats at doses of 2 or 4% of for three 1-hour periods during a 24-hour starvation before ulcer induction. The treatment potentiated ulcer formation, which was accompanied by a reduction of gastric blood flow at the ulcer base and ulcer margin. Smoke exposure alone did not produce any macroscopic injury in the stomach but significantly decreased the basal gastric blood flow in a concentration-dependent manner, which was coupled with an increase in mucosal xanthine oxidase activity. The increment of constitutive NOS activity but not prostaglandin-E2 level was markedly attenuated by cigarette smoke exposure. Tobacco cigarette smoke and subcutaneous nicotine, administered to rats, significantly attenuated the ulcer margin hyperemia in a dose-related manner. Repeated exposure to tobacco cigarette smoke increased ulcer size in the acute and the healing stages. Subcutaneous nicotine also increased the size of ulcers in the acute stage.

Gene Expression

Aqueous extract of the smoke (smoke-bubbled phosphate-buffered saline), in Swiss 3T3 cell culture for 24 hours, produced differential gene expression, mainly antioxidant response genes, genes coding for transcription factor, cell cycle-related genes, and genes-mediators of an inflammatory/immune-regulatory response. Cigarette smoke was administered to the ODS rats at doses of 4 mg/day, S4 or 40 mg/day, S40 of ascorbic acid, and exposed to smoke daily for 25 days. The treatment produced a significant decrease of SOD, MnSOD, catalase and protein disulfide isomerase (PDI) by high-dose ascorbic acid administration, and a nonsignificant decrease of plasma glutathione peroxidase. Cigarette smoke exposure slightly increased gene expression of PDI and catalase, but not significantly. The differently expressed 27 genes in the liver were found by differential display methods. From 27 genes, altered expression of plasma proteinase inhibitor, α-1-inhibitor III and CYP1A2 were confirmed by competitive RT-PCR. Side- stream smoke was administered to male Wistar rats with a single intratracheal instillation of 2 mg of chrysotile or refractory ceramic fiber. The rats were exposed to smoke 5 days/week for 4 weeks. Administration of smoke alone increased IL-1(α) mRNA levels in alveolar macrophages. The smoke stimulated gene expression of inducible NOS in alveolar macrophages and IL-6 and basic fibroblast growth factor in lungs treated with chrysotile; IL-1 (α) in alveolar macrophages and basic fibroblast growth factor in lungs did the same in lungs with refractory ceramic fiber.

Genotoxicity

Cigarette smoke, administered to male ICR mice, produced DNA single-strand breaks measured 15, 30, 60, 120, and 240 minutes after the exposure. Fifteen minutes after the animals were exposed for 1 minute to a sixfold dilution of smoke, the effect appeared in the lungs, stomach, and liver. The damage in the lungs and liver returned to almost control levels by 60 minutes and the stomach by 120 minutes. Kidney, brain, and bone marrow DNA were not damaged. Twelve- or 24-fold smoke dilution did not produce DNA damage. Single oral pretreatment (100 mg/kg) of either ascorbic acid or α-tocopherol acetate 1 hour before inhalation, prevented single-strand breaks in the stomach and liver, while α-tocopherol acetate but not ascorbic acid significantly reduced single-strand breaks in the lung. Five consecutive days of either ascorbic acid or α-tocopherol acetate (100 mg/kg/day) pre-treatment completely prevented single-strand breaks in the lung, stomach, and liver. Smoke condensate, in two hepatoma cell lines, induced a higher frequency of micronuclei in Hepa1c1c7 cells relative to TAOc1BP(r)c1 cells, which express 10-fold less aromatic hydrocarbon receptor (AhR). Smoke condensate, administered intraperitoneally to Ahr+/+ and Ahr–/– mice at doses of 0.5–10 μg/kg/day for 3 days, produced an increase in the incidence of micronucleated reticulocytes in Ahr+/+ mice, and no increase in the null allele animals. The frequency of micronucleated erythrocytes was slightly but significantly higher in Ahr+/+ relative to Ahr–/– mice. Mainstream smoke from TOB-HT or 1R4F cigarettes, administered intranasally to male B6C3/F1 mice at concentrations of 0.16, 0.32, and 0.64 mg total particulate matter/L of air 1 hour/day 5 days/week for 4 weeks, produced an exposure-dependent increase of DNA adducts in lung and heart of animals exposed to 1R4F smoke at all concentrations. The concentration of DNA adducts in lung and heart of TOB-HT cigarette-treated mice was not significantly increased. Aqueous suspension of the acetone extract of the paste-like tobacco preparation, administered orally to mice at single or multiple doses, induced significantly high frequencies of chromosome aberrations, micronuclei, and sister chromatid exchange. Single treatment with different doses revealed a distinct dose- dependent increase of the effect. There was a significant positive correlation between time-course of chronic treatment and frequencies of micronucleated cells. Incidences of chromosome aberrations, micronuclei, and sister chromatid exchange in bone marrow cells after repeated treatment for different periods did not differ significantly from each other and from the respective single treatment data for the same dose.

Aqueous extract of Swedish moist oral snuff, in human V79 lymphocytes, induced sister chromatid exchange and chromosome aberrations, with and without metabolic activation. No induction of point mutations was detected. The methylene chloride extract produced genotoxicity, but no induction of gene mutations in V79 cells was observed. The extract did not induce of micronuclei in mice or of sex-linked recessive lethal mutations in *Drosophila melanogaster*. Smoke, in *Salmonella typhimurium* TA97a, TA100, and TA102, produced a threefold to ninefold increase in the frequency of his+ revertants. Activation by a postmitochondrial fraction from liver of rats pretreated with Aroclor-1254 or methyicholanthrene was required. Fractions from phenobarbital-pretreated or untreated rats had no effect. Vitamins A and E, but not C, inhibited the smoke-induced mutagenesis. Treatment of mice with smoke for 60 minutes/day increased the frequency of micronuclei in polychromatic erythrocytes in bone marrow and in fetal liver, and the number of micronucleated normochromatic erythrocytes in peripheral blood by four- to fivefold. Simultaneous treatment of mice with smoke and Na_2SeO_3 reduced the clastogenic effect of tobacco smoke. Ascorbic acid had no effect on clastogenicity but reduced toxicity as measured by body weight loss. NNK, administered intravenously to pregnant C57Bl mice, diffused through the placenta and reached the fetal tissues.

During the last days of gestation, nasal, pulmonary, and hepatic tissues developed the enzymatic capacity to activate NNK to alkylating species, which bind covalently to cellular macromolecules. Within 4 hours of the injection, a considerable proportion of NNK metabolites present in the fetal tissues were excreted in the amniotic fluid via the fetal urinary tract. Incubation of tissue slices with NNK indicated that the nose, the lung, and the liver of 13-day-old fetuses could reduce NNK to 4-(methylnitrosamino)-1-(3-pyridyl)butan-1-ol (NNA1), but could not activate NNK by á-carbon hydroxylation. Activating enzymes were competent in 18-day-old fetuses, and the activities increased during the first 6 days of life. NNK and NNN, in *Salmonella* TA100, TA7004, 7005, and 7006 at concentrations of 250–2000 mg/plate, produced missense backmutations. NNN was active on TA100 and TA7 004 but inactive in the presence of rat or hamster S9. NNK was mutagenic only in TA7004 strain with rat or hamster S9, but not in TA 100. NNK and NNN, in dark mutant M-169 of *Vibrio fischeri* at concentrations up to 1 mg/mL, were active (Mutatox test). Nicotine, cotinine, *trans*-3-hydroxycotinine, cotinine-*N*-oxide, and nicotine-*N*-oxide were not mutagenic to *Salmonella* TA 100 and TA7004 in the presence or absence of rat or hamster S9. The Mutatox test produced direct mutagenicity for COT, 3HC, and NNO, but not CNO. The latter was mutagenic in the Mutatox test with rat or hamster S9, but only rat S9 was effective for COT, NNO, and 3HC. Inhibitory potentiations of NNN by NIC and COT were observed on strain TA7004 and by NIC on strain TA100. There were no interactions on NNK in the presence of S9 for strain TA7004 or TA100. In contrast, a complex inhibition and enhancement behavior occurred in the Mutatox test for each interaction, but no effects were observed for CNO on NNK without S9, and few for NIC on NNK with hamster S9.

Glutathione Formation Induction

Methyl chloride extract of the leaf, administered intragastrically to rats at a dose of 3 mg/animal daily for 21 months in DMSO solvent, was active. The rats were divided into two groups: one was fed a vitamin A diet, and the other a vitamin A-deficient diet. Treatment with extract increased glutathione levels in both liver and lung in the vitamin-fed rats. In the vitamin-deficient group, hepatic and pulmonary glutathione levels were decreased by treatment with extract.

GSH Peroxidase Activity

Cigarette smoke was administered to the ODS rats at doses of 4 mg/day, S4 or 40 mg/day, S40 of ascorbic acid, and exposed to smoke daily for 25 days. The treatment produced a significant decrease of CuZn-SOD, MnSOD, catalase, and PDI by high-dose ascorbic acid administration, and nonsignificant decrease of plasma GSH peroxidase.

GST Activity

Cigarette smoke, administered intranasally to young male C57BL mice fed 0, 5, and 100 ppm of vitamin E, 20 minutes/day for 8 weeks, produced no effect GST.

GST Inhibition

Methyl chloride extract of the leaf, administered intragastrically to rats at a dose of 3 mg/animal daily for 21 months in DMSO solvent, was active. The rats were divided into two groups: one was fed a vitamin A diet, and the other a vitamin A-deficient diet. In the vitamin-deficient group, the treatment decreased GST levels in liver and lung vs control and vitamin-fed groups.

Glycation Products Formation

Reactive glycation products from an aqueous extract of tobacco and tobacco smoke reacted with proteins to form advanced glycation end products. The end products circulated in high concentrations in the plasma of patients with diabetes or renal insufficiency and have been linked to the accelerated vasculopathy seen in patients with these diseases. Glycotoxins (glycation products) exhibited a specific fluorescence when cross-linked to proteins and were mutagenic. Glycotoxins were transferred to the serum proteins of human smokers. Advanced glycation end products (AGE)-apolipoprotein (apo) B and serum AGE levels in cigarette smokers were significantly higher than those in nonsmokers.

Hematopoiesis Inhibition

Smoke, in bone-marrow culture in a long-term treatment, produced an inhibition of hematopoiesis. Nicotine significantly delayed the onset of hematopoietic loci and reduced their size, the number of long-term culture- initiating cells, nonadherent mature cells, and their progenitors but failed to influence the proliferation of committed hematopoietic progenitors when added into methylcellulose cultures. Exposure to nicotine decreased CD44 surface expression on primary bone marrow-derived fibroblast-like stromal cells and MS-5 stromal cell line but not on hematopoietic cells. Mainstream smoke altered the trafficking of hematopoietic stem/progenitor cells (HSPC) in vivo. Smoke exposure produced an inhibition of HSPC homing into bone marrow. Nicotine and cotinine treatment resulted in reduction of CD44 surface expression on lung micro-vascular endothelial cell line (LEISVO) and bone marrow-derived (STR-12) endothelial cell line. Nicotine increased E-selectin expression on LEISVO cells but not on STR-12 cells.

Heme Oxygenase Expression

Mainstream smoke trapped in phosphate-buffered saline solutions, in Swiss albino 3T3 fibroblasts, produced dose-dependent and transiently elevated expression of heme oxygenase. Heme oxygenase protein and its mRNA were detectable between 1 and 24 hours after exposure to 0.03 puffs (approx 1 cm^3/mL of medium). A nearly 50-fold increase in the amount of heme oxygenase mRNA was determined after 8 hours of exposure, compared to control levels. A decrease of more than 60% in glutathione levels was observed after the exposure. No elevated amounts of heme oxygenase mRNA appeared in smoke treated-cells when cysteine was exogenously added.

Hepatic Enzymes Activity

Tobacco smoke, administered by Hamburg II machine to mice at a dose of eight cigarettes per day for 2, 4, 8, or 31 days, significantly increased HbCO after 4 or 8 days of exposure and decreased after 31 days. The enzymate activities were significantly higher during the period of exposure.

Hepatic Lipid Peroxidation

Aqueous extract of STE activated macrophages with the resultant production of reactive oxygen species, including nitric oxide. Administration to rats at doses of 125–500 mg/kg induced dose-dependent increase in mitochondrial and microsomal lipid peroxidation, enhanced DNA single-strand breaks, and significantly increased the urinary excretion of the lipid metabolites malondialdehyde, formaldehyde, acetaldehyde, and acetone. Extract, administered orally to female Sprague–Dawley rats at a dose of 25 mg/kg daily for 105 days, increased lipid peroxidation 1.4- to 3.3-fold in hepatic mitochondria and microsome. Maximum increase in lipid peroxidation and DNA single-strand breaks occurred between 75 and 90 days of treatment. Maximum increase of urinary excretion of the four lipid metabolites malondialdehyde, formaldehyde, acetaldehyde, and acetone was observed between 60 and 75 days of treatment. Aqueous STD, administered to rats at doses of 125, 250, and 500 mg/kg, produced dose-dependent increase of 1.8, 2.3, and 4.4-fold in mitochondrial and 1.5, 2.1, and 3.6-fold in microsomal lipid peroxidation at doses tested, respectively, relative to control values. At the same three doses of the extract, 1.3, 1.4, and 2.7-fold increases in hepatic DNA single-strand breaks occurred relative to control values. Administration also resulted in significant increases in excretion of urinary metabolites. Urinary excretion of the four lipid metabolites, malondialdehyde, formaldehyde, acetaldehyde, and acetone, were increased at every dose and time point with maximum increase between 12 and 24 hours after treatment. Cigarette smoke, administered intranasally to young male C57BL mice fed 5 and 100 ppm of vitamin E, 20 minutes/day for 8 weeks, increased hepatic lipid peroxidation.

Histological Changes

Mainstream smoke from a 1R4F and 2R4F research cigarette was administered intranasally to rats at doses of 0.06, 0.20, or 0.80

mg wet total particulate matter per liter of air for 1 hour/day, 5 days/ week for 13 weeks. The treatment produced no significant differences between both tobacco types. After 13 weeks of the recovery period, there were no statistically significant differences in histopathological findings observed between the 1R4F and the 2R4F cigarettes. The complete toxicological assessment in this comparative inhalation study of 1R4F and 2R4F cigarettes suggests no overall biologically significant differences between the rats exposed to the two cigarettes. Flue-cured tobacco was administered to rats at doses of 0.06, 0.20, or 0.80 mg wet total particulate matter per liter of air for 1 hour/day, 5 days/ week, for 13 weeks. The only significant difference was increased epithelial hyperplasia of the anterior nasal cavity in males in the high- exposure group for the heat-exchanger cigarette. At the end of the exposure period, subsets of rats from each group were maintained without smoke exposures for an additional 13 week (recovery period). At the end of the recovery period, there were no statistically significant differences in histopathological findings between heat-exchanger-cured tobacco cigarette when compared to direct-fired cured tobacco cigarette.

Hypercholesterolemic Activity

Leaf, administered by inhalation to adults of both sexes at variable concentrations, was active. Serum cholesterol was higher in persons smoking cigarettes in all age groups. The number of cigarettes smoked per day had a direct correlation with the serum total cholesterol, which increased as the number of cigarettes smoked per day increased.

Hyperplasia Induction

Smoke condensate with nonpolar arotinoid, Ro 15-0778, in rodent respiratory epithelia organ culture antagonized the carcinogen-induced hyperplasia and metaplasia. In neonatal rat tracheas and fetal mouse lungs grown in vitro, 3,4-benzpyrene and cigarette smoke condensate induced an increased proliferation of epithelial cells associated with a loss of secretory activity and ciliary function. In explants pretreated with cigarette smoke condensate, Ro 15-0778 reversed the high proliferation rate and restored secretory differentiation and ciliary function. Cigarette smoke condensate, in fetal mouse lung and neonatal rat tracheas organ cultures, induced a striking increase of epithelial mitosis within 12–14 days of treatment. The increase was associated with a loss of secretory activity and of ciliary function. Administration of etretinate with smoke condensate inhibited the increase in cell division and prevented the loss of secretory activity or ciliary function. In explants pretreated with cigarette smoke condensate, etretinate reduced the smoke condensate-induced increase in mitotic activity to normal levels and restored secretory differentiation and ciliary function.

Hypoxic Pulmonary Effect

Cigarette smoke, in isolated rat lungs perfused with blood, produced no change in pulmonary vascular resistance. The hypoxic pulmonary vasoconstriction was significantly enhanced by smoking. Indomethacin, an inhibitor of prostaglandins biosynthesis, administered in the perfusing blood (20 μg/mL) increased hypoxic pulmonary vasoconstriction in the nonsmoking lungs but not in lungs after smoking. Diethylcarbamazine citrate (DEC) (1 mg/mL), an inhibitor of leukotrienes biosynthesis, decreased hypoxic pulmonary vasoconstriction before and after smoking. After perfusion with both indomethacin and DEC, hypoxic pulmonary vasoconstriction also decreased. Results indicated that leukotrienes act as mediators, whereas prostaglandins as modulators in hypoxic pulmonary vasoconstriction and prostaglandins and leukotrienes may play an important role in the increase of hypoxic pulmonary vasoconstriction by cigarette smoking. Cigarette smoke extract, administered intravenously to Wistar rats during hypoxic ventilation, produced significant decrease in microvascular internal diameter of pulmonary arterioles and venules. After pretreatment of animal with smoke extract, much more remarkable pulmonary vasoconstriction was induced by hypoxia than before injection of the extract. During hypoxia, mean pulmonary arterial pressure increased by 13.53% before administration of smoke extract and by 30.57% after administration, respectively. Results indicated that smoke extract can strengthen pulmonary vasoconstriction and hypertension induced by acute alveolar hypoxia.

Immunogenicity

Tobacco was transferred by cholera toxin B (CTB) subunit of *Vibrion cholerae* encoding gene. An aliquot of total protein from the transgenic leaf tissue, administered intradermally to Balb/c (H2K-[d]) mice at a concentration of 5 μg/mL recombinant CTB, produced CTB-specific serum IgG in animals. Macrophages isolated from mice immunized with native or plant-expressed CTB showed enhanced secretion of IL-10. The secretion of lipopolysaccharide-induced IL-12 and TNF-α was inhibited. Results indicated that plant-expressed protein behaved like native CTB regarding effects on T-cell proliferation and cytokine levels. Tobacco transgenized with recombinant FaeG protein (major subunit and adhesion of K88ad fimbriae), administered to mice, produced immunogenicity comparable to that generated with traditional approaches. Tobacco, transgenized with an anti-hepatitis B virus surface antigen mouse IgG-1 monoclonal antibody, was active. Extract of the leaf expressing Norwalk virus-like particles, administered intragastrically to CD1 mice, produced serum and secretory specific IgA. Transgenic tobacco expressing genes encoding *Escherichia coli* heat-labile enterotoxin or *Escherichia coli* heat-labile enterotoxin fusion protein, administered intragastrically to mice, resulted in production of serum and gut mucosal anti-*Escherichia coli* heat-labile enterotoxin immunoglobulins that neutralized the enterotoxin in cell protection assay. Transgenic tobacco leaf expressing recombinant hepatitis B surface antigen produced response qualitatively similar to those obtained by immunizing mice with commercial vaccine. T-cells were obtained from mice primed with the tobacco-derived recombinant hepatitis B surface antigenic peptide that represents part of the a determinant of hepatitis B surface antigen. Tobacco smoke administered to mice for 3 days, 18 or 28 weeks before sheep red blood cell (SRBC) inoculation, produced "shorter-lived" splenomegaly. Mice exposed to smoke for 3 days or 18 weeks produced a reduction in both the magnitude and the duration of the primary immune response as evidenced by the pattern of expansion of splenic white pulp and "RNA-rich" white pulp volumes. Mice exposed to smoke for 28 weeks produced white pulp and "RNA-rich" white pulp volumes similar to those of control mice. Extract of tobacco leaf and tobacco smoke components, administered to mice and rabbits, produced reaginic antibody in mice and precipitating antibody in rabbits. The results indicated that tobacco smoke extracts stimulated immune responses to tobacco leaf antigens in rabbits and mice. The immunogen is apparently not a product of incineration because air passed through unlit cigarettes clearly extracted the antigenic component.

Immunostimulatory Activity

Water extract of the dried leaf, administered to mice at a dose of 0.5%, was active on splenocytes and produced polyclonal AB response. Mouse-tobacco hybrid calli and complete plants generated by somatic cell fusion of mouse spleen cells and tobacco mesophyll protoplasts produced mouse Ig-γ-3-heavy and λ-light chains. Aqueous extract of the STD, in mouse lymphoid cells, produced a significant increase in the proliferation of spleen cells. The polyclonal IgM antibody responses were elevated in smoke-stimulated spleen cell cultures. Similar immuno-stimulatory results were found in the mesenteric lymph node cells. The extract stimulated the spleen cells of the immune defective CBA/N mice. The extract was mitogenic to B- and T-cells in the lipopolysaccharide-resistant C3H/HeJ mice spleen cells. The proliferation of T cells was not accompanied by secretion of IL-2 or expression of IL-2 receptors on T-cells. There was an increase of IL-1 activity in spleen cells. Activation of B- or T-lymphocytes did not result in the elevation of intracellular calcium levels.

Immunosuppressive Activity

Water-soluble condensate of tobacco smoke, administered to C57Bl/6 mice at sublethal doses, inhibited the ability to respond to immunization with sheep erythrocytes by the formation of plaque-forming cells. Spleen cells from water-soluble condensate- treated mice were unable to mount a primary response to SRBC in vitro. There was a decrease in T-lymphocytes in the spleens of treated mice. T-cells from water-soluble condensate-treated mice were unable to cooperate with normal B-cells and macrophages in the response to SRBC. A less marked suppression of B-cell function was noted in condensate-treated mice. Although B-cells from such animals were able to co-operate with normal T-cells and macrophages to give a detectable primary response to SRBC, the response was depressed. Macrophages from water-soluble condensate-treated animals enhanced the response of normal T- and B-cells to SRBC.

Inflammation Induction

Water extract of the leaf, administered intragastrically to mice at a dose of 4.9 g/kg 6 days per week for 5 months, was active. The extract consisted of 15 g *Areca catechu* nut, 10 g tobacco leaf, and either 3 or 10 g lime to which was added water extract of *Piper betle* leaf. Addition of the latter enhanced the effect, whereas the larger dose of lime antagonized the effect. Ether extract of the dried leaf, administered externally to mice at a dose of 10 mL, was inactive.

Insecticidal Activity

Alkaloid fraction and methanol extract of the leaf were active on *Culex pipens* larvae. Fifty or more percent lethality after 24–48 hours was obtained. Acetone extract of the dried leaf, at variable doses, produced 80.6% mortality vs snout moths larva of rice. Water extract of the dried leaf was active on *Phyllocnistis citrella*. Methanol extract of the dried leaf, at a dose of 50 mg/mL, was inactive on *Rhinocephalus appendiculatus*. Inhibition of oviposition was used as a measure of ascaricidal activity. Ethanol (95%) extract of the dried leaf, at a concentration of 50 μg/mL, was inactive on *Rhodnius neglectus*. Methanol extract of the dried root, at a concentration of 50 μg/mL, was active on *Rhipicephalus appendiculatus*. The extract of dried stem was inactive. Inhibition of oviposition was used as a measure of ascaricidal activity.

Interferon-α/β Production

Mainstream and sidestream smoke from Kentucky 2R1 reference cigarette, in murine L-929 cells were administered at the highest doses possible to generate a minimum toxic effect. The dose was then serially diluted to lower doses. The treatment produced viability of exposed cells equivalent to control cell cultures. Addition of polyriboinosinic–polyribocytidylic acid to the cells reduced the IFN production in viable smoke-exposed cells. Aging of smoke by delaying time of exposure of the cells to the smoke or filtration of smoke through activated charcoal has substantially decreased the alteration of IFN production by smoke exposure. 4-aminobiphenyl, aniline-HCl, hydrazine sulfate, and 2-methylquinoline, administered intraperitoneally to mice, induced IFN production at 2, 24, or 48 hours after treatment. Mice treated with 4-aminobiphenyl showed some depression of IFN production 2 hours after treatment. Maximum inhibition of IFN induction was observed 24 hours after treatment and a return to control levels 48 hours after treatment. Mice treated with hydrazine sulfate showed maximum inhibition of IFN induction 24 hours after treatment, but no effects at any other treatment time. Treatment of mice with aniline-HCl resulted in marginal depression of IFN induction 24 hours after treatment. 2-methylquinoline had no effect. 4-aminobiphenyl and aniline-HCl from sidestream cigarette smoke, in mouse embryo fibroblast cell cultures, produced severely reduced levels of α/β IFN after challenge with polyribo-inosinic–polyribocytidylic acid when compared to control cultures. Treatment of additional cell cultures with 2-methylquinoline and intermediate-level component of sidestream tobacco smoke or hydrazine-sulfate also resulted in inhibition of IFN induction with polyriboinosinic acid–polyribocytidylic acid.

IL-12 Activity Stimulation

STD stimulated p40 and p35 promoter activity of the IL-12 (p70), and enhanced IFN-γ-induced p40 and p35 promoter activity. It had no effect on lipopolysaccharide-induced p35 and p40 promoter activity and diminished IFN-γ/lipopolysaccharide-induced p35 promoter activity. The results indicated that tobacco extract stimulation of bioactive IL-12 production is correlated with its effect on both p35 and p40 subunits. Stimulation of IL-12 production can increase the chances of oral inflammatory disease. STD, in cell culture, decreased production of IL-12 p40 and p70 from lipopolysaccharide-stimulated peritoneal macrophages, lipopolysaccharide/IFN γ-stimulated peritoneal and splenic macrophages, and increased production of IL-12 p40 and p70 from IFN γ/CD40-stimulated splenic macrophages or IFN-γ-stimulated peritoneal macrophages. None of the effects resulted from nicotine, rutin, or chlorogenic acid. Nicotine, at a concentration of 100 μg/mL, significantly elevated production of IL-12 p40 and p70 from splenic macrophages stimulate by IFN-γ/lipopolisaccharide.

Intraepithelial Mucosubstance Effect

Diluted mainstream cigarette smoke was administered to F344 rats for 9 days for a 2-week period. After the last exposure, the treatment produced 270% more intraepithelial mucosubstances in the dorsal septum, 58% less intraepithelial mucosubstances in the midseptum, and amounts of intraepithelial mucosubstances in the ventral septum similar to controls. Smoke-exposed rats humanely killed 14 days after exposure still had increased amounts of intraepithelial mucosubstances in the dorsal septal region. There was no effect in regions of squamous metaplasia or amounts of intraepithelial mucosubstances in the midseptal and ventral septal regions that were different from air-exposed controls. Smoke exposure resulted in a significant increase in the unit length-labeling index at 1 day but not 14 days after exposure in the ventral and midseptal regions only.

Irradiation-induced Pneumonitis Suppression

Diluted mainstream tobacco smoke was administered intranasally to rats at a concentration of approx 0.4 mg/L for 1 hour/day, 1–5 days/week for 10 weeks, 3 weeks before irradiation. The treatment produced less inflammation in the alveolar tissue in the irradiated smoke-exposed group than in the irradiated not exposed to smoke group. Mast cells were increased 100-fold in the lung interstitium and 30-fold in the peribronchial area in the irradiated not exposed to smoke group, whereas no increase was found in the irradiated smoke-exposed group or in the controls. The alveolar septa of the irradiated not-exposed group were thickened, with occurrence of inflammatory cells and mast cells, whereas the irradiated smoke-exposed group displayed no difference as compared to controls.

L-Ascorbic Acid Influence

Sidestream cigarette smoke, administered to male Wistar rats for 2 hours daily for 25 days, produced an increase of the excreted amount of L-ascorbic acid in the urine. At the end of the experimental period, the L-ascorbic acid content of the plasma and tissues, liver cytochrome

P450 content, and the activities of drug-metabolizing enzymes in the test group were higher than the control.

Leukocyte Dynamics

The effect of chronic smoking on the microcirculation immediately after exposure to smoke for 2, 4, and 6 weeks and after withholding smoke for 2 weeks from those previously exposed for 4 weeks was investigated. The mean rolling leucocytes at 2, 4, and 6 weeks were 11.10 ± 1.8, 23.7 ± 2.3, 40.2 ± 3.9 ($p < 0.001$). The rolling leukocytes, after smoking for 4 weeks and then having smoke withheld for 2 weeks, was 9.6 ± 1.4. The mean adherent leucocytes were 5 ± 0.7, 7.5 ± 1.1, 12.6 ± 1.8 ($p < 0.001$). The adherent leucocytes, after smoking for 4 weeks and then having smoke withheld for 2 weeks, was 3.5 ± 0.5. The results confirmed those of many previous studies of the adverse effects of cigarette smoking and that those deleterious effects are time-dependent. The reversibility of the deleterious effect of cigarette smoking after cessation of cigarette smoking before facelift or flap reconstruction is at least 2 weeks. This information is also important for clinical management of patients who smoke and are scheduled for face-lift and flap reconstruction. Two weeks without cigarettes is a necessary period for successful elective plastic surgery.

Lipemic Activity

Cigarette smoke extract, in macrophages with LDL culture at a dose of 100 μg/mL, stimulated cholesteryl oleate synthesis approximately equal to 12.5-fold. Enhancement in cholesteryl ester synthesis was dependent on the concentration of smoke-modified LDL and exhibited saturation kinetics. There was extensive fragmentation of apo B. This LDL modification depended on the incubation time and concentration of the smoke extract. Superoxide dismutase inhibited LDL modification by 52%, suggesting that superoxide anion is involved. The results indicated that smoke extract alters LDL into a form recognized and incorporated by macrophages. Environmental tobacco smoke, administered to rats, produced an increase in the rate of LDL accumulation. LDL accumulation was primarily dependent on LDL interaction with environmental tobacco smoke–plasma rather than the interaction of environmental tobacco smoke–plasma with the artery wall.

Lung Aryl Hydrocarbon Hydroxylase Activity

Cigarette smoke, administered intranasally to young male C57BL mice fed 5 and 100 ppm of vitamin E, 20 minutes/day for 8 weeks, produced no effect on hepatic aryl hydrocarbon hydroxylase. All of the mice on the vitamin E-free diet showed reduced lung aryl hydrocarbon hydroxylase activity. Lung aryl hydrocarbon hydroxylase activity was increased in all of the smoke- exposed mice.

Lymphocyte Viability and Proliferation

Acetaldehyde, benzene, butyraldehyde, isoprene, styrene, and toluene in mouse lymphocytes cell culture for 3 hours produced no effect on either viability or proliferation. Formaldehyde, catechol, acrylonitrile, propionaldehyde, and hydroquinone significantly inhibited T-lymphocyte and B-lymphocyte proliferation, inhibitory concentration $(IC)_{50}$ 1.19×10^{-5} *M* to 8.20×10^{-4} *M*. Acrolein and crotonaldehyde inhibited T-cell and B-cell proliferation and acted on viability with IC_{50} 2.06×10^{5} *M* to 4.26×10^{-5} *M*. Mixtures of acrolein, formaldehyde, and propionaldehyde or crotonaldehyde interactive effects at 0.5 and $1 \times IC_{50}$ were observed.

Malignant Cell Transformation

Smoke of cured leaf, administered to leaf-cutter ants at an undiluted concentration, was active on primary spermatocytes.

MAO Inhibition

2-Naphthylamine from smoke, in cell culture, inhibited mouse brain MAO A and B by mixed competitive- and noncompetitive-type inhibition.

Mastocytoma Induction

Cigarette smoke condensate suspensions ("tars") from different cigarettes, administered to female CD-1 mice, produced cutaneous mastocytomas accompanied by diffuse dermal mast cell infiltration[NT145]. Cigarette smoke condensate suspensions ("tar"), administered to CAF1/J and ARS-HA (ICR) female mice on long-term application, produced a significant incidence of cutaneous mastocytomas. The skin mastocytomas were constantly accompanied by diffuse dermal mast cell infiltration, which was also seen in the tumor-free skin of the "tar"-treated mice.

Metabolizing Enzymes Induction

Masheri, a pyrolyzed tobacco product, administered orally to Swiss mice, Sprague–Dawley rats, and Syrian golden hamsters at a dose of 10% diet for 20 months, produced a significant induction of CYP and benzo(a) pyrene hydroxylase in proximal and distal parts of the three species. GSH and GST were depleted on masheri treatment in all three species only in proximal and distal parts of the intestine.

Metaplasia Induction

Smoke condensate with nonpolar arotinoid, Ro 15-0778, in rodent respiratory epithelia organ culture, antagonized the carcinogen-induced hyperplasia and metaplasia. In neonatal rat tracheas and fetal mouse lungs grown in vitro, 3,4-benzpyrene and cigarette smoke condensate induced an increased proliferation of epithelial cells associated with a loss of secretory activity and ciliary function. In explants pretreated with cigarette smoke condensate, Ro 15-0778 reversed the high proliferation rate and restored secretory differentiation and ciliary function. Cigarette smoke condensate, in fetal mouse lung and neonatal rat tracheas organ cultures, induced a striking increase of epithelial mitosis within 12–14 days of treatment. The increase was associated with a loss of secretory activity and of ciliary function. Administration of etretinate with smoke condensate inhibited the increase in cell division and prevented the loss of secretory activity or ciliary function. In explants pretreated with cigarette smoke condensate, etretinate reduced the smoke condensate-induced increase in mitotic activity to normal levels and restored secretory differentiation and ciliary function.

Metaplasia Induction

Water extract of the leaf, administered intragastrically to mice at a dose of 4.9 g/kg, was active. The extract consisted of 15.0 g *Areca catechu* nut, 10 g tobacco leaf, and either 3 or 10 g lime to which was added water extract of *Piper betle* leaf. Cigarette smoke, administered to rats at a dose of 250 mg total particulate matter/m^3 6 hours/day for 5 days, increased the number of small mucous cells in the respiratory epithelium of the nasal septum in the early stages of squamous differentiation, but they were gradually replaced by squamous metaplastic cells. At 5 days after the withdrawal of cigarette smoke exposure, the morphology of the midseptal epithelium returned to that of a pseudostratified mucociliary epithelium and the epithelia lining the maxilloturbinates to that of a transitional epithelium.

Mitochondrial ATPase Inhibition

Smoke extract, in the mouse brain mitochondria culture in the presence or absence of vitamin C for 60 minutes, inhibited mitochondrial ATPase and cytochrome C oxidase activities in a dose-dependent manner. The effect of extract on mitochondria swelling response to calcium stimulation was dependent on calcium concentrations. The extract treatment induced mitochondrial inner membrane damage and vacuolization of the matrix, whereas the outer mitochondrial membrane was preserved. Nicotine produced no significant damage.

Mitogenic Activity

Water extract of the dried leaf, administered to mice at a concentration of 0.05%, was active on lymphocytes from mesenteric lymph node and lymphocytes B and T. Intracellular Ca^{2+} level was unchanged. The extract was active on splenocytes. There was an effect in cells from strains irresponsible to lipopolysaccharide.

Mitotic Effect

STE, administered to the buccal mucosa of 15 female HMT rats, 6 months of age, weekly for 1 year, produced hyperorthokeratosis, acanthosis, numerous binucleate spinous cells, and subepithelial connective tissue hyalinization. Verrucous carcinoma and squamous cell carcinoma were not seen. Karyotyping revealed that lymphocytes of tobacco-treated, as well as control rats, had normal chromosome number and morphology. However, approx 25% of buccal epithelial cells of the tobacco-treated rats were tetraploid and 5% octaploid, compared with only 11% tetraploid and no octaploid in the controls. Results indicated that the mitotic process could be disturbed by tobacco treatment.

Molluscicidal Activity

Water extract of the dried leaf, at a concentration of 168 ppm, produced equivocal effect on *Lymnaea luteola*.

Morphologic and Pathological Changes

Cigarette smoke inhalation and hydrocortisone acetate (HCA), administered to C57BL/6 male mice, induced in the lungs a marked reduction of pulmonary macrophage population that is normally elevated by smoke inhalation, an accumulation of surfactant and flocculent material in alveoli, a decrease in alveolar space surrounded by normal septal tissue, and an increase in hypertrophied alveolar parenchyma. Concomitant with altered lung morphology, lung volume and gas diffusing capacity were significantly compromised. Smoke inhalation or HCA administration alone had no ill effects. Tobacco application (snuff water extract or smoking tar condensate) and herpes simplex virus (HSV)-1 inoculation, administered to mice for 2 months, produced epithelial dysplasia and other histomorphologic changes (hyperkeratosis, increased granular cell layer thickness, acanthosis, and increased inflammatory cell infiltration) in a significant number of animals. Tobacco or HSV-1 when administered alone did not induce dysplasia in the epithelium of labial mucosa. The result indicated that HSV-1 and tobacco could possibly act synergistically in the development of precancerous oral lesions and oral cancer. Fresh smoke, administered to Balb/c mice at a daily equivalent of 30 high-tar filtered cigarettes for up to 95 weeks, produced an induction or production of significant numbers of malignant tumors of several types. Significant histopathological changes that consisted mainly of interstitial pneumonia and focal low-grade emphysema were observed. Alveolar and bronchiolar spaces in the lungs of cigarette smokers contained numerous macrophages with pigmented cytoplasmic granules resulting from increased numbers of lysosomes and phagolysosomes and "smokers' inclusions" in the interstitial and alveolar macrophages of cigarette users.

The inclusions have been referred to as "needle-shaped" and "fiber-like." Thin sectioning techniques impart varying lengths to the inclusions, suggesting that they have a disc, or platelet, configuration. Surgically resected lung tissue contained varying numbers of hexagonal plate-like particles that had features consistent with those of the aluminum silicate kaolinite, and energy-dispersive X-ray spectrometry confirmed the presence of these two elements. The origin of aluminum silicate inclusions in pulmonary macrophages has yet to be determined, although preliminary evidence strongly suggests that they were derived from inhaled tobacco smoke. A mixture of mainstream and sidestream cigarette smoke, administered by whole-body exposure to Sprague–Dawley rats for 28 days, produced dramatic alterations of DNA adducts in bronchoalveolar lavaged cells, tracheal epithelium, lung, and heart. Oxidative damage to pulmonary DNA, hemoglobin adducts of 4-aminobiphenyl and benzo(a)pyrene-7,8-diol-9, 10-epoxide, micronucleated and polynucleated alveolar macrophages, and micronucleated polychromatic erythrocytes in bone marrow were also observed. Mainstream smoke from an Eclipse and 1R4F reference cigarettes, administered to Sprague–Dawley rats of each gender at concentrations of 0, 0.16, 0.32, or 0.64 mg wet total particulate matter/liter air for 1 hour/day, 5 days/week for 13 weeks, produced a decrease of respiratory rate at all concentrations of 1R4F smoke and at the high concentration of Eclipse smoke. Tidal volume was depressed and minute volume was lower for all smoke-exposed rats.

Carboxyhemoglobin and serum nicotine were directly related to the exposure concentrations of carbon monoxide and nicotine in an exposure-dependent manner. Body weights were slightly lower in smoke-exposed rats. The only treatment- related effect found in organ weights was an increase in heart weight in females in the Eclipse high-concentration exposure group. Nasal epithelial hyperplasia and ventral laryngeal squamous metaplasia were noted after exposure to either the 1R4F or Eclipse smoke. The degree of change was less in Eclipse smoke-exposed rats. Lung macrophages were increased to a similar extent in the Eclipse and 1R4F smoke-exposed groups. Brown/gold pigmented macrophages were detected in the lungs of rats exposed to 1R4F smoke, but not those exposed to Eclipse smoke.

mRNA Expression

Cigarette smoke was administered to intact animals and animals with ulcers at concentrations of 2 or 4%. The treatment significantly reduced the thickness of the mucous secreting layer and gastric mucosal ornithine decarboxylase activity in animals with or without ulcers. The extract significantly reduced mucus synthesis and ornithine decarboxylase activity but not its mRNA expression in MKN-28 cells.

Murine CD4 T-cell Costimulatory Counterreceptors

STD, in splenic mononuclear cells at $1{:}10^2$ to $1{:}10^3$ dilutions or 1–100 μg/ mL nicotine for 48 and 72 hours of stimulation with anti-CD3, produced an increase of percentage and intensity of CTLA-4 expression and a decrease of CD28 expression during an exposure to a $1{:}10^2$ dilution. Exposure to nicotine decreased the percentage of CD4+ T-cells expressing both CD28 and CTLA-4 and decreased the intensity of CD28 expression. Responding T-cells exposed to nicotine produced significantly less Th1 cytokines, IL-2 and IFN-γ, but significantly more Th2 cytokines, IL-4, and IL-10. Cytokine specific mRNA expression was only slightly affected by the exposure to nicotine.

Murine Embryopathy

NNK, administered intraperitoneally to pregnant CD-1 mice during organogenesis at a dose of 100 mg/kg, produced open eye and one case of a cleft palate in 3 of 374 fetuses, which were not observed in 160 controls. With phenobarbital plus NNK, two fetuses had a cleft palate, two had exencephaly and one had a kinky tail, although phenobarbital controls showed no anomalies. NNK-initiated fetal postpartum lethality was enhanced by phenobarbital pretreatment. There were no fetal skeletal anomalies or alterations in resorptions or fetal body weight in any group. In embryo culture, gestational day 9.5 embryos exposed to and 10 μM of NNK had decreased yolk sac diameter, crown- rump length and somite development, and 100 μM of NNK decreased anterior neuropore closure and crown-rump length ($p < 0.05$). Embryos exposed to 100 μM of NNK were assessed for K-*ras* codon 12 mutations and none was detected.

Mutagenic Activity

Smoke of cured leaf, in broth culture was active on *Salmonella typhimurium*. Environmental cigarette smoke, administered by whole-body to p53 mutant (UL53-3 x A/J)F_1 mice of both genders for up to 9.5 months, produced similar oxidative DNA damage in lung and heart of both mutants and wild-type littermate controls, proliferation of the bronchial epithelium, and levels of p53 oncoprotein as assessed after exposure for 28 days. Smoke- exposed mutant mice underwent a lower induction of apoptosis in bronchial epithelium, a greater formation of DNA adducts in lung and heart, and a more intense cytogenic damage. At the end of experiment, DNA adducts were not repaired in either wild-type or mutant mice after discontinuing exposure to smoke for 1 week. A weak but significant increase of lung tumor incidence and multiplicity was induced in p53 mutant mice after exposure to smoke for either 5 months. No tumorigenic effect was observed in their wild-type controls, carrying a 99.9% A/J background and 5% FVB genome. Leaf, administered intragastrically to rats at a dose of 150 mg/animal twice daily 5 days-week for 15 weeks, was active on natural-killer (NK) cells. Activity of peripheral NK cells was assayed by lysis of YAC-1 lymphoma cells. Smoke condensate, in combination with a 2 Gy dose of ã-rays, in C3H 10T1/2 cell system, induced a toxicity and transforming response that was largely additive in nature. Similar additive modes of interaction were observed when smoke condensate was combined with He4 ions at a dose that was equivalent in cell death to that used for γ-rays. Smoke, at a concentration of 240 cm^3 for 1 or 5 minutes, was active on *Salmonella typhimurium* TA98 in an S9 mix-type-dependent manner. Black and brown masheri, a pyrolyzed tobacco product, were active on *Salmonella typhimurium* TA98 with metabolic activation and on V79 Chinese hamster cells producing 8-azaguanine resistant mutations. Both tobacco varieties induced statistically significant increase in micronuclei formations compared to the solvent controls and structural chromosomal aberrations in bone marrow cells of mice. Smoke, at concentrations of 120–480 cm^3 in a 16-1 glass chamber, for 1–10 minutes of exposure and activated by S9 mix, induced a threefold to ninefold increase of spontaneous His+ reversion mutation rate in *Salmonella typhimurium* TA98, but not TA97a, TA100 and TA102. Smoke, administrated to BDF1 mice at a dose of 600 cm^3, two exposures of 30 minutes each, produced a twofold dose-dependent elevation of the number of micronucleated polychromated erythrocytes in bone marrow. No cumulative effect was detected when mice were treated with tobacco smoke for 2 to 28 consecutive days.

The effect observed 24 hours after tobacco-smoke exposure was abolished 48 hours later. Tobacco smoke (180 or 360 cm^3), passed through the culture medium (with or without S9 mix) of human peripheral lymphocytes, did not increase the spontaneous rate of UDS. Masheri extract was highly mutagenic in the presence of an exogenous metabolic system in the Ames test and in the micronucleus test, in a dose-dependent manner. It also induced 8-azaguanine-resistant mutants in Chinese hamster V79 cells. Dermal administration produced weak carcinogenic effect in Swiss nude mice. The saliva of masheri users produced high levels of NNN (14–43 ppb) and NPYR (2.2–8.3 ppb). The condensates collected after pipe smoking of a natural tobacco and a cavendish type tobacco produced an increase of the number of revertants induced with cavendish type tobacco on *Salmonella typhimurium* TA100 and TA98 in the presence of S9 activation in both strains compared to the natural tobacco.

The increase in the number of revertants (approximately three times) was found when the tobacco was smoked after paper wrapping "savers". Alcoholic extract of the chewing tobacco, in *Salmonella typhimurium* TA98 culture, was active after activation with S9 mix. Extract induced 8-azaguanine-resistant mutation in V79 Chinese hamster cells. Urine concentrates from smokers, chewers, and nonsmokers, in *Salmonella typhimurium* TA1538 with metabolic activation by S9, produced an effect by cigarette and bidi smokers' urine, whereas nonsmokers' urine was devoid of mutagenic effects. Urine of tobacco chewers produced variation in its mutagenic potential. Bidi smokers' urine showed maximum activity. Bidi (Indian cigarette) smoke condensate, produced frame-shift mutations in *Salmonella typhimurium* TA98 and TA1538, and induced 8-azaguanine-resistant mutations in V79 Chinese hamster cells in the presence of S9 mixture. Administration to Swiss mice induced elevated frequencies of micro-nucleated erythrocytes in the bone marrow. A crude alcoholic extract of tobacco containing *N*-nitrosonornicotine and NNK, in histidine-deficient *Salmonella typhimurium* TA98 in the presence of 9000 × g supernatant fraction, was active.

NK-cell Activity

Cigarette smoke, administered to mice preimmunized with a sublethal infection of influenza virus daily for 36 weeks, mounted a secondary immune response of normal height on subsequent challenge with the homologous virus strain. The response was less specific than that elicited in control mice, with high titers of cross-reacting antibody by hemagglutination inhibition to the following strain in the same antigenic series. Return of antibody to the previous level in the antigenic series was not observed. Snuff, administered orally to male adult rats for 15 weeks, significantly decreased NK-cell activity in peripheral blood against murine NK-cell-sensitive target cells (YAC-1 lymphoma).

Nervous System Development

Water-soluble substances of cigarette smoke and combined effect of smoke and high temperature was investigated in pregnant rats at the 8th–11th day of gestation. The treatment produced changes of all indices correlated with embryonic nervous system development and morphological differentiation, with an apparent dose-effect relationship ($p < 0.01$).

Neuronal Acetylcholine Receptor Blockade

Cembranoids ([4R]-2,7,11-cembratriene-4-6-diol and its diastereoisomer 4S), in SH-EP1-h$\alpha 4\beta 2$ cell line heterologously expressing human $\alpha 4\beta 2$-nicotine acetylcholine receptors (nAChRs), the SH-SY5Y neuroblastoma line naturally expressing human ganglionic $\alpha 3\beta 4$-nAChRs, and the TE671/RD cell line naturally expressing embryonic muscle $\alpha 1\beta 1\gamma\delta$-nAChRs, blocked carbamylcholine-induced (86)Rb(+) flux with IC_{50} in the low micromolar range. Tobacco ([4R]-2,7, 11-cembratriene-4-6-diol and its diastereoisomer 4S) cembranoids blocked binding of the noncompetitive inhibitor [^{3}H]tenocyclidine to nAChRs from *Torpedo californica* electric organ. IC_{50} values were in the submicromolar to low-micromolar range, with (4R)-2,7,11-cembratriene-4-6-diol displaying an order of magnitude higher potency than its diastereoisomer, 4S. Presynaptic nAChRs mediated a calcium influx that enhanced the release of both glutamate and γ-aminobutyric acid (GABA). Fura-2 detection of calcium in single mossy fiber presynaptic terminals indicated that nAChRs directly mediated a calcium influx. In hippocampal neurons in primary culture, both spontaneous vesicular release and evoked release of glutamate and GABA were enhanced by nicotine. The nicotinic current displayed rapid desensitization kinetics, and the response to nicotine was inhibited by α-bungarotoxin and methylcaconitine, suggesting that nAChRs containing the α-7 subunit mediated the effect. Modulation of synaptic activity by presynaptic calcium influx may represent a physiological role of acetylcholine in the brain, as well as a mechanism of action of nicotine.

Neuroprotective Effect

2,3,6-trimethyl-1,4-naphthoquinone, administered to C57 BL/6 mice, protected against MPTP Parkinson's disease-mediated depletion of neostriatal dopamine levels and lowered the brain MAO activity. Smoke, administered to *N*-methyl-4-phenyl-1,2,3,6-tetrahydropyridine-treated C57 black mice, produced no protective effect. Mainstream of cigarette smoke from 15 Kentucky 2R1F research cigarettes (28.6 mg tar, 1.74 mg nicotine/cigarette), administered to kainic acid-treated rats for 10 minutes daily, 6 days/week, for 4 weeks, indicated pre-exposure to smoke significantly reduced the seizures, mortality, and severe loss of cells in regions CA1 and CA3 of the hippocampus after kainic acid administration and attenuated the kainic acid-induced increased *Fos*-related antigen immunoreactivity in the hippocampus.

In contrast, pretreatment with central nicotinic antagonist, mecamylamine (2 or 10 mg/kg, intraperitoneally) blocked the neuroprotective effects mediated by smoke in a dose-dependent manner. Results indicated that smoke exposure provided neuroprotection against the kainic acid insulted via nicotinic receptor activation. 1,2,3,4-tetrahydro-β-carboline (THβC), an endogenous or environmental neurotoxic factor putatively involved in the development of Parkinson's disease, reacted in vitro with some components of cigarette smoke. Significant differences in the recovery of some of THβC-derivatives were obtained for Burley and Bright tobacco. Several of the reported compounds showed reversible and competitive MAO-A inhibitory properties.

The detection of some of these compounds in rat brain after chronic administration of THβC and a solution of cigarette smoke proved that the reported interactions also occur in vivo. 1,2,3,4-tetrahydroisoquinoline (TIQ) and some components of tobacco smoke were investigated for their ability to inhibit rat brain MAO. 1-Cyano-TIQ (1CTIQ), *N*-(1'-cyanoethyl)-TIQ (CETIQ), *N*-(1'-cyanopropyl)-TIQ (CPTIQ), and *N*-(1'-cyanobutyl)-TIQ (CBTIQ) acted as competitive inhibitors for both MAO-A and MAO-B. K_i values ranged from 16.4 to 37.6 μM. *N*-(Cyanomethyl)-TIQ (CMTIQ) was not to be an inhibitor ($K_i > 100.0\ \mu M$). 1,2,3,4-TIQ, a presumed proneutrotoxin linked with Parkinson's disease, interacted with some components of cigarette smoke.

The in vitro formation of these compounds under physiological conditions occurred rapidly and with a high yield. Significant differences in the recovery of the different compounds were obtained for Burley tobacco compared to Bright tobacco. After chronic administration of TIQ and a solution of cigarette smoke to rats, the presence of some of these compounds was also detected in the brain.

Nicorandil Kinetics

Cigarette smoke, administered to rats at a dose of 10 mg/kg for 8 minutes, produced the maximum nicorandil plasma levels in the rats inhaling standard cigarette and nicotine-less cigarette smoke 4.7 and 4.9 μg/mL, respectively, after 1–2 hours, compared to the controls. Nicorandil plasma level reached the maximum (7.6 μg/mL) after an hour and then decreased gradually.

NOS Activity

Cigarette smoke or smoke extract, administered to ulcerated rats once daily for 3 days, produced concomitant and dose-dependent reduction of angiogenesis and constitutive NOS activity. The same treatments also delayed ulcer healing. Sidestream smoke was administered to male Wistar rats with a single intratracheal instillation of 2 mg of chrysotile or refractory ceramic fiber. The rats were exposed to smoke 5 days/week for 4 weeks. Administration of smoke alone increased IL-1[α] mRNA levels in alveolar macrophages. The smoke stimulated the gene expression of inducible NOS in alveolar macrophages and IL-6 and basic fibroblast growth factor in lungs treated with chrysotile.

Non-Hodgkin's Lymphoma

The 1450 cases of non-Hodgkin's lymphoma (NHL) and 1779 healthy controls from 11 Italian areas with different demographic and productive characteristics were included in a study, corresponding to approx 7 million residents. Odds ratios (ORs) adjusted for age, gender, residence area, educational level, and type of interview were estimated by unconditional logistic regression model. A statistically significant association (OR = 1.4, 95% confidence interval [CI] 1.1–1.7) was found for blond tobacco exposure and NHL risk. A dose-response relationship was limited to men younger than 52 years. Subjects starting smoking at an early age showed a higher risk in men younger than 65 years, whereas no clear trend was evident for the other age and gender subgroups. The analysis by Working Formulation categories showed the highest risks for follicular lymphoma in blond (OR = 2.1, 95% CI 1.4–3.2) and mixed (OR = 1.8, 95% CI 1.1–3) tobacco smokers and for large cell within the other Working Formulation group (OR = 1.6, 95% CI 1.1–2.4) only for blond tobacco.

NF-κB Inhibition

Aqueous extract of mainstream smoke (smoke-bubbled phosphate-buffered saline), in Swiss 3T3 cells, decreased DNA binding of NF-κB during the first 2 hours of exposure and increased more than twofold over controls after 4–6 hours of exposure. There was lack of phosphorylation and degradation of IκB-α and a significant increase in thioredoxin reductase mRNA after 2–6 hours of exposure. Results indicated that the activity of NF-κB in smoke-treated cells was subject mainly to a redox-controlled mechanism dependent on the availability of reduced thioredoxin rather than being controlled by its normal regulator, IκB-α.

Oral Tumorigenic Effect

Snuff, administered to mice in the diet at concentrations of 25% gradually decreasing to 5% in a 14-month study, did not increase the tumor incidence. Administration to rats at a concentration of 5% for 18 months produced no result. Administration to hamsters at a dose of 20% for 2 years, produced forestomach tumors. Snuff, administered orally to hamster, produced a high incidence of squamous cell carcinomas. No carcinogenic activity was observed when snuff was inserted into the cheek pouch of the hamster or spread over the oral mucosa. This negative result was obtained in numerous experiments whether snuff was applied once only and left in place for several months or inserted repeatedly for up to 2 years. In the rat, a few tumors were observed when snuff was inserted into the artificial lip canal[NT052]. Tobacco smoke tar condensate or water-extract of snuff was administered dermally to mice in upper lips inoculated with latent HSV, for 2–3 months. Tar condensate induced reactivation of latent HSV in the ganglia of 10–20% of the animals, but snuff extract did not. The infectious virus was also detected in the lips after chronic application of tar condensate in 10% of the animals. Three months' exposure to tobacco produced epithelial dysplasia and other changes in a significant number of latent HSV-infected mice. Tobacco alone did not induce dysplasia in the labial epithelium of uninfected mice. Nicotine, administered by surgically created canals in the mandibular lips of male Sprague–Dawley rats twice daily for 6 weeks, decreased the thromboxane B2 levels in nicotine treated tissues. Within the nicotine group, thromboxane B2 concentrations were lower at the nicotine site compared to the posterior site (18.3 ± 5.4 pg/mg). There was also a trend toward reduced 6-keto-PGF1α in the nicotine-treated tissues compared to saline-exposed sites. These alterations in cyclo-oxygenase metabolites were not accompanied by changes in epithelial proliferation or histologic parameters. 12(S)-hydroxyeicosate traenoic acid and leukotriene B4 were not affected by nicotine.

Ornithine Decarboxylase Activity

Cigarette smoke was administered to intact animals and animals with ulcers at concentrations of 2 or 4%. The treatment significantly reduced the thickness of the mucous secreting layer and gastric mucosal ornithine decarboxylase activity in animals with or without ulcers. The extract significantly reduced mucus synthesis and ornithine decarboxylase activity but not its mRNA expression in MKN-28 cells. Cigarette smoke and its extract, in human MKN-28 cells, markedly decreased mucus synthesis in vivo and in vitro and suppressed ornithine decarboxylase activity.

Ovarian Toxicity

Cigarette smoke, administered to pregnant C57BL/6 and DBA/2 inbred strain mice during days 1–18 of pregnancy, did not affect the number of primordial follicles in the ovaries of the mothers. Counts were significantly decreased (31%) in the ovaries of DBA/2 offspring, and not significantly (20%) decreased in the ovaries of C57BL/6 offspring.

Oxidative Stress

Smoke, administered to mice for 10 weeks, produced an increase in the levels of lipid peroxidation in the heart, catalase activity, and a decrease in glutathione level. Superoxide dismutase and heat-stable lactate dehydrogenase in serum activities were not affected. Smoke, administered to mice, produced a significant decrease in total antioxidant capacity in bronchoalveolar lavage fluid and significant changes in oxidized glutathione, ascorbic acid, protein thiols, and 8 epi-PGF(2α). Treatment with smoke induced a 50% decrease in the inhibitory activity of human recombinant SLPI. Peroxynitride formed by smoke in aqueous solutions, in Swiss 3T3 cells, sustained c-*fos* expression was obtained for smoke-bubbled PBS, peroxynitrite itself, and a compound known to stoichiometrically release superoxide and nitric oxide (NO) (3-morpholino-sydnonimine [SIN-1]). c-*fos* expression in cells exposed to aqueous smoke fractions was inhibited by either the superoxide-scavenging enzyme superoxide dismutase, in combination with catalase, or the NO-scavenger oxyhemoglobin (HbO_2). Activation of guanylate cyclase in rat lung cells was observed only when bubbling was performed with filtered smoke and with whole smoke in the presence of SOD/catalase. STD, in macrophage J774A.1 cell culture, produced an increase of lactate dehydrogenase in concentration- and time-dependent manner. The addition of 250 μg/mL dose produced 2.9-fold increase in the release of lactate dehydrogenase. Tobacco smoke condensate, in rat pulmonary microvascular endothelial cell culture at a concentration of 20 μg/mL, significantly upregulated xanthine dehydrogenase/oxidase activity after 24 hours of exposure. Longer exposure (1 week) to a lower concentration of smoke (2 μg/mL) also produced an increase in xanthine dehydrogenase/oxidase activity. Unlike hypoxia, smoke treatment did not alter the phosphorylation of xanthine dehydrogenase/oxidase, but increased XO mRNA expression and the xanthine dehydrogenase/oxidase gene promoter activity. Actinomycin D blocked the activation of xanthine dehydrogenase/oxidase by smoke concentrate.

Gas-phase cigarette smoke, in rat lung tissue, produced infiltration of the terminal bronchioles by lymphocytes in the peribronchiolar region and a mild to moderate degree of emphysema in the alveolar spaces. The terminal bronchioles also produced marked lipid peroxidation, dilatation, and peribronchiolar fibrosis. The expression of inducible NOS, NF-κB, mitogen-activated protein kinases (MEK1, ERK2), phosphotyrosine protein, and c-*fos* was increased in the terminal bronchioles but protein kinase C (PKC), MEKK-1, c-*jun*, p38 and c-*myc* was not changed. Smoke, administered to rats once only or daily for 1,2,7, or 28 days, did not change NOS-1 gene expression and protein levels. Levels of NOS-2 expression was twofold higher in smokers at day 1 and decreased to control values during 1 month with daily smoke exposure, whereas protein levels did not change. NOS-2 was diffusely expressed in the lung parenchyma, airways, and vessels. NOS-3 expression was increased approx 35% after 2 days of smoke exposure and remained increased to 28 days, whereas protein levels were increased by approx 60% at day 7 and remained elevated. NOS-3 was strongly expressed in vascular endothelium. Protein distribution was identical to mRNA tissue distribution, and these distributions were not changed by smoke.

p53 Mutations

Four head and neck squamous cell carcinomas from three patients using snuff and one patient not using snuff showed p53 mutations in tumors resected from two of three patients using snuff. No p53 mutations were observed in the tumor from the patient not using snuff. No K-*ras* (codons 12 and 13) or H-*ras* (codon 12) mutations were found in any of the tumors. Smoke condensate, in sarcoma derived from Rat-1 cells transformed by smoke condensate-treated human fetal lung DNA, produced overexpression of p53 and contributed to the initiation of human lung carcinogenesis.

Pancreatic Effect

Cigarette smoke, administered to anesthetized rats alone or in combination with iv ethanol infusion, reduced pancreatic blood flow temporarily and increased leukocyte–endothelium interaction (roller $p < 0.001$, sticker $p < 0.01$ vs baseline). Cigarette smoke potentiated the impairment of pancreatic capillary perfusion caused by ethanol, and both the number of rolling leukocytes and myeloperoxidase activity levels were increased compared with ethanol or nicotine administration alone. Tobacco-specific nitrosamines, administered to rats, induced pancreatic acinar cell and ductal cell neoplasms. One of the tumors had a mixed ductal-squamous-islet cell components.

Periodontal Disease

A total of 1085 people who smoke were examined for periodontal status. There was a significant dose-effect relationship between the exposure to tobacco smoke and the extent of periodontal disease assessed as attachment loss and tooth loss. There was a gene–environmental interaction. Subjects bearing at least one copy of the variant allele 2 at positions IL-1A-889 and IL-1B+3954 had an enhanced smoking-associated periodontitis as compared with their IL-1 genotype-negative counterparts. Snuff, at a dose of 500 mg (1% nicotine), induced a rapid increase gingival blood flow that was higher than the increase in blood pressure, indicating an active vasodilatation, partly blocked by infraorbital nerve block anesthesia and more so by the superficial mucosal anesthesia. Piroxicam and dexchlorpheniramine had no effect. In 240 patients, smoking had a 2.7 times greater probability to have established periodontal disease in smokers and 2.3 times in former smokers compared to nonsmokers, independent of age, sex, and plaque index. Among cases, probing depth, gingival recession, and clinical attachment level were greater in smokers than in former smokers or nonsmokers. The plaque index did not show differences. Bleeding on probing was less evident in smokers than in nonsmokers. There was a dose-dependent relationship between cigarette consumption and the probability of having advanced periodontal disease.

Plasma Lipid Profile

Tobacco was administered to 184 patients with head and neck cancer, 153 patients with oral precancerous conditions, and 52 controls. The treatment produced a significant decrease in plasma total cholesterol and HDL cholesterol in patients with cancer and oral precancerous conditions, compared to controls. The plasma very low-density lipoproteins (VLDL) and triglycerides levels were significantly lower in patients with cancer compared to the patients with oral precancerous conditions and controls. The results indicated that the tobacco habituates showed lower plasma lipid levels than the nonhabituates.

Polymorphonuclear Cells

High-tar (16 mg tar/cigarette) filtered cigarettes, administered to Balb/c and C57 Black mice for up to 32 weeks, produced in Balb/c mice less phagocytic and degradative capacities of polymorphonuclear cells than those of C57 Black mice. Heat inactivation of complement within the serum reduced the differences between smoke-exposed animals and age-matched controls in both strains of mice. The treatment effects were seen at all times tested from 3 days to 32 weeks of tobacco smoke exposure. The results indicated that fc-receptor site activity of PMN cells was not significantly affected by smoke exposure but that complement interactions within the phagocytic process are significantly suppressed.

Prenatal Influence

In 589 10-year-old children who have been observed from their gestation to tobacco smoke, half were females and 52% were African-American. During pregnancy, 52.6% of the mothers were smokers, 59.7% were smokers when their children were 10 years old. Six percent of the children (37/589) reported ever smoking cigarettes. Maternal smoking was significantly associated with an increased risk of the child's tobacco experimentation. Offspring exposed to more than half a pack per day during gestation had a 5.5-fold increased risk for early experimentation. Prenatal tobacco exposure had a direct and significant effect on the child's smoking and predicted child anxiety/depression and externalizing behaviors. Maternal current smoking had no significant effect. Cell-free amniotic acid from groups of smokers and nonsmokers showed a presence of NNAL in 11/21 (52.4%) of smokers and in 2/30 (6.7%) of nonsmokers. There was not convincing evidence of NNAL-Gluc in the amniotic acid. A total of 40 very-low-birth-weight infants (750–1500 g) of smoking during pregnancy (50%) and alcohol-consuming (42%) mothers were hospitalized and ventilated for respiratory distress syndrome. At birth, infants of mothers who smoked and consumed alcohol during pregnancy had significantly higher blood docosahexaenoic acid (DHA) than infants of nonsmoking and nondrinking mothers. Mothers of 91 cases and 321 population controls matched for age, sex, and residence and smoking during pregnancy produced higher an odds ratio of 1.7 (95% CI 0.8, 3.8) of brain tumor in their children, although this was not statistically significant. Among nonsmoking mothers, the risk for light and heavy exposure to passive smoking was 1.7 (0.8, 3.6) and 2.2 (1.1, 4.5), respectively, and a statistically significant dose–response relationship was found. Maternal use of masheri tobacco in pregnancy was associated with low birth-weight of the offspring, lower birth- weights in girls than in boys, and decreased male:female ratio in live newborns. Mainstream and sidestream smoke, administered by inhalation to hamsters, retarded transport of preimplantation embryos through the hamster oviduct. Oviductal muscle contraction rate decreased significantly during a single exposure of animals to either mainstream or sidestream smoke, and contraction rate failed to return to initial control values during a 25-minute recovery period. Both preimplantation embryo transport and muscle contraction were more sensitive to sidestream than mainstream smoke.

Progesterone Production Inhibition

Nicotine, cotinine, and anabasine, together, or an aqueous extract of cigarette smoke, in MA-10 Leydig cells, produced a dose- dependent inhibition of progesterone and 20-α-dihydroprogesterone synthesis. The number of cells in the treated dishes was less than the controls. Growth of MA-10 cells was inhibited. Cotinine, anabasine, a combination of nicotine, cotinine and anabasine, or an aqueous extract of cigarette smoke in human granulose cells produced an inhibition of progesterone synthesis. The alkaloids and some extract decreased the DNA content of the culture dish. Both cotinine and anabasine slightly stimulated the synthesis of normalized estradiol. Nicotine, combination of the three alkaloids, and cigarette smoke extract had no significant influence on estradiol production.

Prostaglandins Formation

Aqueous cigarette tar extracts, in rat pulmonary alveolar macrophages, increased cyclo-oxygenase activity threefold above the initial activity within 2 hours of incubation and gradually decreased below the initial activity after 8 hours of incubation. Accumulated levels of prostaglandin-2 increased dramatically after 12 hours of incubation. Release of arachidonic acid from the cells was dramatically increased in cells incubated with smoke extracts in parallel to prostaglandins accumulation.

Prostate Tumorigenic Effect

Tobacco smoke was administered to rats with implanted bilaterally with Dunning R3327 tumor fragments at 10 weeks of age and smoke exposed for an hour each day, 5 days a week, for 9 or 20 weeks. The treatment produced only minor changes in the growth rates of both the control and the irradiated tumors. At the cellular level, smoking produced a small, but significant, increase in the fraction of tumor cells relative to controls. The main difference observed was in the mast cell numbers. Smoking produced a fourfold increase in mast-cell density. The combination of smoking and irradiation resulted in a 10-fold intermediate increase.

Protein Synthesis Stimulation

Smoke of the leaf, administered to rats at variable doses, was active. The effect of a combination of hashish and cigarette smoke on brain proteins and catecholamines was measured. This was compared with the effects of animals under stress.

Pulmonary Arterial Effects

Cigarette smoke, administered to 2- and 3-month-old rats at a dose of one cigarette 10 times a day, increased significantly the volume fractions of the fibroblasts, the collagenous bundles, and the elastic laminae of the pulmonary arteries. The volume fractions of smooth muscle cells and the remainder were decreased significantly in both groups compared to controls. An increase in the stiffness of the pulmonary arteries was found in both the 2- and 3-month smoke-exposed rats.

Pulmonary Effect

Smoke, administered to mice at a dose of two cigarettes daily, 5 days/ week for 2–4 months, decreased the number of dendritic cells in the lung tissue and reduced the percentage of B7.1-expressing dendritic cells. Inoculation with 2×10^8 pfu of a replication-deficient adenovirus three times 2 weeks apart during the last month of tobacco exposure, prevented the expansion and maximal activation of CD4 T-cells and reduced the number of both activated CD4 and CD8 T-cells. Smoke exposure shifted the activated CD4:CD8 T-cells ratio from 3 to 1.5, and decreased serum adenovirus-specific pan IgG, IgG1, and IgG2a levels. NNK, in pulmonary cells, produced promutagenic adduct *O*-6-methylguanine. Administration of doses from 0.1 to 50 mg/kg increased the number of adducts (3- to 30-fold) in Clara cells those detected in type II cells and whole lung. Very low rates of repair of this adduct were detected in Clara cells, whereas efficient adduct removal occurred in type II cells. There was a strong correlation between the concentration of *O*-6-methylguanine in Clara cells and tumor incidence in the Fischer rat with NNK doses from 0.03 to 50 mg/kg. No differences in adduct concentration between type II and Clara cells from A/J mice were observed under conditions resulting in pulmonary tumor formation. Activation of the K-*ras* gene was detected in lung tumors from A/J mice. An early proliferative lesions observed in both mice and rats involved the alveolar areas. Ultrastructural examination of these lesions and adenomas revealed morphologic features characteristic of the type II cell. Masheri (pyrolyzed tobacco), administered orally to Swiss mice, Sprague–Dawley rats, and Syrian golden hamsters at a dose of 10% of the diet for 20 months, produced significant increase in activities of phase I activating enzymes and a remarkable decrease in the phase II detoxification system in most extrahepatic tissues of the treated animals of the three species. Cigarette smoke and hydrocortisone acetate (HCA), administered to C57BL/6J male mice, induced marked abnormalities in lungs, high congestion with surfactant and flocculent material in alveoli, prominent alveolar collapse, and septal hypertrophy. Results indicated that the genesis of abnormal conditions that resemble pulmonary alveolar proteinosis is potentiated by cumulative effects of different treatments (i.e., smoke, HCA, and stress), most significant being the interaction between cigarette smoke and the steroid. Whole cigarette smoke from reference cigarettes administration produced the prompt (maximal activity was 6 hours), but fairly weak (similar to twofold), induction of murine pulmonary microsomal monooxygenase activity not unequivocally linked to the Ah locus. This activity can be detected by using as substrates either benzo(a)pyrene or ethoxyresorufin and can be inhibited by treatment with cycloheximide or actinomycin D. Whole-smoke condensate and fractions induced pulmonary monooxygenase activity, inhibited benzo(a)pyrene metabolism in vitro, were metabolized to forms mutagenic to *Salmonella typhimurium* TA153 and TA98, transformed C3H 10T1/2 cells in vitro, and enhanced the carcinogenicity of benzo(a)-pyrene in murine pulmonary tissue. A potentially important observation was that whereas hepatic tissue is capable of activating whole cigarette smoke condensate to mutagenic forms in vitro, murine pulmonary tissue does not seem capable of such activation. Although these pulmonary-derived tissue homogenates had significant aryl hydrocarbon hydroxylase activity and can metabolize Aflatoxin B1,2-aminofluorene, and 7,8-dihydro-7,8-dihydroxybenzo(a)pyrene to mutagenic forms, these homogenates failed to activate both cigarette smoke condensate and the promutagen 6-aminochrysene. Smoke, administered to Sprague–Dawley rats at a dose of seven cigarettes/day for 5 days/week during a total period of 12 months, produced early abnormalities in pulmonary function, with the forced expiratory volume in one second/forced vital capacity (FEV1/FVC) ratio, showing an acceleration of ageing effect, particularly between 4 and 8 months of exposure.

Pulmonary Macrophage Mobilization

Smoke, administered to smoke-exposed and subsequently halothane anesthetized normal and to C57BL/6 mice, produced airway cilia shorter in stature and fewer in number. There was also disorientation of ciliary basal bodies. Airway macrophages were larger in size and contained more lysosomes and inclusions than phagocytes in airways of all other animals. Smoke inhalation alone produced a significant increase in the number of lung parenchymal macrophages when compared to the number of cells in sham- treated and control animals. The total macrophage population was significantly greater in lungs of smoke-exposed mice 48 hours after anesthesia than in lungs of smoke-exposed mice not subjected to halothane anesthesia and to those of sham-treated and control animals. Airway macrophage numbers were significantly elevated in smoke-exposed mice 48 hours after halothane when compared to those of all other groups, but the number of parenchymal macrophages decreased in lungs.

Pulmonary Surfactant Activity

Smoke from Kentucky cigarette, administered intranasally to female Sprague–Dawley rats twice daily for 60 weeks, produced no difference in total phospholipids content of the bronchoalveolar lavage fluids and the lung tissues among the groups. Desaturated phosphatidylcholine levels in the bronchoalveolar lavage fluids were significantly decreased. The lung tissue desaturated phosphatidylcholine content in smoke-treated rats was not significantly different compared to controls. Phospholipids profile analysis did not reveal any significant differences among other major constituents of surfactant from control and smoke-treated rats.

Renal Damages

Cigarette smoke condensate was administered to the oral mucosa of subtotally nephrectomized Sprague–Dawley rats, and to sham-operated Sprague–Dawley rats daily for 12 weeks. The treatment increased the indices of structural renal damage in the nephrectomized group. The cigarette smoke condensate increased the indices of glomerulosclerosis and tubulointerstitial damage in nephrectomized, but not sham-operated rats. This increase was completely prevented by renal denervation. Urinary albumin excretion went in parallel with the indices of glomerulosclerosis and tubulointerstitial damage and urinary endothelin-1 excretion was significantly increased in the nephrectomized animals.

Serotonin Uptake

Cigarette smoke, administered to mice, increased serotonin concentration as a function of the frequency of exposure. Serotonin uptake by the skin was maximally increased after two 8-minute exposures. MAO activity for serotonin, but not for tyramine, was decreased by cigarette smoke exposure of more than 8 minutes.

Sister Chromatid Exchange Inhibition

Water extract of the dried leaf, in cell culture at a concentration of 20 μL/mL, was active on Chinese hamster ovary cells. There was an elevation in frequency in cultures treated with 20 μL/mL of growth media. Seed, administered orally to adults, was active in chewers with oral cancer or oral submucosal fibrosis and healthy chewers, compared to control. An average of 6 quids of tobacco leaf, *Areca* nut and lime were used daily. Whole Kentucky reference 3A1 and American Blend cigarettes smoke were administered intranasally to B6C3F1 mice at a dose of 10% v/v 1,4,9, and 18 exposures/day (one exposure equal to one cigarette smoke) 5 days/week for 2 weeks. The treatment increased bone marrow sister chromatid exchange for both cigarette types in a dose-dependent manner. There was no effect on bone marrow cell replication kinetics. Whole cigarette smoke, administered intranasally to B6C3F1/Cum mice daily for 1 to 46 weeks, produced a twofold increase in sister chromatid exchange over sham-exposed control mice. In animals exposed either chronically or for 1 week to either type of smoke, the increase in sister chromatid exchange persisted for at least 1 week after cessation of smoke exposure. Smoke, administered to young adult male mice, elevated pulmonary macrophage population. Pulmonary macrophages (free, attached, and septal or interstitial) divided only rarely. During the marked progressive increase in the labeled macrophage population in the lungs, the number of silver grains over the nuclei of labeled macrophages did not become significantly diluted. Results indicated that the markedly elevated macrophage population resulted from the immigration of cells from bone marrow rather than *in situ* division of resident macrophages. Cigarette smoke, administered to adult male mice for 42 to 82 days, induced increased DNA activity in pulmonary tissue. No such induction was noted in the liver or spleen. An increase in DNA activity reflected a marked increase in the number of labeled pulmonary macrophages. At times, more than 50% of the total pool of labeled cells were identifiable as macrophages.

Skin Tumorigenic Effect

Aqueous extract of bidi tobacco, administered dermally according skin tumorigenesis protocol to hairless S/RV Cri-ba mice, did not exhibit carcinogenesis and effectively promoted skin papilloma formation in 7,12-dimethylbenz[a] -anthracene-initiated mice. An increase in papilloma yield above the control was noted only after 30 weeks of promotion. At week 40 weeks of promotion with 5 mg and 50 mg extract, it was significantly higher than that in the control mice. Mild epidermal hyperplasia, increase in mitotic activity and dermal thickness induced by a single application of extract persisted on multiple treatment and correlated well with its tumor-promoting activity. Brown and black varieties of pyrolyzed tobacco (masheri) extracts were administered dermally to Swiss mice and Swiss bare mice. In Swiss mice, there was no tumorigenic effect but a marginal synergistic effect of 7,12-dimethylbenz[a]anthracene (DMBA). The effect of black masheri extract was observed when DMBA was used as an initiator. In Swiss bare mice, black masheri extract induced tumors in 20–35% of the animals the two doses tested. In an initiation/promotion protocol with DMBA as an initiator, induction of tumors in 50–52% of the Swiss bare mice and a slight synergistic effect of black masheri extract were observed with a low dose of DMBA, suggesting a synergistic effect.

Small Airway Remodeling

Airway remodeling is usually attributed to the effects of cigarette smoke-induced inflammation in the airway wall, but little is known about its pathogenesis. Cigarette smoke, in rat tracheal explants at 24 hours after smoke exposure, produced a dose-dependent increase in gene expression of procollagen and a significant increase in tissue hydroxyproline, a measure of collagen content. Greater increases in procollagen gene expression were found with repeated smoke exposures. Results indicated that cigarette smoke can directly induce airway remodeling, specifically airway wall fibrosis, probably through active oxygen species-dependent transactivation of the epidermal growth factor receptor and subsequent NF-κB activation. Smoke-evoked inflammatory cells were not required for this process. Cigarette smoke, in rat bronchioles previously exposed to smoke in vivo, induced a small but consistent degree of contraction of the airways in vitro, which could be reduced by an endothelin receptor antagonist in the animals which had no previous smoke exposure in vivo, and reduced by the oxidant scavengers SOD or catalase in the animals with previous smoke exposure. Results indicated that cigarette smoke induced acute small airways constriction through both endothelin release and direct oxidant effects. Mainstream cigarette smoke, administered to rats at a dose of 250 mg total particulate matter/m^3 air for 6 hours/day, 5 days/week, for 2 weeks, increased the type II epithelial BrdU labeling index (LI). The axial airway and terminal bronchiolar LIs were enhanced only in the pump-labeled group. In the pump-labeled rats, the type II LI elevation was greater than the LI elevation in conducting airways, suggesting that the parenchyma may have been injured more than the conducting airways. The exposure did not increase the total number of muco-substances-containing cells or the total number of axial airway epithelial cells, but there was a phenotype change in the mucosubstance cells. Neutral mucosubstance cells (periodic acid-Schiff-positive) were significantly decreased, while acid mucosubstance cells (Alcian blue-positive) were slightly increased by smoke exposure. Either cell replication and differentiation or differentiation alone may have changed the phenotype in the smoke cell population.

Spermatozoal Effect

Filtered and non- filtered cigarette smoke, streamed at a rate of 100 mL/second into chamber with washed human spermatozoa, produced a dramatic drop in sperm motility, which caused sperm immobilization in about 15 minutes. This effect showed dose-response relationship with the amounts streamed or with the time of exposition and was almost the same for filtered or unfiltered smoke. Semen of 119 tobacco chewers and 218 smokers selected from an idiopathically hypofertile population, produced some decrease in the ejaculating volume, density and total count compared to non-smokers, but statistically insignificant. No difference was found in motility and morphology. Sperm of 103 smokers produced a significant decrease of density and motility compared with nonsmokers. Seventy-five percent of smokers vs 26% of nonsmokers had a sperm density under 40×10^6 sperm/mL. Morphologic abnormalities, particularly bicephali, did not differ significantly. Cigarette smoke, at doses of 10 cigarettes per day (57 cases), 11–20 (115 cases) or more than 20 (25 cases), produced significantly poorer sperm density, lower viability, motility and morphology in smokers. These parameters were worse in the heavy smoking groups. The ejaculate content of seminal vesicles, prostate gland, and epididymis of 29 smokers and 25 chewers produced a significant decrease in vesicular and prostatic parameters in smokers compared to nonusers of tobacco, whereas these parameters were unchanged in chewers. The activity of α-1,4-glucosidase was significantly lowered in both types of tobacco users. Serum levels of estradiol, prolactin, and total testosterone in 50 heavy smokers (median 23.5 cigarettes/day), produced higher levels of estradiol and prolactin, but not testosterone.

Sudden Infant Death Syndrome

Water- soluble smoke extract, in cell culture supernatants of mouse fibroblasts (L-929 cell line), produced an increase in TNF-α from respiratory syncytial virus-infected cells. It decreased TNF-α from cells incubated with toxic shock syndrome toxin. Incubation with cigarette smoke extract decreased the NO production from respiratory syncytial virus-infected cells and increased the NO production from cells incubated with toxic shock syndrome toxin. Monocytes from a minority of individuals demonstrated extreme TNF-α responses and/or very high or very low NO. The proportion of samples in which extreme responses with a very high TNF-α and very low NO were detected was increased in the presence of the three agents to 20% compared with 0% observed with toxic shock syndrome toxin. One to 4% was observed with cigarette smoke extract or respiratory syncytial virus.

Symphatomimetic Activity

Water extract of the dried leaf, administered intravenously to cats at doses of 0.05 and 10–20 mg/kg, enhanced the contractile response of the cat nictitating membrane evoked by preganglionic cervical sympathetic nerve stimulation. At higher doses it produced contractions and nictitating membrane contraction without nerve stimulation.

Systemic Inflammatory Cytokine Production

Sidestream smoke was administered to mice at 60 and 120 minutes per day, 5 days/week for a 16 weeks. The treatment produced a significant increase in the pro-inflammatory cytokines, IL-6, TNF-α, and IL-1β in 120-minutes smoke exposed mice. A decrease of the stroke volume, cardiac and hepatic antioxidant vitamin E levels, and heart pathology and increased peripheral arterial resistance were observed in 120-minute smoke-exposed mice. Hepatic lipid peroxides were increased on 60-minute smoke exposure.

T-cells Influence

SSTE and nicotine were incubated in splenic mononuclear cells at concentrations of 1:10^2 or 1:10^3 dilutions of STD, or 10 or 100 μg/mL nicotine, during 4 days of stimulation with anti-CD3. The treatment sustained expression of IL-2, IFN-γ, IL-10, and IL-4 cytokine mRNA at 100 μg/mL nicotine. STD did not exhibit residual expression of cytokine mRNA. Restimulated STD exhibited maximum IL-2, IL-4, IFN-γ, and IL-10 mRNA at 48 hours. STE, in splenic mononuclear cell culture at 1:10^2 to 1:10^4 dilutions, increased IL-2 production and decreased IL-10 at 1:10^2 dilution. IFN-γ production was decreased at all concentrations. STE did not alter IL-4 production.

P-benzoquinone, a thiol-reactive benzene derivative from cigarette tar, was incubated in human peripheral blood mononuclear cells at a concentration of 10 μM. The treatment inhibited mitogen-induced IL-2 production by 76 $\pm$ 7% without affecting lymphocyte/macrophage agglutination or blast transformation. The effect of p-benzoquinone appeared to be specific for IL-2 production, since *de novo* induction of the IL-2 receptor α-chain (CD25) and intercellular adhesion molecule-1 (CD54) and upregulation of LFA-1 α/β (CD11a and CD18) were unaffected. Smoke, administered by inhalation to mice, inhibited the antigen- specific T-cell proliferative response of lung-associated lymph nodes. Cell-mixing experiments demonstrated that the defect in smoke exposed mice resulted from an abnormality in T-lymphocyte function. The activity of antigen-presenting cells was similar in smoke-exposed and sham-smoke-exposed control animals. Diluted, mainstream cigarette smoke was administered to rats for up to 30 months or nicotine at a concentration of 1 mg/kg body weight/24 hours via mini-osmotic pumps for 4 weeks. The treatment decreased antigen-mediated T-cells proliferation and constitutive activation of protein tyrosine kinase and phospholipase C-γ1 activities. Spleen cells from smoke-exposed and nicotine-treated animals have depleted inositol-1,4,5-trisphosphate-sensitive Ca^{2+} stores and a decreased ability to raise intracellular Ca^{2+} levels in response to T-cell antigen receptor ligation. The results indicated that chronic smoking affects T-cell anergy by impairing the antigen receptor-mediated signal transduction pathways and depleting the inositol-1,4,5-trisphosphate-sensitive Ca(2+) stores. Moreover, nicotine may account for or contribute to the immunosuppressive properties of cigarette smoke.

Teratogenic Effect

Stem, administered orally to pigs, produced 80% congenital deformities if eaten between days 10 and 30 of pregnancy. STE was administered to 65 pregnant CD-1 mice at the following doses: equivalent to 8 mg/kg nicotine (group ST), ethanol 1.8 g/kg (EtOH), a combination of ST and EtOH in the same dosages, or D-glucose (controls and ST alone) three times daily on 6–15 days of gestation.

The mean maternal plasma drug levels were: nicotine, 321 ng/mL and ethanol, 0.105 g%. No significant differences were observed in maternal weight gain, litter size, or in the incidence of resorptions, deaths, and/or malformations. Fetal weights were reduced in the three treatment groups, with the greatest reduction (13% decrease) recorded in the ST group, and a 7% decrease in the ST/EtOH group. Placentas of the ST group weighed significantly less than controls. Ossification of the fetal skeleton, observed in 10 sites, was affected to the greatest extent in the ST group, followed by the EtOH and ST/EtOH groups. Craniofacial measurements were significantly affected in all three treatment groups, compared to controls. Aqueous extract of STE was administered intragastrically to CD-1 mice at doses: ST/D-1 and ST/D-2, equivalent of 12 and 20 mg/nicotine kg body wt, respectively, three times daily for 5 weeks: 2 weeks before conception, during conception, and during gestational days 0–17. The weight gain was not significantly affected. The mean maternal plasma nicotine level for the low-dosage group was 363 ng/mL and 481 ng/mL for the high-dosage group. Maternal lethality for the low and high doses were at 9.6% and 28.2%, respectively. No significant differences were found between control and ST/D-1 maternal and/or fetal values. In ST/D-2 group, fetal weights were reduced by 5.4%; decreased ossification in femur measurements and in 9 of 10 characteristics measured; the frequency of resorptions was twofold higher than in controls; and the frequency of deaths and malformations was not affected. The low dose produced a negligible effect on the CD-1 mouse fetus. The high dose demonstrated growth retardation, increased embryotoxicity, and a significant decrease in ossification. Aqueous extract of the STE, administered intragastrically to CD-1 mice at doses of: 1× extract equivalent to 4 mg/mL nicotine/kg body weight; 3 × 12 mg nicotine, and 5 × 20 mg nicotine, three times daily for gestation days 1–17, produced no significant effect on the weight gain in comparison to treated control. Difference was significant in comparison to untreated controls. Placental weighs were unaffected by the extract. The lowest extract dosage produced a negligible effect on the CD-1 mouse and the fetus. The highest dose demonstrated embryotoxicity, growth retardation, few malformations, and maternal toxicity. The intermediate dose showed a range of effects between the highest and lowest doses to both the fetus and the mother. NNK was administered intraperitoneally to A/J, C3B6F1, and Swiss out- bred Cr:NIH(S) mice at a dose of 100 mg/ kg on days 14,16, and 18 of gestation to A/ J and C3H/He mice and on days 15,17, and 19 of gestation to the Swiss mice. There was significant incidences of tumors in the lungs of A/J progeny and in the livers of male C3B6F1 and Swiss progeny. A lung tumor incidence in male offspring of treated A/J mice vs control was not statistical significant but was significantly greater in progeny A/J mice of both sexes compared to controls. The incidence of liver tumors in the male C3B6F1 mice exposed transplacentally to NNK was 40%, compared with 17% in controls. No effect of postnatal sodium barbital or PCB was observed on trans- placental NNK tumorigenicity in C3B6F1 mice. The combined incidence of liver carcinoma in male mice in all NNK-treated groups was significantly greater than in controls. In male Swiss mice exposed transplacentally to NNK, the incidence of liver tumors was 5%, compared to 0% in controls, and postnatal treatment with PCB on day 56 produced a significant increase in the incidence of NNK-induced liver tumors.

The combined incidence of liver tumors in the male offspring of the Swiss mice treated with NNK, with or without PCB, was 10%, compared to 0% in controls. Aqueous extract of STE was administered intragastrically to pregnant Sprague–Dawley rats three times daily on gestational days 6–18 at doses equivalent to 1.33 mg nicotine/kg body weight (STD- 1) or 6 mg nicotine/kg body weight (STD-2). The treatments reduced the weight gain of mothers, but fetal weights were reduced in the STD-2 group only. Placental weights, litter size, resorptions, deaths, and malformations were not significantly affected. Skeletal examinations revealed several dose-related differences between the smoke extract-treated and control groups.

In the STD-1 group, reductions in ossification were seen in the nasal and femur width measurements only. In the STD-2 group, reductions in ossification were seen in femur length and width, in the number of ossification centers in the forelimb, and in the maxillary, mandibular, and nasal bone measurements. STE was administered continuously via Alzet osmotic mini-pumps to CD-1 mice at doses of 3.2 mg/mL (dosage I) and 6.4 mg/mL (dosage II) for gestational days 7–14 and 6–13. The treatment produced plasma nicotine levels in the range of 29.4 $\pm$ 4.8 ng/mL to 44.3 $\pm$ 16 ng/mL for dosage I and in the range of 34.6 $\pm$ 10.9 ng/mL to 75.5 $\pm$ 19.9 ng/mL for the dosage II. Dosage I produced a tendency toward weight reduction, an increase in the incidence of hemorrhages and super-numerary ribs, and significant delay in ossification of the supraoccipital bone, the sacrococcygeal vertebrae, and the bones of the forefoot and hindfoot. Dosage II produced a significant (8.6%) weight reduction from normal and an increase in fetal deaths. There were no significant differences between placental weights. Weights of mothers were significantly reduced only at the higher doses.

Testosterone Effect

Cigarette smoke diluted with 90% air, administered to 12 male adult rats for 2 hours/day for 60 days, produced significant decrease of the mean plasma testosterone level. The mean plasma luteinizing hormone and follicle-stimulating hormone levels of the two groups did not change significantly after exposure. Histological examination of the testes showed fewer Leydig cells and degeneration of the remaining cells. The results indicated that the decrease in plasma testosterone levels induced by exposure to smoke was not associated with changes in plasma gonadotrophin levels. The decrease in testosterone levels may be related to the toxic effects of smoke on Leydig cells.

Tourette's Syndrome

Administration of nicotine (either 2 mg nicotine gum or 7 mg transdermal nicotine patch) potentiated the therapeutic properties of neuroleptics in treating patients with Tourette's syndrome, and a single patch may be effective for a variable number of days. These findings suggest that transdermal nicotine could serve as an effective adjunct to neuroleptic therapy.

Toxicity

Fresh tobacco smoke, administered to high leukemic AKR strain of mice at low levels for 1 day, produced significantly different mortality profiles associated with the sex of the animals and the age at which smoke exposure commenced. Females were more susceptible and died sooner than males, where a significant proportion of animals survived longer than age-matched controls. This prolongation of life appeared to result from a failure of the leukemic state to be mobilized in the smoke-exposed males. Exposure of both females and males to the smoke did not induce significant detectable immunological reactivity against the leukemic cells for the parameters tested. This possibly resulted from a significant enhancement of suppressor activity in the serum of the chronically exposed animals that also occurs in age-matched control animals. Smoke of cured leaf, administered by inhalation to adults, was active. Leaves, on agar plate at a concentration of 50%, produced changes in cell morphology of the monkey kidney cells. Leaf, administered orally to adults, was active. Ultrastructure abnormalities included discontinuous and fragmented basement membrane, reduction in hemidesmosomes, and widened intercellular spaces of the esophageal mucosa. Leaf, administered by inhalation to male adults at a dose of 15 g/day, was active vs various male sex gland functions in tobacco smokers. An 8-month-old infant ate two cigarette butts. On presentation 2.5 hours later, she was very lethargic and had depressed respiration, subsequently became somnolent with difficulty breathing. Urine toxicity screen was positive only for nicotine. She recovered with supportive treatment. Fresh leaf, administered externally to adults, was active vs green tobacco sickness in tobacco workers. The exposure of 47 patients to wet leaves resulted in nicotine toxicity, specifically nausea, vomiting, weakness, and dizziness. Mean time from exposure to onset of symptoms was 10 hours. Smoke extract, in the mouse brain mitochondria culture in the presence or absence of vitamin C for 60 minutes, inhibited mitochondrial ATPase and cytochrome C oxidase activities in a dose- dependent manner. The effect of extract on mitochondria swelling response to calcium stimulation was dependent on calcium concentrations. The extract treatment induced mitochondrial inner membrane damage and vacuolization of the matrix, whereas the outer mitochondrial membrane was preserved. Nicotine produced no significant damage. Nicotine, preincubated with normal human oral keratinocytes, altered the ligand-binding kinetics of their nicotinic acetylcholine receptors. It produced transcriptional and translational changes and changed the mRNA and protein levels of the cell cycle and cell differentiation markers Ki-67, PCNA, p21, cyclin D1, p53, filaggrin, loricrin, and cytokeratins 1 and 10. Environmental cigarette smoke or drinking water containing equivalent concentrations of nicotine that are pathophysiologically relevant, administered to rats and mice for 3 weeks, produced changes of the nicotinic acetylcholine receptors and the cell cycle and cell differentiation genes similar to those found in vitro.

Transglutaminase Activity

Water-soluble extract of gas-phase cigarette smoke was incubated with mouse bone marrow-derived macrophage culture containing both tissue- type transglutaminase and factor XIII-associated transglutaminase for 15 minutes at 3 7°C. A dose-dependent decrease in tissue- type (thrombin-independent) transglutaminase activity was produced, as compared to control cells. Factor XIII (zymogen) was not inactivated after incubation of macrophages with smoke extracts. Smoke exposure had no effect on cell viability or adherence. The results indicated that bone marrow-derived macrophages contain factor XIII and tissue-type transglutaminase. Gas-phase cigarette smoke can inactivate tissue transglutaminase within viable murine bone marrow-derived macrophages but cannot inactive zymogenic factor XIII.

TNF-α production

Water-soluble cigarette smoke extract and/or respiratory syncytial virus (RSV), in cell culture, stimulated TNF-α release from monocytes by both RSV infection and smoke extract and an additive effect was observed. There was a decrease in NO release, significant only with smoke extract or a combination of smoke extract and RSV infection. Nicotine decreased both TNF-α and NO responses. The proportion of extreme responses with very high TNF-α and very low NO in the presence of both RSV and smoke extract increased to 20% compared with 5% observed with smoke extract or RSV alone. Aqueous extract of the STE, in macrophage J774-A1 cell line and mouse splenocyte cultures at low concentrations, enhanced the production of both TNF-α and IL-1β and mitogen-induced murine splenocyte proliferation. Tobacco smoke, in alveolar macrophages of C57BL/6 mice in vivo, produced a significant decrease in the production of TNF-α by alveolar macrophages with the stimulation of lipopolysaccharide. In vitro exposure of alveolar macrophages to tobacco smoke water-soluble extracts decreased the production of TNF-α up to 93% of control with stimulation of lipopolysaccharide without any decrease in cellular viability. Smoke was administered to rats at a concentration of 10 mg/m^3 of cigarette smoke for 8 hours daily and the recovered alveolar macrophages were incubated with chrysotile or ceramic fibers for 24 hours. TNF of alveolar macrophages that were not stimulated produced no difference between treated and control rats. In alveolar macrophages stimulated by chrysotile or ceramic fibers, production of TNF in smoke-exposed rats was higher than that of controls.

Ulcer Healing Inhibition

Cigarette smoke or smoke extract, administered to ulcerated rats once daily for 3 days, produced concomitant and dose-dependent reduction of angiogenesis and constitutive NOS activity. The same treatments also delayed ulcer healing. Results indicated that cigarette smoke and its extract repressed the processes of new blood vessel formation and NOS activity during tissue repair in the gastric mucosa.

Urinary Metabolites

Different biomarkers of NNK exposure and metabolism, including the urinary metabolite NNAL and the presumed detoxification product [4-(methylnitrosamino)-1-(3-pyridyl)but-1-yl]-β-*O*-D-glucosiduronic acid

(NNAL-Gluc) were examined along with questionnaire data on lifestyle habits and diet in a metabolic epidemiological study of 34 black and 27 white healthy smokers. The results demonstrated that urinary NNAL-Gluc–NNAL ratios, a likely indicator of NNAL glucuronidation and detoxification, were significantly greater in whites than in blacks ($p < 0.02$). In addition, two phenotypes were apparent by probit analysis representing poor (ratio < 6) and extensive (ratio ≥ 6) glucuronidation groups. The proportion of blacks falling into the former, potentially high-risk group was significantly greater than that of whites ($p < 0.05$). The absolute levels of urinary NNAL, NNAL-Gluc, and cotinine were also greater in blacks than in whites when adjusted for the number of cigarettes smoked. None of the observed racial differences could be explained by dissimilarities in exposure or other sociodemographic or dietary factors. Also, it is unlikely that the dissimilarities are due to racial differences in preference for mentholated cigarettes, because chronic administration of menthol to NNK-treated rats did not result in either increases in urinary total NNAL or decreases in NNAL-Gluc–NNAL ratios. Altogether, these results suggest that racial differences in NNAL glucuronidation, a putative detoxification pathway for NNK, may explain in part the observed differences in cancer risk.

Vasoconstriction Activity

Cigarette smoke, in isolated rat lungs perfused with blood, produced no change in pulmonary vascular resistance. The hypoxic pulmonary vasoconstriction was significantly enhanced by smoking. Indomethacin, an inhibitor of prostaglandins biosynthesis, administered in the perfusing blood, increased hypoxic pulmonary vasoconstriction in the nonsmoking lungs but not in lungs after smoking. Diethylcarbamazine citrate (1 mg/mL), an inhibitor of leukotrienes biosynthesis, decreased hypoxic pulmonary vasoconstriction before and after smoking. After perfusion with both indomethacin and diethylcarbamazine citrate, hypoxic pulmonary vasoconstriction also decreased. Results indicated that leukotrienes act as mediators whereas prostaglandins as modulators in hypoxic pulmonary vasoconstriction, and prostaglandins and leukotrienes may play an important role in the increase of hypoxic pulmonary vasoconstriction by cigarette smoking. Cigarette smoke extract, administered intravenously to Wistar rats during hypoxic ventilation, produced significant decrease in microvascular internal diameter of pulmonary arterioles and venules. After pretreatment of animal with smoke extract, much more remarkable pulmonary vasoconstriction was induced by hypoxia than before injection of the extract. During hypoxia, mean pulmonary arterial pressure increased by 13.53% before administration of smoke extract and by 30.57% after administration, respectively. Results indicated that smoke extract can strengthen pulmonary vasoconstriction and hypertension induced by acute alveolar hypoxia.

Viral Stimulant

Smoke of the leaf, administered by inhalation to mice at variable doses twice daily for 5 days followed by 2 nontreatment days, dosed for two to six cycles, was inactive on encephalomyocarditis virus, Herpes virus type 1, and influenza virus APR-8.

Vitamins A and C Concentrations Influence

Methyl chloride extract of the leaf, administered intragastrically to rats at a dose of 3 mg/animal daily for 21 months in DMSO solvent, was active. Rats were divided or two groups: one was fed a vitamin A diet, and one a vitamin A-deficient diet. Vitamin C levels in both liver and plasma elevated significantly in extract-treated animals, while vitamin A level decreased. NNN and NNK, administered to Swiss and Balb/c male mice, produced a significant decrease in liver vitamin A levels by both nitrosamines. NNK treatment also produced a decrease in the levels of vitamin A in plasma.

Weight Loss

Smoke of the leaf, administered by inhalation to mice at variable doses twice daily for 5 days followed by 2 nontreatment days, dosed for two to six cycles, was active. Tobacco smoke, administered to 3-week-old Wistar rats on a vitamin E-depleted or normal diet, for 4 weeks, significantly suppressed body weight increases, particularly in the vitamin E-depleted group.

2

Camellia Sinensis

Camellia sinensis is an evergreen tree or shrub of the *Theaceae* family that grows to 10–15 m high in the wild, and 0.6–1.5 m under cultivation. The leaves are short- stalked, light green, coriaceous, alternate, elliptic-obovate or lanceolate, with serrate margin, glabrous, or sometimes pubescent beneath, varying in length from 5 to 30 cm, and about 4 cm wide. Young leaves are pubescent. Mature leaves are bright green in color, leathery, and smooth.

Flowers are white, fragrant, 2.5–4 cm in diameter, solitary or in clusters of two to four. They have numerous stamens with yellow anthers and produces brownish-red, one- to four-lobed capsules. Each lobe contains one to three spherical or flattened brown seeds. There are numerous varieties and races of tea. There are three main groups of the cultivated forms: China, Assam, and hybrid tea, differing in form. *Camellia sinensis assamica*, the source of much of the commercial tea crop of Ceylon is a tree that, unpruned, may attain a height of 15 m and has proportionally longer, thinner leaves than typical species.

Origin and Distribution

The cultivation and enjoyment of tea are recorded in Chinese literature of 2700 BC and in Japan about 1100. Through the Arabs, tea reached Europe about 1550. Native to Assam, Burma, and the Chinese province of Yunnan, it is highly regarded in southern Asia and planted in India, southern Russia, East Africa, Java, Ceylon, Sumatra, Argentina, and Turkey. China, India, Indonesia, and Japan produced about a half of the total world production.

Fig. 2.1. Camellia sinensis – flowering branch.

Traditional Uses

India. Decoctions of the dried and fresh buds and leaves are taken orally for headache and fever. Powder or decoction of the dried leaf is applied to teeth to prevent tooth decay. Fresh leaf juice is taken orally for abortion, and as a contraceptive and hemostatic.

Mexico. Hot water extract of the leaf is taken orally by nursing mothers to increase milk production.

Turkey. Leaves are taken orally to treat diarrhea.

China. Hot water extract of the dried leaf is taken orally as a sedative, an antihypertensive, and anti-inflammatory.

Guatemala. Hot water extract of the dried leaf is used as eyewash for conjunctivitis.

Kenya. Water extract of the dried leaf is applied ophthalmically to treat corneal opacities. The infusion is used for chalzion and conjunctivitis.

Thailand. Hot water extract of the dried leaf is taken orally as a cardiotonic and neurotonic. Hot water extract of the dried seed is taken orally as an antifungal.

Medicinal Values

Antibacterial Activity

Alcohol extract of black tea, assayed on *Salmonella typhi* and *Salmonella paratyphi* A, was active on all strains of Salmonella paratyphi A, and only 42.19% of Salmonella typhi strains were inhibited by the extract. Hot water extract of the dried entire plant and the tannin fraction, on agar plate, were active on *Escherichia coli*, *Pseudomonas aeruginosa*, and *Staphylococcus aureus*.

Anticancer Activity

Catechin, administered to pheochromocytoma cells in cell culture, was active. The cells were incubated with different concentrations of catechin at short-term (2 days) and long-term (7 days) in Dulbecco's modified Eagle medium. The activity of superoxide dismutase was measured and its mRNA assayed by Northern blotting. After incubation for 2 days, catechin significantly increased the activity of copper/zinc superoxide dismutase. However, it did not produce significant effect at 7 days. The magnesium superoxide dismutase activity produced significant changes in both short- and long-term treatment groups. The amount of mRNA also showed similar changes.

Anticarcinogenic Activity

The anti-carcinogenic activity of tea phenols has been demonstrated in rats and mice transplantable tumors, carcinogen-induced tumors in digestive organs, mammary glands, hepatocarcinomas, lung cancers, skin tumors, leukemia, tumor promotion, and metastasis. The mechanisms of this effect indicated that the inhibition of tumors may be the result of both extracellular and intracellular mechanisms indicating the modulation of metabolism, blocking or suppression, modulation of DNA replication and repair effects, promotion, inhibition of invasion and metastasis, and induction of novel mechanisms. The association of green tea and cancer has been investigated in 8552 Japanese women 40 years of age. After 9 years of follow-up study,

384 cases of cancer were identified. There was a negative association between cancer incidence and green tea consumption, especially among females consuming more than 10 cups of tea a day. A slow down in increases of cancer incidence with age was observed among females who consumed more than 10 cups daily. Tea, taken by lung cancer patients at a dose of two or more cups per day, reduced the risk by 95%. The protected effect was more evident among Kreyberg I tumors (squamous cell and small cells) and among light smokers. The green tea polyphenols, epigallocatechin-*3*-gallate, applied topically to human skin, prevented penetration of ultraviolet (UV) radiation. This was demonstrated by the absence of immunostaining for cyclobutane pyrimidine dimers in the reticular dermis. Topical administration to the skin of mice inhibited UVB-induced infiltration of CDIIb$^+$ cells. The treatment also results in reduction of the UVB-induced immuno-regulatory cytokine interleukin (IL)-10 in the skin and draining lymph nodes, and an elevated amount of IL- 12 in draining lymph nodes. Green tea extract, in human umbilical vein endothelial cells, did not affect cell viability but significantly reduced cell proliferation dose-dependently and produced a dose-dependent accumulation of cells in the gastrointestinal phase. The decrease of the expression of vascular endothelial growth factor receptors fms-like tyrosine kinase and fetal liver kinase-I/kinase insert domain containing receptor in the cell culture by the extract was detected with immunohistochemical and Western blotting methods. Green and black tea, administered orally to hairless mice in the absence of any chemical initiators or promoters, resulted in significantly fewer skin papillomas and tumors induced by UVA and UVB light. Black tea however, provided better protection against UVB-induced tumors than green tea. Black tea consumption was associated with a reduction in the number of sunburn cells in the epidermis of mice 24 hours after irradiation, although there was no effect of green tea. Other indices of early damage such as necrotic cells or mitotic figures were not affected. Neutrophil infiltration as a measure of skin redness was slightly lowered by tea consumption in the UVB group. Epigallocatechin-3-gallate, in cell culture, activated proMMP-2 in U-87 glioblastoma cells in the presence of concanavalin A or cytochalasin D, two potent activators of MT1-MMP, resulted in proMMP-2 activation that was correlated with the cell surface proteolytic processing of Mt1-MMP to it's inactive 43 kDa form. Addition of epigallocatechin-3-gallate strongly inhibited the MT1-MMP-driven migration in the cells. The treatment of cells with non-cytotoxic doses of epigallocatechin-3-gallate significantly reduced the amount of secreted

pro MMP-2, and led to a concomitant increase in intracellular levels of that protein. The effect was similar to that observed using well-characterized secretion inhibitors such as brefeldin A and manumycin, indicative that epigallocatechin could also potentially act on intracellular secretory pathways. Green tea polyphenols, at a dose of 30 mg/mL, inhibited the photolabeling of P-glycoprotein (P-gp) by 75% and increased the accumulation of rhodamine-123 in the multidrug-resistant cell line CH(R)C5. This result indicated that green tea polyphenols interact with P-gp and inhibited its transport activity. The modulation of P-gp was a reversible process. Epigallocatechin-3-gallate potentiates the cytotoxicity of vinblastine in CH(R)C5 cells. The inhibitory effect on P-gp was also observed in human Caco-2 cells.

Anticataract Activity

Tea, administered in culture to enucleated rat lens, reduced the incidence of selenite cataract in vivo. The rat lenses were randomly divided into normal, control and treated groups and incubated for 24 hours at 37°C. Oxidative stress was induced by sodium selenite in the culture medium of the two groups (except the normal group). The medium of the treated group was additionally supplemented with tea extract. After incubation, lenses were subjected to glutathione and malondialdehyde estimation. Enzyme activity of superoxide dismutase, catalase, and glutathione peroxidase were also measured in different sets of the experiment. In vivo cataract was induced in 9-day-old rat pups of both control and treated groups by a single subcutaneous injection of sodium selenite. The treated pups were injected with tea extract intraperitoneally prior to selenite challenge and continued for 2 consecutive days thereafter. Cataract incidence was evaluated on 16 postnatal days by slit lamp examination. There was positive modulation of biochemical parameters in the organ culture study. The results indicated that tea act primarily by preserving the antioxidant defense system.

Antifungal Activity

Ethanol (50%) extract of the entire plant, in broth culture at a concentration of 1 mg/mL, was inactive on *Aspergillus fumigatus* and *Trichophyton mentagrophytes*. Hot water extract of the leaf on agar plate at a concentration of 1.0% was active on *Alternaria tenuis*, *Pythium aphanidermatum*, and *Rhizopus stolonifer*. Saponin fraction of the leaf on agar plate was active on *Microsporum audonini*, minimum inhibitory concentration (MIC) 10 mg/mL; *Epidermophyton floccosum* and *Trichophyton mentagrophytes*, MICs 25 μg/ mL.

Antihypercholesterolemic Activity

Tea supplemented with vitamin E, administered to male Syrian hamsters, reduced plasma low-density lipoprotein (LDL) cholesterol concentrations, LDL oxidation, and early atherosclerosis compared to the consumption of tea alone by the hamsters. The antioxidant action of vitamin E is through the incorporation of vitamin E into the LDL molecule. The hamsters were fed a semi- purified hypercholesterolemic diet containing 12% coconut oil, 3% sunflower oil, and 0.2% cholesterol (control), control and 0.625% tea, control and 1.25% tea or control and 0.044% tocopherol acetate for 10 weeks. The hamsters fed the vitamin E diet compared to the different concentrations of tea significantly lower plasma LDL cholesterol concentrations, –18% ($p < 0.007$), –17% ($p < 0.02$), and –24% ($p < 0.0001$), respectively. Aortic fatty streak areas were reduced in the vitamin E diet group compared to the control, –36% ($p < 0.04$) and low tea –45% ($p < 0.01$) diets. Lag phase of conjugated diene production was greater in the vitamin E diet compared to the control, low tea, and high tea diets, 41% ($p < 0.0004$), 40% ($p < 0.0004$), and 39% ($p < 0.0008$), respectively. Rate of conjugated diene production was reduced in the vitamin E diet compared to the control, low tea, and high tea diets, –63% ($p < 0.002$), –57% ($p < 0.005$), and –59% ($p < 0.02$), respectively. Infusion of black tea leaves was taken by 31 men (ages 47 ± 14) and 34 females (ages 35 ± 13) in a 4-week study. Six mugs of tea were taken daily vs placebo (water, caffeine, milk, and sugar) and blood lipids, bowel habit, and blood pressure measured during a run-in period and at the end weeks 2, 3, and 4 of the test period.

Compliance was established by adding a known amount of *p*-aminobenzoic acid to selected tea bags and then measure it excretion in the urine. Mean serum cholesterol values during run-in, placebo and on tea drinking were 5.67 ± 1.05, 5.76 ± 1.11, and 5.69 ± 1.09 mmol/L ($p = 0.16$). There were also no significant changes in diet, LDL-cholesterol, high-density lipoprotein (HDL) cholesterol, triacylglycerols, and blood pressure in the tea intervention period compared with placebo. Stool consistency was softened with tea compared with the placebo, and no other differences were observed in bowel habit. The results were unchanged within 15 "non-compliers" whose *p*-aminobenzoic acid excretion indicated that fewer than six tea bags had been used, were excluded from the analysis, and when differenced between run-in and tea periods were considered separately for those who were given tea first or second.

Anti-inflammatory Effect

Epigallocatechin-3-gallate was shown to mimic its anti-inflammatory effects in modulating the IL-I β-induced activation of mitogen activated protein kinase in human chondrocytes. It inhibited the IL-I β-induced phosphorylation of c-Jun N-terminal kinase (JNK) isoforms, accumulation of phosphoc-Jun and DNA-binding activity of AP-1 in osteoarthritis chondrocytes, IL-I β but not epigallocatechin-3-gallate, and induced the expression of JNK p46 without modulating the expression of JNK p54 in osteoarthritis chondrocytes. In immune complex kinase assays, epigallocatechin-3-gallate completely blocked the substrate phosphorylating activity of JNK but not p38-mitogen activated protein kinase (MAPK). Epigallocatechin-3-gallate had no inhibitory effect on the activation of extracellular signal- regulated kinase p44/p42 (ERKp44/p42) or p38-MAPK in chondrocytes. Epigallocatechin-3-gallate did not alter the total nonphosphorylated levels of either p38- MAPK or ERKp44/p42 in osteoarthritis chondrocytes. Epigallocatechin-3-gallate administered to primary human osteoarthritis chondrocytes at a concentration of 100 μM in cell Culture, inhibited the IL-I β-induced production of nitric oxide by interfering with the activation of nuclear factor (NF)κB. Tea, in culture with bovine nasal and meta-carpophalangeal cartilage and human nondiseased osteoarthritis and rheumatoid cartilage with and without reagents known to accelerate cartilage matrix breakdown, produced chondroprotective effect that may be beneficial for the arthritis patient by reducing inflammation and the slowing of cartilage breakdown. Individual catechins were added to the cultures and the amount of released proteoglycan and type II collagen were measured by metachromatic assay and inhibition enzyme-linked immunosorbent assay (ELISA), respectively. Possible nonspecific or toxic effects of the catechins were assessed by lactate output and proteoglycan synthesis. Catechins, particularly those containing a gallate ester, were effective at micromolar concentrations at inhibiting proteoglycan and type II collagen breakdown.

Antimutagenic Activity

The anticarcinogenic activity of tea phenols has been demonstrated in rats and mice, transplantable tumors, carcinogen-induced tumors in digestive organs, mammary glands, hepatocarcinomas, lung cancers, skin tumors, leukemia, tumor promotion, and metastasis. The mechanisms of this effect indicated that the inhibition of tumors maybe the result of both extracellular and intracellular mechanisms indicting the modulation of metabolism, blocking or suppression, modulation of

DNA replication and repair effects, promotion, inhibition of invasion and Metastasis, and induction of novel mechanisms. Green and black teas, administered orally to human adults, were effective. Between 60 and 180 minutes after the teas were administered, the antimutagenic active compounds were recovered from the jejunal compartment by means of dialysis. The dialysate appeared to inhibit the mutagenicity of the food mutagen 2-amino-3,8-dimethylimidazo[4,5-f]quinoxaline on *Salmonella typhimurium*. The maximum inhibition was measured at 2 hours after administration and was comparable for black and green teas. The maximum inhibition observed with black tea was reduced by 22, 42, and 78% in the presence of whole milk, semi-skimmed milk, and skimmed milk, respectively. Whole milk and skimmed milk abolished the antimutagenic activity of green tea by more than 90% and semi- skimmed milk by more than 60%. When a homogenized breakfast was taken with black tea, the antimutagenic activity was eliminated. When tea and mutagen 2-amino-3,8-dimethylimidazo[4,5-f]quinoxaline were added to the system, 2-amino-3,8-dimethylimidazo[4,5-f]quinoxaline mutagenicity was efficiently inhibited, with green tea showing a slightly stronger antimutagenic activity than black tea. The addition of milk had only a small inhibiting effect on the antimutagenicity. The antimutagenic activity corresponded with reduction in antioxidant capacity and with a decrease of concentration of catechin, epigallocatechin gallate, and epigallocatechin. Chinese white tea, tested on rat liver S9 in assay for methoxyresorufin *O*-demethylase, inhibited methoxyresorufin *O*-demethylase activity and attenuated the mutagenic activity of 3 -methylimidazo [4,5-f] quinoline (IQ) in absence of S9. Nine of the major constituents found in green and white teas were mixed to produce artificial teas according to their relative levels in white and green teas. The complete tea exhibited higher antimutagenic potency compared with the corresponding artificial tea. Green and black tea polyphenols, applied to the surfaces of ground beef before cooking, inhibited the formation of the mutagens in a dose-related fashion. Green or black tea polyphenols sharply decreased the mutagenicity of a number of aryl- and heterocyclic amines, of aflatoxin B_1, benzo[a]pyrene, 1, 2-dibromoethane, and more selectively of 2-nitropropane, all involving an induced rat liver S9 fraction. Good inhibition was found with two nitrosamines that required a hamster S9 fraction for biochemical activation. No effect was found with 1-nitropyrene and with the direct-acting (no S9) 2-chloro-4-methylthiobutanoic acid. Hot water extract on the leaf was evaluated in cell cultures on various systems vs decaffeinated and caffeinated teas. On mouse mammary gland vs decaffeinated and caffeinated teas, ICs_{50} were 10 mg/mL and 10 μg/mL on CA-A427, IC_{50} 27 mg/mL and 31 μg/mL, and on epithelial cells, IC_{50} 0.01 ng/mL and 0.3 ng/mL. Hot water extract of the leaf, on agar plate at a concentration of 1 mg/ plate, was active on Salmonella typhimurium TA98 vs 2-amino-3-methylimidazo [4,5-f]quinoline-induced mutagenesis and produced weak activity vs benzo[a]pyrene-induced mutagenesis. Infusion of the leaf, on agar plate at a concentration of 0.7 mg/plate, was active on *Salmonella typhimurium* TA98 and TA100 vs 2-amino-3-methylimidazo [4,5-f]quinoline-; 3-amino-1,4-dimethyl-5H-pyrid[4,3-b]indole(Trp-1); aflatoxin B1-; 2-amino-6-methyl-dipyrido[1,2-A:3,2-d] imidazole-, and benzo[a]pyrene-induced carcinogenesis. Infusion of the leaf, on agar plate at a concentration of 50 mg/plate, was active on *Salmonella typhimurium* TA98 vs 2-amino-3-methylimidazo[4,5-f]quinoline-; 2-amino-3,4-dimethyl-imidazo[4,5-f]quinoline-;2-amino-3,8-dimethylimidazo[4,5 -f]quinoxaline-; 2-amino-1-methyl-6-phenylimidazo [4,5-b]-pyridine-;2-amino-3,7,8-trimethylimidazo[4,5-f]quinoxaline-; 2-amino-3,4,7,8-tetramethyl-3H-imidazo-[4,5-f]quinoxaline-inoxaline-; 3-amino-1,4-dimethyl-5 H-pyrid[4,3-b] indole (Trp-P-I)- and 3-amino-1-methyl-5H-pyrido [4,3-b] indole-induced mutagenesis. Metabolic activation was required for positive results.

Anti-neoplastic Effect

Green tea, administered orally at a dose of 6 g per day in six doses to 42 patients who were asymptomatic and had manifested, progressive prostate specific antigen elevation with hormone therapy, produced limited antineoplastic activity. Continued use of lute inizing hormone-releasing hormone agonist was permitted. However, patients were ineligible if they had received other treatments for their disease in the preceding 4 weeks or if they had received a long-acting antiandrogen therapy in the preceding 6 weeks. The patients were monitored monthly for response and toxicity. Tumor response, defined as a decline of 50% or greater in the baseline prostate-specific antigen (PSA) value, occurred in a single patient, or 2% of The cohort (95% confidence interval [CI], 1–14%). This one response was not sustained beyond 2 months. At the end of the first month, the median change in the PSA value from baseline for the cohort increased by 43%. Infusion of the leaf, administered in the drinking of female mice at a concentration of 1.25%, was active vs UV radiation-induced papillomas and tumors[CS172]. Leaves in the drinking water of female mice at a dose of 0.6% reduced lung tumor multiplicity and volume in 4-(methyl-nitrosamine)-1-(3-pyridyl)-1-butanone (NNK) treated mice.

Antioxidative Effect

Tea, administered orally to rats, decreased the thiobarbituric acid reactive substances (TBARS) contents in urine and lowered the esterified and total cholesterol contents in plasma as compared with a control group. TBARS contents in liver, plasma, and cholesterol levels in the liver were not affected. The lower plasma cholesterol concentration could not be explained by increased fecal excretion of cholesterol or bile acids. On the other hand, a relationship between decreased plasma cholesterol and significantly higher acetate concentrations in the cecum, colon, and portal blood of rats was assumed. Copper absorption was significantly increased while iron absorption was not affected. Epigallocatechin gallate, tea polyphenols, and tea extract were added to human plasma and lipid peroxidation induced by the water-soluble radical generator 2,2'-azobis (2-amidinopropane) dihydrochloride. Following a lag phase, lipid peroxidation was initiated and it occurred at a rate that was lower in a dose that was lowered in a dose-dependent manner by the polyphenols. Similarly, epigallocatechin gallate and the extract added to plasma strongly inhibited 2,2′-azobis(2-amidinopropane) dihydrochloride-induced lipid peroxidation. The lag phase preceding detectable lipid peroxidation was the result of the antioxidant activity of endogenous ascorbate, which was more effective at inhibiting lipid peroxidation than the tea polyphenols and was not spared by these compounds. When eight volunteers consumed the equivalent of six cups of tea, the resistance of their plasma to lipid peroxidation did not increase over a period of 3 hours. Black tea leaves, administered to human red blood cells, was effective against damage by oxidative stress induced by inducers such as phenylhydrazine, Cu^{2+}-ascorbic acid, and xanthine/xanthine oxidase systems. Lipid peroxidation of pure erythrocyte membrane and of whole red blood cell was completely prevented by black tea extract. Similarly, the tea provided total protection against degradation of membrane proteins. Membrane fluidity studies as monitored by the fluorescent probe 1,6-diphenyl-hexa-1,3,5-triene showed considerable disorganization of its architecture that could be restored back to normal on addition of black tea or free catechins. The tea extract in comparison to free catechin seemed to be a better protecting agent against various types of oxidative stress. Ethanol/water (7:3) extract of green tea, tested on 2,2-azino-di-3-ethylbenzthiazoline sulphonate, produced antioxidant activity compared with that of ascorbic acid (10 mmol/L). The Nonpolyphenolic fraction of residual green tea (after hot water extraction) produced a significant suppression against hydroperoxide generation from oxidized linoleic acid in a dose-dependent manner. Using silica gel TLC plate, chlorophylls a and b, pheophytins a and b, β-carotene, and lutein were isolated. All of these constituents exhibited significant antioxidant activites, the ranks of suppressive activity against hydroperoxide generation were chlorophyll a > lutein > pheophytin a > chlorophyll b > b-carotene > pheophytin b.

Antiproliferative Activity

Green tea fractions, tested on human stomach cancer (MK-1) cells, indicated six active flavan-3-ols, epicatechin, epigallocatechin, epigallocatechin gallate, gallocatechin, epicatechin gallate, and gallocatechin gallate. Among the six active flavan-3-ols, epigallocatechin gallate and gallocatechin gallate produced the highest activity. Epigallocatechin, gallocatechin, and epicatechin gallate followed next, and the activity of epicatechin was lowest. This suggests that the presence of the three adjacent hydroxyl groups (pyrogallol or galloyl group) in the molecule would be a key factor for enhancing the activity.

Antiviral Activity

Epigallocatechin-3-gallate, administered to Hep2 cells in culture, produced a therapeutic index of 22 and an IC_{50} of 25 μM. The agent was the most effective when added to the cells during the transition from the early to the late phase of viral infection suggesting that the polyphenol inhibits one or more late steps in virus infection. Ethanol (50%) extract of the entire plant, in broth culture at a concentration of 50 μg/mL, was inactive on Raniket and Vaccinia viruses. Hot water extract of the leaf in cell culture was active on Coxsackie A9, B1, B2, B3, B4, and B6 viruses, Echo type 9 virus, herpes simplex virus, poliovirus III, vaccinia virus, and REO type 1 virus.

Anti-yeast Activity

Ethanol (50%) extract of the entire plant, in broth culture at a concentration of 1 mg/mL, was inactive on *Candida albicans*, *Cryptococcus neoformans*, and *Sporotrichum schenckii*. Ethanol extract of the leaf on agar plate produced MIC 9.3 mg/mL on *Candida albicans*.

Cytochrome P50 Expression

Fresh leaves of green, black, and decaffeinated black tea enhanced lauric acid hydroxylation. The decaffeinated black tea produced no significant effect. Green tea and black tea but not decaffeinated black tea, stimulated the *O*-dealkylations of methoxy-, ethoxy-, and pentoxy-resorufin indicating upregulation of cytochrome P50 (CYP) 1A and CYP2B. Immunoblot analysis revealed that green and black tea, but not decaffeinated black tea, elevated the hepatic CYP1A2 apoprotein levels. Hepatic microsomes from green and black tea-treated rats, but not those from the decaffeinated black tea-treated rats, were more effective than controls in converting IQ into mutagenic species in the Ames test.

Dental Enamel Erosion

Herbal tea and conventional black tea, tested on teeth, resulted in erosion of dental enamel. After exposure to tea, sequential profilometric tracings of the specimens were taken, superimposed, and the degree of enamel loss calculated as the area of disparity between the tracings before and after exposure. Tooth surface loss resulted from herbal tea (mean 0.05 mm^2) was significantly greater than that which resulted from exposure to conventional black tea (0.01 mm^2), and water (0.00 mm^2). Tannin, catechin, caffeine, and tocopherol, tested in vitro on tooth enamel, demonstrated that these components possess the property of increasing the acid resistance of tooth enamel. The effects increased dramatically when the components were used in combination with fluoride. A mixture of tannic acid and fluoride showed the highest inhibitory effect (98%) on calcium release to an acid solution. Tannin in combination with fluoride inhibited the formation of artificial enamel lesions in comparison with acidulated phosphate fluoride (APF) as determined by electron probe microanalysis, polarized-light microscopy, and Vickers microhardness measurement.

DNA Effect

Green tea extract, in cell culture at a dose of 10 mg/L corresponding to 15 mmol/L EGCg for 24 hours, did not protect Jurkat cells against H_2O_2-induced DNA damage. The DNA damage, evaluated by the Comet assay, was dose-dependent. However, it reached plateau at 75 mmol/L of H_2O_2 without any protective effect exerted by the extract. The DNA repair process, completed within 2 hours, was unaffected by supplementation.

Fluoride Retention

Tea, used as a mouth rinse, demonstrated strong avidity of enamel for tea and salivary pellicle components. Thirty-four percent of the fluoride was retained in the oral cavity. Differences in retention at the tooth surface in the presence and absence of an acquired pellicle were not statistically significant at incisor or molar sites. Fluoride from tea showed strong binding to enamel particles, which was only partially dissociated by solutions of ionic strength considerably greater than that of saliva.

Gastrointestinal Effect

Green tea, administered to rats fasted for 3 days, reverted to normal the mucosal and villous atrophy induced by fasting. Black tea ingestion had no effect. Ingestion of black tea, green tea, and vitamin E before fasting protected the intestinal mucosa against atrophy. Characterization of melanin extracted from tea leaves proved similarity of the original compound to standard melanin. The Langmuir adsorption isotherms for gadolinium (Gd) binding were obtained using melanin. Melanin-Gd preparation demonstrated low acute toxicity. LD_{50} for the preparation was in a range of 1.25–1.50 g/kg in mice. Magnetic resonance imaging (MRI) properties of melanin itself and melanin-Gd complexes have been estimated. Gadolinium-free melanin fractions possess slighter relaxivity compared with its complexes. The relaxivity of lower molecular weight fraction was 2 times higher than relaxivity of Gd(DTPA) standard. Postcontrast images demonstrated that oral administration of melanin complexes in concentration of 0.1 m*M* provides essential enhancement to longitudinal relaxation times (T[1])-weighted spin echo image. The required contrast and delineation of the stomach wall demonstrated uniform enhancement of MRI with proposed melanin complex.

Hypocholesterolemic Effect

Green tea, in human HepG2 cell culture, increased both LDL receptor-binding activity and protein. The ethyl acetate extract, containing 70% (w/w) catechins, also increased LDL receptor-binding activity, protein, and mRNA, indicating that the effect was at the receptor level of gene transcription and that the catechins were the active constituents. The mechanism by which green tea upregulated the LDL receptor was investigated. Green tea decreased the cell cholesterol concentration (–30%) and increased the conversion of the sterol-regulated element binding protein (SREBP-1) from the inactive precursor form to the active transcription-factor form. Consistent with this, the mRNA of 3-hydroxy-3-methylglutaryl coenzyme-A reductase, the rate limiting enzyme in cholesterol synthesis, was also increased by green tea.

Immunomodulatory Effect

To determine the effects of tea on transplant-related immune function in vitro lymphocyte proliferation tests using phytohemagglutinin, mixed lymphocytes culture assay, IL-2, and IL-10 production from mixed lymphocyte proliferation were performed. Tea had immunosuppressive effects and decreased alloresponsiveness in the culture. The immunosuppressive effect of tea was mediated through a decrease in IL-2 production. Tea, assayed in cell culture, enhanced neopterin production in unstimulated peripheral mononuclear cells, whereas an effective reduction of neopterin formation in cells stimulated with concanavalin A, phytohemagglutinin or interferon (IFN)-γ was observed. Theaflavins potently suppressed IL-2 secretion, IL-2 gene expression, and the activation of NF-κB in murine spleens enriched for CD4(+) T-cells. Theaflavins also inhibited the induction of IFN-γ mRNA. However, the expression of the T(H2) cytokines IL-4 and IL-5, which lack functional NF-κB sites within their promoters was unexpectedly suppressed by theaflavins as well.

Insulin-enhancing Effect

Tea, as normally consumed, was shown to increase insulin activity more than 15-fold in vitro in an epididymal fat cell assay. The majority of the insulin-potentiating activity for green and oolong teas was owing to epigallocatechin gallate. For black tea, the activity was present in addition to epigallocatechin gallate, tannins, theaflavins, and other undefined compounds. Several known compounds found in tea were shown to enhance insulin with the greatest activity due to epigallocatechin gallate followed by epicatechin gallate, tannins, and theaflavins. Caffeine, catechin, and epicatechin displayed insignificant insulin-enhancing activities. Addition of lemon to the tea did not affect the insulin-potentiating activity. Addition of 5 g of 2% milk per cup decreased the insulin-potentiating activity one-third, and addition of 50 g of milk per cup decreased the insulin-potentiating activity approx 90%. Non-dairy creamers and soymilk also decreased the insulin-potentiating activity.

Iron Absorption

Tea, administered by gastric intubation to rats, did not affect iron absorption when tea was consumed for 3 days but when delivered in tea the absorption was decreased. Rats maintained on a commercial diet were fasted overnight with free access to water and then gavaged with 1 mL of ^{59}Fe labeled FeCl3 (0.1 m*M* or 1 m*M*) and lactulose (0.5 *M*) in water or black tea. Iron absorption was estimated from Fe retention. Intestinal permeability was evaluated by lactulose excretion in the urine. Iron absorption was lower with given with tea at both iron concentrations but tea did not affect lactulose excretion.

Lipid Peroxidation Activity

Solubilized green tea, administered orally to rats for 5 weeks, reduced lipid peroxidation products. The treatment produced increased activity of glutathione (GSH) peroxidase and GSH reductase, increased content of reduced GSH, a marked decrease in lipid hydroperoxides and malondialdehyde in the liver, an increase in the concentration of vitamin A by about 40%. A minor change in the measured parameters was observed in the blood serum. GSH content increased slightly, whereas the index of the total antioxidant status increased significantly. In contrast, the lipid peroxidation products, particularly malondialdehyde, was significantly diminished. In the central nervous tissue, the activity of superoxide dismutase and glutathione peroxidase decreased, whereas the activity of GSH reductase and catalase increased after drinking green tea. Moreover, the level of lipid hydroperoxides, 4-hydroksynonenal, and malondialdehyde decreased significantly[CS036].

Neuromuscular-blocking Action

Thearubigin fraction of black tea was investigated for neuromuscular-blocking action of botulinum neurotoxin types A, B, and E in the mouse phrenic nerve-diaphragm preparations. On binding, A (1.5 n*M*), B (6 n*M*), and E (5 n*M*) abolished indirect twitches within 50, 90, and 90 minutes, respectively. Thearubigin fraction mixed with each toxin protected against the neuromuscular-blocking action of botulinum neurotoxin types A, B, and E by binding with the toxins.

Oral Submucousal Fibrosis Effect

Tea, administered orally to 39 patients with oral submucous fibrosis, indicated that the treatment was effective for patients with abnormal hemorheology. The patients were divided into control and experimental groups. The control group included 22 oral submucous fibrosis patients who were treated by oral administration of vitamins A and D, vitamin B complex, and vitamin E. The experimental group included 17 patients who were treated with vitamins and tea pigment after their examination of hemorheology. The results showed that 7 of 12 patients in the experimental group with abnormal hemorheology had average 7.9 mm improvement on the open degree (58.3%), and the open degree of the other five patients whose hemorheology was normal only increased 2 mm (20%). The therapeutical results of the experimental group (58.3%) were significantly better than that of the control group (13.6%) ($p < 0.005$).

P-glycoprotein Activity

Green tea polyphenols (30 μg/mL) inhibited the photo-labeling of P-gp by 75% and increased the accumulation of rhodamine-123 threefold in a multidrug-resistant cell line CH(R)C5, indicating that the polyphenols interact with P-gp and inhibit its transport activity. The modulation of P-gp transport by polyphenols was a reversible process.

Photoprotection Effect

Tea extracts, administered topically, produced a dose- dependent inhibition of the erythema response evoked by UV radiation. The (–)-epigallocatechin-3-gallate and (–)-epicatechin-3 -gallate polyphenolic fractions were most efficient at inhibiting erythema, whereas (–)-epigallocatechin and (–)-epicatechin had little effect. On histological examination, skin treated with the extracts reduced the number of sunburn cells and protected epidermal Langerhans cells from UV damage. The extract also reduced damage that formed after UV radiation. Green tea polyphenols, applied topically to the human skin, prevented UVB-induced cyclobutane pyrimidine dimers, which are considered to be mediators of UVB-induced immune suppression and skin cancer induction. The treatment, prior to exposure to UVB, protected against UVB-induced local as well as systemic immune suppression in laboratory animals. Additionally, treatment of mouse skin inhibited UVB-induced infiltration of CD11b cells. CD11b is a cell-surface marker for activated macrophages and neutrophils, which are associated with induction of UVB-induced suppression of contact hypersensitivity responses. The treatment also resulted in reduction of the UVB-induced immunoregulatory cytokine IL-10 in skin as well as in draining lymph nodes, and an elevated amount of IL-12 in draining lymph nodes.

Protease Inhibition

Epigallocatechin-3-gallate, in cell culture at a concentration of 100 μM, reduced virus yield by 2 orders of magnitude producing an IC_{50} of 25 μM and a therapeutic index of 22 in Hep2 cells. The agent was the most effective when added to the cells during the transition from the early to the late phase of viral infection, suggesting that it inhibited one or more late steps in virus infection. One of these steps appears to be virus assembly, because the titer of infectious virus and the production of physical particles were much more affected than the synthesis of virus proteins. Another step might be the maturation cleavages carried out by adenain. When tested on adenain, epigallocatechin-3-gallate produced an IC_{50} of 109 μM.

Radical Scavenging Activity

Green tea, evaluated using the 1,1-diphenyl-2-picrylhydrazyl radical, indicated that the galloyl moiety showed more potent activity. The contribution of the pyrogallol moiety in the B-ring to the scavenging activity seemed to be less than that of the galloyl moiety.

Tetanus Toxin Protection

Thearubigin fraction of black tea was investigated for neuromuscular-blocking action on tetanus toxin in the mouse phrenic nerve-diaphragm preparations and on binding of this toxin to the synaptosomal membrane preparations of rat cerebral cortices. Tetanus toxin (4 μg/mL) abolished indirect twitches in the mouse phrenic nerve-diaphragm preparations within 150 minutes. Thearubigin fraction mixed with tetanus toxin blocked the inhibitory effect of the toxin.

Toxicity

Green tea, administered orally at a dose of 6 g per day in six doses to 42 patients who were asymptomatic and had manifested, progressive prostate specific antigen elevation with hormone therapy, produced grade 1 or 2 toxicity in 69% of the patients and included nausea, emesis, insomia, fatigue, diarrhea, abdominal pain, and confusion. However, six episodes of grade 3 toxicity and one episode of grade 4 toxicity also occurred, with the latter manifesting as severe confusion.

Toxicity Assessment

Ethanol (50%) extract of the entire plant, administered intraperitoneally to mice produced lethal dose $(LD)_{50}$ 316 mg/kg. Ethanol (95%) extract of the leaf, administered by gastric intubation to mice, produced LD_{50} 10 g/kg. Intraperitoneal administration produced CD_{90} 0.7 g/kg.

3

Larrea Tridentata

Larrea tridentata is a member of the caltrop family Zygophyllaceae. It is a native, drought-tolerant, evergreen shrub slowly growing to 2–4 m tall and 1.8 m wide, with numerous flexible stems projecting at an angle from its base. The bush is a group of four to 12 plants that shoot up from one plant in all directions. The root system consists of a shallow taproot and several lateral secondary roots, each approx 3 m in length and 20–35 cm deep. The taproot extends to a depth of approx 80 cm. The leaves are thick, waxy, resinous, 12–25 mm long, alternate leaves with two leaflets, pointed, yellow-green in color, covered with a varnish; darker and aromatic after rainfall. These leaves grow directly from the branches of the bush. The bush may lose some of leaves during extreme drought. Yellow flowers are solitary and axillary, numerous, up to 2 cm wide, mostly bloom from February to August, some individuals maintain flowers year-round. Fruits are small, reddish-white, globose, consisting of five united, indehiscent, one-seeded carpels that may or may not break apart after maturing. Each carpel is densely covered by long, gray or white trichomes.

Origin and Distribution

Larrea tridentata occurs throughout the Mojave, Sonoran, and Chihuahuan Deserts. Its distribution extends from southern California northeast through southern Nevada to the southwest corner of Utah and southeast through southern Arizona and New Mexico to western Texas and north-central Mexico. It is known to attain ages of several thousand years; some clones may be the earth's oldest living organisms. The age of the largest clone in Johnson Valley, California, is estimated at 9400 years; one estimated the average longevity to be 1250 years at a study site in Dateland, CA, and 625 years at a San Luis site. *Larrea tridentata* commonly grows on gentle well-drained slopes, plains, valley floors, and sand dunes and in arroyos at elevations up to 1515 m and occurs on calcareous, sandy, and alluvial soils with a layer of caliche. It can survive without any added water. Often the most abundant shrub, even forming pure stands.

Traditional Uses

Mexico. Decoction of the bark and dried branches is taken orally as an abortive and for diabetes. Decoction of the dried root is taken orally by pregnant humans as an abortive and for diabetes. Infusion of the shade-dried entire plant is taken orally to treat infectious diseases. Decoction of the dried leaf is taken orally for treatment of diabetes. Hot water extract of the dried leaf is taken orally as a blood purifier; to treat kidney problems, urinary tract infections, and frigidity; for gallstones, rheumatism and arthritis, diabetes, wounds, and skin injuries, displacement of the womb, and paralysis; and to dissolve tumors.

United States. Hot water extract of the dried leaf is taken orally as a stimulating expectorant and tonic, for tuberculosis, and is drank by Indians of the Southwest for bowel cramps, as a diuretic, and for venereal disease. Hot water extract of the dried leaf is used externally for wound healing. Hot water extract of the dried plant is taken orally for cancer. Effects described are from multicomponent reaction.

Medicinal Values

Alkaline Phosphatase Stimulation

Extract of the leaf, administered orally to adults, was active. Patients with subacute hepatic necrosis had negative workup, except for consumption of 15 tablets of the herbal extract per day for 4 months.

Anti-amoebic Activity

The resin of *Larrea* produced inhibitory activity at a concentration of 1 ppm on *Entamoeba invadens* PZ axenic cultures. The nordihydroguaiaretic acid activity was observed at 10^{-6} to 10^{-8} concentrations.

Antibacterial Activity

Methylene chloride extract of the dried aerial parts of the plants, on agar plate at a concentration of 1 g/mL, was active on *Bacillus subtilis*. Methanol extract of the shade-dried plant, on agar plate at a concentration of 0.6 mg/mL, was inactive on *Staphylococcus aureus*. A concentration of 10 mg/mL was inactive on *Escherichia coli* and *Pseudomonas aureuginosa*.

Antidiabetic Activity

Masoprocol, a compound derived from *Larrea*, administered orally to streptozotocin-induced diabetic rats at a dose of 0.83 mmol/kg body weight twice daily for 4 days, lowered glucose concentrations an average of 35% compared with vehicle (14.2 ± 1.1 vs 21.7 ± 1.0 mmol/L, $p < 0.001$). The animals were fed a 20% fat diet for 2 weeks before iv injection with streptozotocin (STZ, 0.19 mmol/kg). Diabetic animals (glucose 16–33 mmol/L) were treated with vehicle, metformin (0.83 mmol/kg), or masoprocol. Masoprocol decreased triglyceride level 80% compared with vehicle; nonesterified fatty acids and glycerol concentration by approx 65%, in comparison to vehicle. Adipocytes isolated from normal animals, treated with masoprocol (30 μmol/L) had higher basal and insulin-stimulated glucose clearance than adipocytes treated with vehicle ($p < 0.05$). Oral administration of masoprocol to two mouse models for type 2 diabetes reduced plasma glucose concentration approx 8 mmol/L in male C57BL/ks-db/db or C57BL/6J-ob/ob mice. The decline in plasma glucose concentration after masoprocol treatment in the mice was achieved without any change in plasma insulin concentration. Oral glucose tolerance improved, and the ability of insulin to lower plasma glucose concentrations was accentuated in masoprocol-treated db/db mice.

Antifungal Activity

Ethanol and methanol (41.5–100%) extracts prepared from 6 g of dried leaf and stem powders were active on *Aspergillus flavus*, *Aspergillus niger*, *Penicillium chrysogenum*, *Penicillium expansum*, *Fusarium poae*, and *Fusarium moniliforme*.

Antihypertriglyceridemic Activity

Masoprocol (nordihydroguaiaretic acid), administered orally to rodent models of type 2 diabetes at a dose range of 10 to 80 mg/kg twice daily for 4 to 8 days, decreased serum glucose and triglyceride levels. Masoprocol, at a dose of 40 or 80 mg/kg twice daily, significantly reduced hepatic triglyceride secretion ($p < 0.01$) and liver triglyceride content ($p < 0.001$), whereas lower doses of masoprocol decreased serum triglyceride without an apparent reduction in hepatic triglyceride secretion. The adipose tissue hormone-sensitive lipase was decreased, while adipose tissue lipoprotein lipase activity was increased in masoprocol-treated rats.

Anti-implantation Effect

Chloroform extracts of the dried leaf, twig, and stem, administered intragastrically to pregnant rats at a dose of 0.58 g/kg for 10 days, were active. The phenolic fraction, at a dose of 0.52 g/kg and methanol extract at a dose of 0.70 g/kg, were active. Water extract, at a dose of 1 g/kg and petroleum ether extract at a dose of 0.38 g/kg, were inactive.

Anti-tumor Activity

Water extract of the dried root, administered intraperitoneally to mice at a dose of 400 mg/kg, was inactive on Leuk (friend virus-solid) and Leuk-L1210. A dose of 500 mg/kg was inactive on sarcoma 180(ASC).

Antiviral Activity

Chloroform/methanol extract (1:1) of the dried leaf, in cell culture, was active on HIV-1 virus. TAT transactivation was inhibited. Ethanol acetate soluble fraction of the dried leaf, in cell culture at a concentration of 0.75 μg/ mL, was active on HIV virus vs HIV cytopathic effect.

Cytotoxic Activity

Water extract of the dried root, in cell culture, was inactive on CA-9KB, ED_{50} greater than 0.1 μg/mL. The methanol extract was active on Leuk-P388, ED_{50} 0.57 μg/mL.

Detoxification Activity

Phenolic resin, in increasing levels, was mixed with alfalfa pellets and fed to wood rats. Three detoxification pathways and urine pH, which are related to detoxification of allelochemicals, were measured. The excretion rate of two-phase II detoxification conjugates, glucuronides, and sulfides increased with increasing resin intake, whereas excretion of hippuric acid was independent of resin intake. Urine pH declined with increasing resin ingestion. The results indicated that a wood rat's tolerance to resin intake is related to the capacity for amination, sulfation, or pH regulation.

Gene Expression Inhibition

Chloroform/methanol extract (1:1) of the dried leaf, in cell culture, was active on hepatoma-Cos-7, IC_{50} 600.0 μg/mL vs TAT-dependent activation of HIV promoter bioassay.

Hepatotoxic Activity

The leaf, taken orally by a female adult, was active. A patient consumed 15 tablets of the leaf per day for 4 months. Approximately 1 year after stopping consumption, liver enzymes returned to normal and fatigue was no longer a complaint. Infusion of the dried leaf, taken orally by a female adult at variable doses, was active. The 60-year-old woman who took *Larrea tridentata* for 10 months developed severe hepatitis for which no other cause could be found. Despite aggressive supportive therapy, the patient's condition deteriorated and required orthotropic liver transplantation. Dried leaves, administered orally to adults at variable doses, were active. A public warning has been issued by the US Centers for Disease Control based on reports of liver toxicity after use of *Larrea tridentata* tea. Dried leaves, administered orally to adults of both sexes at variable doses, were active. The plant, administered orally to adults at variable doses, was active. Dried leaves, administered orally to adults at variable doses, were active. One case of hepatotoxicity induced by *Larrea tridentata* taken as a nutritional supplement was reported. Thirteen patients were identified for whom *Larrea tridentata* tincture for internal use was prescribed. Additionally, 20 female and three male patients were identified from whom an extract of *Larrea tridentata* in castor oil for topical use was prescribed. None of the patients had history of liver disease. In all of the cases, *Larrea tridentata* was given as either part of a complex herbal formula individualized for each patient containing less than 10% *Larrea tridentata* tincture or an extract in castor oil for topical use. The four patients with complete before and after blood chemistry panels and complete blood counts had no indication of liver damage from use of *Larrea tridentata*. This included one patient who was taking medications with significant potential for hepatotoxicity. No patient showed any sign of organ damage during the follow-up period. Nordihydroguaiaretic acid is a lignan found in high amounts (up to 10% by dry weight) in the leaves and twigs of *Larrea tridentata*. It has been shown to reduce cystic nephropathy in the rats, but no reports have been made concerning the hepatotoxic potential of the compound. *Larrea*-containing medications induce hepatotoxicity and nephrotoxicity in humans. Intraperitoneal administration of nordihydroguaiaretic acid produced LD_{50} 75 mg/kg. Administration is associated with a time- and dose-dependent increase in serum alanine aminotransferase levels, which suggest liver damage. Freshly isolated mouse hepatocytes are more sensitive to nordihydroguaiaretic acid than human melanoma cells. Glucuronidation was identified as a potential detoxification mechanism for nordihydroguaiaretic acid.

Insecticide Activity

Acetone extracts of the dried leaf, dried root, and dried stem, at a low concentration, were inactive on *Culex quinquefasciatus*. Water extract of the dried leaf, administered intravenously, produced weak activity on *Periplaneta americana*.

Pigmented Cholelithiasis Prophylaxis

Powdered hydroalcoholic extract of the leaf was administered to Syrian golden hamster (ChCM). The extract was added to the lithogenic diet (basic diet plus 25 000 IU of vitamin A) at the 4% level for 70 days. The results indicated that the *Larrea*-fed group did not develop pigment cholelithiasis, whereas the group that received the lithogenic diet alone developed cholelithiasis in 63% of the cases. It is suggested that the active principle present in the leaves of *Larrea*, are responsible for the prevention is nordihydroguaiaretic acid, a potent antioxidant. However, the hamsters that received the diet containing *Larrea* showed serious signs of toxicity and pathological changes, such as marked reduction of growth, pronounced irritability and aggressiveness, and a marked hypoplasia of testicular and accessory sex glands.

Plant Growth Inhibitor

Water extract of the aerial parts, administered externally, was toxic to tomato plant seedlings.

Prothrombin Time Increased

Extract of the leaf, administered orally to adults, was active. Patients with subacute hepatic necrosis had negative workups, except for their consumption of 15 tablets of the herbal extract daily for 4 months.

Skin Cancer Chemoprevention

Topically applied nordihydroguaiaretic acid prevented phorbol ester promotion of tumors in mouse skin, indicating that nordihydroguaiaretic acid may be a candidate drug for the chemoprevention of skin cancer. Nordihydroguaiaretic acid, investigated as a potential inhibitory agent for ultraviolet-B (UVB)-induced signaling pathways in the human keratinocyte cell line HaCaT, significantly inhibited UVB-induced c-fos and activator protein-1 transactivation. It also inhibited the activity of phosphatidylinositol 3-kinase, a UVB-inducible enzyme that contributes c-fos expression and activator protein-1 transactivation by inhibiting the phosphatidylinositol 3-kinase signaling pathway.

4

CANNABIS SATIVA

Cannabis sativa is an annual herb of the *Moraceae* family that grows to 5 m tall. It is usually erect; stems variable, with resinous pubescence, angular, sometimes hollow, especially above the first pairs of true leaves; basal leaves opposite, the upper leaves alternate, stipulate, long petiolate, palmate, with 3–11, rarely single, lanceolate, serrate, acuminate leaflets up to 10 cm long, 1.5 cm broad. Flowers are monoecious or dioecious, the male in axillary and terminal panicles, apetalous, with five yellowish petals and five poricidal stamens; the female flowers germinate in the axils and terminally, with one single-ovulate ovary. Fruit is brown, shining achene, variously marked or plain, tightly embraces the seed with its fleshy endosperm and curved embryo; late summer to early fall; year-round in tropics. Drug-producing selections grow better and produce more drugs in the tropics; oil- and fiber-producing plants thrive better in the temperate and subtropical areas. The form of the plant and the yield of fiber from it vary according to climate and particular variety. Varieties cultivated for their fibers have long stalks, branch very little, and yield only small quantities of seed. Oil seed varieties are small, mature early, and produce large quantities of seed. Varieties grown for the drugs are small, much branched with smaller dark-green leaves. Between these three main types of plants are numerous varieties that differ from the main one in height, extent of branching, and other characteristics.

ORIGIN AND DISTRIBUTION

Native to Central Asia and long cultivated in Asia, Europe, and China. Now a widespread tropical, temperate, and subarctic cultivar.

Cannabis sativa has been cultivated for more than 4500 years for different purposes, such as fiber, oil, or narcotics. The oldest use of hemp is for fiber, and later the seeds were used for culinary purposes. Plants yielding the drug were discovered in India, cultivated for medicinal purposes as early as 900 BC. In medieval times, it was brought to North Africa, where currently it is cultivated exclusively for hashish or kif.

Traditional Uses

Afghanistan. Hot water extract of the resin is taken orally to induce abortion.

China. Hot water extract of the inflorescence is taken orally for wasting diseases, to clear the blood, to cool the temperature, to relieve fluxes, for rheumatism, to discharge pus, and to stupefy and produce hallucinations. The seed is taken orally as an emmenagogue. Decoction of the seed is taken orally as an anodyne, an emmenagogue, a febrifuge, for migraine, and for cancer. It is taken orally as a hallucinogen and externally for rheumatism.

Guatemala. The leaves are used externally to relieve muscular pains.

India. Hot water extract of the dried entire plant is taken orally as a narcotic and to relieve pain of dysmenorrhea. Hot water extract of the dried flower and leaf is taken orally for dyspepsia and gonorrhea and as a nerve stimulant. Hot water extract of the inflorescence of female plants is taken orally as an abortifacient. Hot water extract of the leaf is taken orally to relieve menstrual pain. For cuts, boils, and blisters, leaf paste is applied topically for 4 days. Hot water extract of the bark is taken orally for hydrocele and other inflammation. Extract of the leaves is used as an insect repellant. Hot water extract of the seed is taken orally as an emmenagogue. The powdered seed is taken orally as an aid in conception. One gram of seeds is powdered, then mixed with water, and given to women in the morning before breakfast for 7 days after menstruation. The use of pepper and cane sugar is avoided. Paste of dried leaves is applied over the anus in the morning and evening for piles. The dried leaf juice is used externally on cuts and piles and taken orally as an anthelmintic. To eliminate cough, bronchitis, and other respiratory ailments, a half tablespoonful of powdered dried leaves is mixed with an equal amount of honey and taken orally three times daily. Seed oil is used externally for burns. The oil is extracted by roasting the seeds. Seeds are taken orally for diabetes, hysteria, and sleeplessness. The aerial parts are smoked to

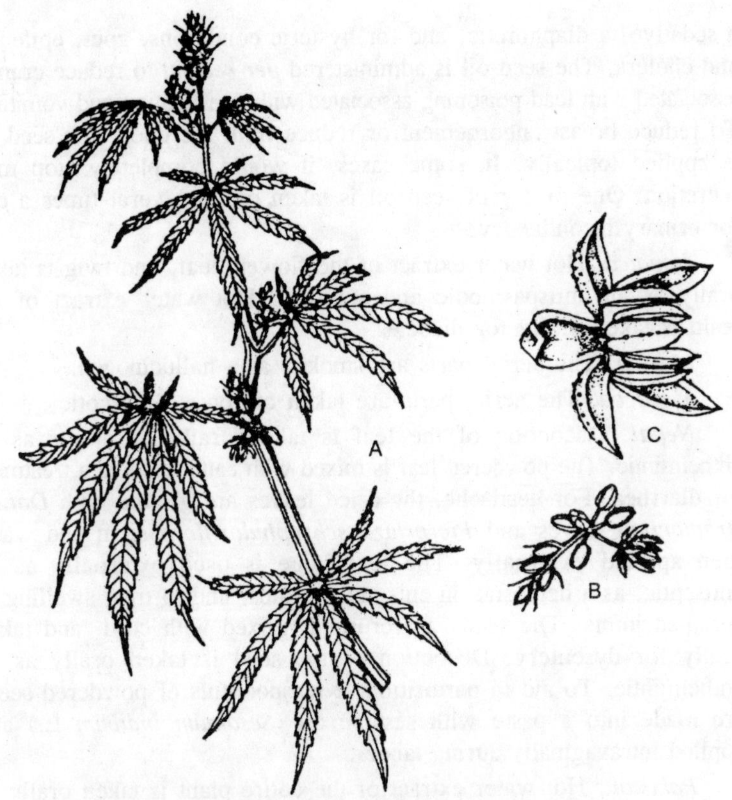

Fig. 4.1. Cannabis sativa. A–Male shoot; B and C–Flowers.

decrease nausea and vomiting induced by anticancer drugs. Hot water extract of the aerial parts is taken orally by males as an aphrodisiac. The dried aerial parts are smoked by women to increase their amorous prowess. The fresh leaves are taken orally for hemorrhoids. Hot water extract of the dried leaf and seed is taken orally for stomach troubles and indigestion. Fresh leaf juice is administered intraural to treat earache. The fruit is used externally for skin diseases. The unripe fruit is taken orally to induce sleep.

Iran. Fluidextract of the dried flowering top or the dried fruit is taken orally for abdominal pain associated with indigestion, for pain associated with cancer, for rheumatoid arthritis, for gastric cramps or neuralgia, for coughing, and as a hypnotic. Fluidextract of the dried fruit is taken orally for whooping cough, as a hypnotic, and a tranquilizer. The dried seed is taken orally as a diuretic. An infusion is taken orally as an analgesic in rheumatism or rheumatoid arthritis, a sedative, a diaphoretic, and for hysteric conditions, gout, epilepsy, and cholera. The seed oil is administered *per rectum* to reduce cramps associated with lead poisoning associated with constipation and vomiting. To reduce breast engorgement or reduce milk secretion, the seed oil is applied topically. In some cases, it would completely stop milk secretion. One to 2 g of seed oil is taken orally several times a day for urinary incontinency.

Jamaica. Hot water extract of the flower, leaf, and twig is taken orally as an antispasmodic and anodyne. Hot water extract of the resin is taken orally for diabetes.

Mexico. The aerial parts are smoked as a hallucinogen.

Morocco. The aerial parts are taken orally as a narcotic.

Nepal. Decoction of the leaf is taken orally by adults as an anthelmintic. The powdered leaf is mixed with cattle feed as a treatment for diarrhea. For headache, the dried leaves are ground with *Datura stramonium* leaves and *Picrorhiza schrophulariflora* stem and water then applied externally. The leaf juice is used externally as an antiseptic, as a hemostat on cuts and wounds, and to treat swelling of sprained joints. The seeds are crushed, mixed with curd, and taken orally for dysentery. Decoction of the seed is taken orally as an anthelmintic. To aid in parturition, 2 teaspoonfuls of powdered seeds are made into a paste with sesame oil (*Sesamum indicum* L.) and applied intravaginally during labor.

Pakistan. Hot water extract of the entire plant is taken orally as a parturifacient. Infusion of the leaf is taken orally for general weakness.

Saudi Arabia. The aerial parts, mixed with honey, sugar, and nutmeg, are taken orally as a psychotropic.

Senegal. The seed is taken orally as an emmenagogue.

South Africa. Hot water extract of the entire plant is taken orally for asthma[CS107]. Hot water extracts of the root and seed are taken orally to induce abortion, labor, and menstruation.

United States. Fluidextract of the inflorescence is taken orally as a narcotic, antispasmodic, analgesic, and aphrodisiac. Hot water extract of the flowering top is taken orally as a potent antispasmodic, anodyne, and narcotic. One teaspoon of plant material is steeped in 2 cups of boiling water, and 1 tablespoonful is taken two to four times a day. The dried aerial parts are smoked by both sexes as an aphrodisiac.

Vietnam. The seeds are taken orally as an emmenagogue.

West Indies. Hot water extract of the entire plant is taken orally as an antispasmodic.

Yugoslavia. Hot water extract of the seed is taken orally for diabetes.

Zimbabwe. Hot water extract of the aerial parts is taken orally as a treatment for malaria.

Medicinal Values

Abortifacient Activity

Alcohol extract of the dried leaf, administered intragastrically to pregnant rats at a dose of 125 mg/kg, produced teratogenic effects. Water extract of the dried leaf, administered intragastrically to pregnant rats at variable dosage levels on days 6–15 of pregnancy was active.

Acute Cardiovascular Fatalities

Six cases of possible acute cardiovascular death in young adults were reported where very recent cannabis ingestion was documented by the presence of Δ-9-tetrahydrocannabinol (Δ-9-THC) in postmortem blood samples. A broad toxicological blood analysis could not reveal other drugs.

Acute Panic Reaction (Koro)

Koro, an acute panic reaction related to the perception of penile retraction, was once considered limited to specific cultures. Over 70 American men responded by telephone to report negative reactions to cannabis. Three of them (Caucasians aged 22–26 years with considerable experience with cannabis) spontaneously mentioned experiencing symptoms of Koro after smoking cannabis. All three cases occurred after the participants had heard about cannabis-induced Koro and used the drug in a novel setting or atypical way. Two of the men had body dysmorphia, which may have contributed to symptoms. All three decreased their cannabis consumption after the Koro experience. Several factors may have interacted to create the symptoms. These include previous knowledge of cannabis-induced Koro, the use of cannabis in a way that might heighten a panic reaction, and poor body image.

Adverse Effects

A causal role of acute cannabis intoxication in motor vehicle and other accidents has been shown by the presence of measurable levels of Δ-9-THC in the blood of drivers in the absence of alcohol or other drugs, by surveys of driving under the influence of cannabis, and by significantly higher accident culpability risk of drivers using cannabis. Evidence demonstrated that cannabis dependence, both behavioral and physical, occurred in about 7–10% of regular users, and that early onset of use—especially of weekly or daily use—is a strong predictor of future dependence. Cognitive impairments of various types are readily demonstrable during acute cannabis intoxication, but there is no suitable evidence yet available to permit a decision as to whether long-lasting or permanent functional losses can result from chronic heavy use in adults. The gender effects on progression to treatment entry and on the frequency, severity, and related complications of the *Diagnostic and Statistical Manual of Mental Disorders*, 3rd edition revised drug and alcohol dependence among 271 substance-dependent patients (mean age: 32.6 years; 156 women) was studied. There was no gender difference among patients in the age at onset of regular use of any substance. Women experienced fewer years of regular use of opioids and cannabis and fewer years of regular alcohol drinking before entering treatment. Although the severity of drug and alcohol dependence did not differ by gender, women reported more severe psychiatric, medical, and employment complications. In a 3-day, double-blind, randomized, counterbalanced study, the behavioral, cognitive, and endocrine effects of 2.5 and 5 mg intravenous Δ-9-THC were characterized in 22 healthy individuals, who had been exposed to cannabis but had never been diagnosed with a cannabis abuse disorder. Prospective safety data at 1,3, and 6 months post-study was also analyzed. Δ-9-THC produced schizophrenia-like positive and negative symptoms, altered perception, increased anxiety and plasma cortisol, euphoria, disrupted immediate and delayed word recall, sparing recognition recall, impaired performance on tests of distractibility, verbal fluency, and working memory, but did not impair orientation. This study examined the behavioral and neurochemical (cannabinoid *CB1* receptor gene expression) changes induced by spontaneous cannabinoid withdrawal in mice. Cessation of CP-55,940 treatment in tolerant mice induced a spontaneous time-dependent behavioral withdrawal syndrome consisting of marked increases (140%) in motor activity, number of rearings (170%), decreases in grooming (57%), wet-dog shakes (73%), and rubbing behaviors (74%) on day 1, progressively reaching values similar to vehicle-treated mice on day 3. This spontaneous cannabinoid withdrawal resulted in *CB1* gene expression up-regulation (20–30%) in caudate-putamen, ventromedial hypothalamic nucleus, central amygdaloid nucleus, and CA1, whereas in the CA3 field of hippocampus, a significant decrease (15–20%) was detected.

Alcohol Interaction

The complementary DNA and genomic sequences encoding G protein-coupled cannabinoid receptors (CB1 and CB2) from several species were cloned. This has facilitated discoveries of endogenous ligands (endocannabinoids). Two fatty acid derivatives characterized to be arachidonylethanolamide and 2-arachidonylglycerol isolated from both nervous and peripheral tissues mimicked the pharmacological and behavioral effects of Δ-9-THC. The down-regulation of CB1 receptor function and its signal transduction by chronic alcohol was demonstrated. The observed down-regulation of CB1 receptor-binding and its signal transduction resulted from the persistent stimulation of receptors by the endogenous CB1 receptor agonists arachidonylethanolamide and 2-arachidonylglycerol, whose synthesis is increased by chronic alcohol treatment. The deletion of CB1 receptor has been shown to block voluntary alcohol intake in mice.

Allergenic Effect

An "All India Coordinated Project on Aeroallergens and Human Health" was undertaken to discover the quantitative and qualitative prevalence of aerosols at 18 different centers in the country. Predominant airborne pollens were *Holoptelea*, *Poaceae*, *Asteraceae*, *Eucalyptus*, *Casuarina*, *Putanjiva*, *Cassia*, *Quercus*, *Cocos*, *Pinus*, *Cedrus*, *Ailanthus*, *Cheno/Amaranth*, *Cyperus*, *Argemone*, *Xanthium*, *Parthenium*, and others. Clinical and immunological evaluations revealed some allergenically important taxa. Allergenically important pollens were *Prosopis juliflora*, *Ricinus communis*, *Morus*, *Mallotus*, *Alnus*, *Querecus*, *Cedrus*, *Argemone*, *Amaranthus*, *Chenopodium*, *Holoptelea*, *Brassica*, *Cocos*, *Cannabis*, *Parthenium*, *Cassia*, and grasses. In the multitest routine skin-test battery, 78 of 127 patients tested (61%) were cannabis-test positive. Thirty of the 78 patients were randomly selected to determine if they had allergic rhinitis and/or asthma symptoms during the cannabis pollination period. By history, 22 (73%) claimed respiratory symptoms in July through September. All 22 of these subjects were also skin test-positive to weeds pollinating during the same period as cannabis (ragweed, pigweed, cocklebur, Russian thistle, marsh elder, or kochia).

Amnesic Syndrome

A 26-year-old woman suffered disseminated intravascular coagulation (DIC) and a brief respiratory arrest following recreational use of 3,4-methylenedioxymethamphetamine (MDMA, or "ecstasy") together with amyl nitrate, lysergic acid (LSD), cannabis, and alcohol. She was left with residual cognitive and physical deficits, particularly severe anterograde memory disorder, mental slowness, severe ataxia, and dysarthria. Follow-up investigations have shown that these have persisted, although there has been some improvement in verbal recognition memory and in social functioning. Magnetic resonance imaging and quantified positron emission tomography investigations revealed severe cerebellar atrophy and hypometabolism accounting for the ataxia and dysarthria; thalamic, retrosplenial, and left medial temporal hypometabolism to which the anterograde amnesia can be attributed. There was some degree of frontotemporal–parietal hypometabolism, possibly accounting for the cognitive slowness. The putative relationship of these abnormalities to the direct and indirect effects of MDMA toxicity, hypoxia, and ischemia was considered.

Amyotrophic Lateral Sclerosis

One hundred thirty one respondents with amyotrophic lateral sclerosis—13 of whom reported using cannabis in the last 12 months—were examined. The results indicated that cannabis might be moderately effective at reducing symptoms of appetite loss, depression, pain, spasticity, and drooling. Cannabis was reported ineffective in reducing difficulties with speech and swallowing, and sexual dysfunction. The longest relief was reported for depression (approx 2–3 hours).

Analgesic Activity

Ethanol (50%) extract of the entire plant, administered intraperitoneally to mice at a dose of 250 mg/kg, was active vs tail pressure method. Flavonoid fraction of the leaf, administered intraperitoneally to mice, was active. The inflorescence, administered orally to male rats, produced weak activity vs paw pressure test, effective dose $(ED)_{50}$ 35.5 mg/kg and hot plate method, ED_{50} 53 mg/kg. Petroleum ether and ethanol (95%) extracts of the dried aerial parts, administered intragastrically to mice, was active vs phenylbenzoquinone-induced writhing, inhibitory concentration $(IC)_{50}$ 0.013 mg/kg and 0.045 mg/kg, respectively.

Analgesic Effect

Ajulemic acid (AJA, CT-3, or IP-751), administered to healthy human adults and patients with chronic neuropathic pain, demonstrated a complete absence of psychotropic actions. It proved to be more effective than placebo in reducing this type of pain as measured by the visual analog scale. Signs of dependency were not observed after withdrawal at the end of the 1-week treatment period. Forty women undergoing elective abdominal hysterectomy were investigated in a randomized, double-blind, placebo-controlled, single-dose trial. Randomization took place when postoperative patient-controlled analgesia was discontinued on the second postoperative day. When patients requested further analgesia, they received a single, identical capsule of either 5 mg of oral Δ-9-THC ($n = 20$) or placebo ($n = 20$) in a double-blind fashion. The primary outcome measure was summed pain intensity difference (SPID) at 6 hours after administration of the study medication derived from visual analog pain scores on movement and at rest. Secondary outcome measures were time-to-rescue medication and adverse effects of study medication. Mean (standard deviation [SD]) visual analog scale pain scores before medication in the placebo and Δ-9-THC groups were 6.3(2.6) and 6.4(1.3) cm on movement, and 3.2(1.9) and 3.3(0.9) at rest, respectively. There were no significant differences in mean (95% confidence interval [CI] of the difference) SPID at 6 hours between the groups (placebo 7.9, Δ-9-THC 4.3[−1.8 to 9] cm per hour on movement; placebo 8.8, Δ-9-THC 4.9[−0.2 to 8.1] cm per hour at rest) and time to rescue analgesia (placebo 217, Δ-9-THC 163[−22 to 130] minutes). Increased awareness of surroundings was reported more frequently in patients receiving Δ-9-THC (40 vs 5%, $p = 0.04$). There were no other significant differences with respect to adverse events. THC, morphine, and a THC–morphine combination were administered to 12 healthy subjects using experimental pain models (heat, cold, pressure, and single and repeated transcutaneous electrical stimulation). THC (20 mg), morphine (30 mg), THC–morphine (20 mg THC + 30 mg morphine), or placebo were given orally as single dose. Reaction time, side effects (visual analog scales), and vital functions were monitored. For the pharmacokinetic profiling, blood samples were collected. THC did not significantly reduce pain. In the cold and heat tests, it even produced hyperalgesia, which was completely neutralized by THC–morphine. A slight additive analgesic effect was observed for THC–morphine in the electrical stimulation test. No analgesic effect resulted in the pressure and heat test, with neither THC nor THC–morphine.

Psychotropic and somatic side effects (sleepiness, euphoria, anxiety, confusion, nausea, dizziness, etc.) were common, but usually mild. Three cannabis-based extracts Δ-9-THC, cannabidiol [CBD], and a 1:1 mixture of them both) were given over a 12-week period in a randomized, double-blind, placebo-controlled, crossover trial. Extracts, which contained THC, proved most effective in symptom control. Regimens for the use of the sublingual spray emerged and a wide range of dosing requirements was observed. Side effects were common, reflecting a learning curve for both patient and study team. These were generally acceptable and little different to those seen when other psychoactive agents are used for chronic pain. Over a 6-week period 209 chronic noncancer pain patients were studied. Seventy-two (35%)

subjects reported ever having used cannabis. Thirty-two (15%) subjects reported having used cannabis for pain relief (pain users), and 20 (10%) subjects were currently using cannabis for pain relief. Thirty-eight subjects denied using cannabis for pain relief (recreational users). Compared with nonusers, pain users were significantly younger ($p = 0.001$) and were more likely to be tobacco users ($p = 0.0001$). The largest group of patients using cannabis had pain caused by trauma and/or surgery (51%), and the site of pain was predominantly neck/upper body and myofascial (68 and 65%, respectively). The median duration of pain was similar in both pain users and recreational users (8 vs 7 years; $p = 0.7$). There was a wide range of amounts and frequency of cannabis use. Of the 32 subjects who used cannabis for pain, 17 (53%) used four puffs or less at each dosing interval, eight (25%) smoked a whole cannabis cigarette (joint), and four (12%) smoked more than one joint. Seven (22%) of these subjects used cannabis more than once daily, five (16%) used it daily, eight (25%) used it weekly, and nine (28%) used it rarely. Pain, sleep, and mood were most frequently reported as improving with cannabis use, and "high" and dry mouths were the most commonly reported side effects. Patients with chronic pain completed a questionnaire about the type of cannabis used, the mode of administration, the amount used and the frequency of use, and their perception of the effectiveness of cannabis on a set of pain-associated symptoms and side effects. Fifteen patients (10 males) were interviewed (median age, 49.5 years; range, 24-68 years). All patients smoked herbal cannabis for therapeutic reasons (median duration of use, 6 years; range, 2 weeks–37 years). Seven patients only smoked at night (median dose eight puffs, range two to eight puffs), and eight patients used cannabis mainly during the day (median dose of three puffs; range, two to eight puffs); the median frequency of use was four times per day (range, 1 to 16 times/day). Twelve patients reported improvement in pain and mood, whereas 11 reported improvement in sleep. Eight patients reported a "high;" six denied a "high." Tolerance to cannabis was not reported. THC was administered to six patients with chronic pain at doses 5–20 mg/day. A sufficient pain relief had been achieved in three patients. The other three suffered from intolerable side effects, such as nausea, dizziness, and sedation without a reduction of pain intensity. In these cases, the treatment was continued with other analgesics.

Ankylosing Spondylitis

Ankylosing spondylitis is a systemic disorder occurring in genetically predisposed individuals. The disease course appears to be characterized by bouts of partial remission and flares. There were 214 patients questioned (169 men, 45 women; average disease duration, 25 years; age of disease onset, 22 years). The main symptoms of flare were pain (all groups), immobility (90%), fatigue (80%), and emotional symptoms, such as depression, withdrawal, and anger, (75%). All of patients experienced between one and five localized flares per year. Fifty-five percent of the groups contained patients ($n = 85$) who experienced a generalized flare. The main perceived triggers of flare were stress (80%) and "overdoing it" (50%). Patients reported that a flare might last anywhere from a few days to a few weeks and relief from flare were by analgesic injections (including opiates), relaxation, sleep, and cannabis. Three-quarters of the groups agreed that there was no long-term effect on the ankylosing sponylitis following a flare.

Anti-arthritic Effect

Oral administration of AJA, a cannabinoid acid devoid of psychoactivity, reduced joint tissue damage in rats with adjuvant arthritis. Peripheral blood monocytes (PBM) and synovial fluid monocytes (SFM) were isolated from healthy subjects and patients with inflammatory arthritis, respectively, treated with AJA (0–30 m*M*) in vitro, and then stimulated with lipopolysaccharide. Cells were harvested for messenger RNA (mRNA), and supernatants were collected for cytokine assay. Addition of AJA to PBM and SFM in vitro reduced both steady-state levels of interleukin-1γ (IL-1γ) mRNA and secretion of IL-1γ in a concentration-dependent manner. Suppression was maximal (50.4%) at 10 m*M* AJA ($p < 0.05$ vs untreated controls, $n = 7$). AJA did not influence tumor necrosis factor-α (TNF-α) gene expression in or secretion from PBM.

Anticonvulsant Activity

Ethanol (95%) extract of the entire plant, administered subcutaneously to male mice and rats at a dose of 2–4 mL/kg, was active vs metrazole and electroshock, respectively. A dose of 4 mL/kg was inactive vs strychnine convulsions in mice. The entire plant, smoked by 29 patients with epilepsy under the age of 30 years, was active. It must be noted that in some species, cannabinoids can precipitate epileptic seizures. Tincture of the resin, administered intraperitoneally to mice at a dose of 25 mg/kg, produced 80% protection vs pentylenetetrazole convulsions.

Anti-emetic Activity

In a qualitative study of self-care in pregnancy, birth, and lactation within a nonrandom sample of 27 women in British Columbia, Canada,

20 women (74%) experienced pregnancy- induced nausea. Ten of these women used antiemetic herbal remedies, which included ginger, peppermint, and cannabis. Only ginger has been subjected to clinical trials among pregnant women, although the three herbs were clinically effective against nausea and vomiting in other contexts, such as chemotherapy-induced nausea and post-operative nausea. CBD, a major non- psychoactive cannabinoid administered by oral infusion to rats with nausea elicited by lithium chloride, and with conditioned nausea elicited by a flavor paired with lithium chloride, was active[CS382]. Oral nabilone, oral dronabinol (THC), and intramuscular levonantradol were administered to 1366 patients. Cannabinoids were more effective antiemetics than prochlorperazine, metoclopramide, chlorpromazine, thiethylperazine, haloperidol, domperidone, or alizapride. Relative risk was 1.38 (95% CI 1.18–1.62), number-needed-to-treat (NNT) was 6 for complete control of nausea; relative risk was 1.28 (CI 1.08–1.51), NNT 8 for complete control of vomiting. Cannabinoids were not more effective in patients receiving very low or very high emetogenic chemotherapy. In crossover trials, patients preferred cannabinoids for future chemotherapy cycles: relative risk 2.39 (2.05–2.78), NNT 3. Some potentially beneficial side effects occurred more often with cannabinoids: "high" 10.6 (6.86–16.5), NNT 3; sedation or drowsiness 1.66 (1.46–1.89), NNT 5; euphoria 12.5 (3–52.1), NNT 7. Harmful side effects also occurred more often with cannabinoids: dizziness 2.97 (2.31– 3.83), NNT 3; dysphoria or depression 8.06 (3.38–19.2), NNT 8; hallucinations 6.10 (2.41–15.4), NNT 17; paranoia 8.58 (6.38–11.5), NNT 20; and arterial hypotension 2.23 (1.75–2.83), NNT 7. Patients given cannabinoids were more likely to withdraw because of side effects (relative risk 4.67 [3.07–7.09]; NNT 11).

Antifungal Activity

Ethanol (50%) extract of the dried leaf was active on *Rhizoctonia solani*, mycelial inhibition was 65.99%. Water extract of the fresh leaf on agar plate at a concentration of 1:1 was active on *Fusarium oxysporum*. The water extract also produced strong activity on *Ustilago maydis* and *Ustilago nuda*. Water extract of the fresh shoot on agar plate was inactive on *Helminthosporium turcicum*.

Antiglaucomic Activity

Water extract of the dried entire plant, administered intravenously to Rhesus monkeys and rabbits at a dose of 0.01 μg/animal, was active. The intraocular pressure rose for 24 hours postinjection, then fell for 3 days. A dose of 25 μg/animal, administered intravenously to rabbits, was also active. The effect was not influenced by atropine, scopolamine, methysergide, haloperidol, chlorpromazine, spironolactone, yohimbine or dexamethasone. Partial inhibition was seen when galactose, glucose or mannose were administered intravenously, concurrently. Water extract of the dried leaf and stem, applied opthalmically to rabbits was active.

Antigonadotropin Effect

Ethanol (80%) extract of the dried aerial parts, administered intragastrically to male langurs at a dose of 14 mg/kg daily for 90 days produced equivocal effect.

Anti-inflammatory Activity

Petroleum ether and ethanol (95%) extracts of the dried aerial parts, applied externally on mice at a dose of 100 μg/ear, was active vs tissue plasminogen activator-induced erythema of the ear. CBD was administered orally to rats at doses of 5–40 mg/kg daily for 3 days after the onset of acute inflammation induced by intraplantar injection of 0.1 mL carrageenan (1% w/v in saline). CBD had a time- and dose-dependent antihyperalgesic effect after a single injection. Edema following carrageenan peaked at 3 hours and lasted 72 hours. A single dose of CBD reduced edema in a dose-dependent fashion and subsequent daily doses produced further time- and dose-related reductions. There were decreases in prostaglandin E2 (PGE2) plasma levels, tissue cyclooxygenase activity, production of oxygen-derived free radicals, and nitric oxide ([NO], nitrite/nitrate content) after three doses of CBD. The effect on NO seemed to depend on a lower expression of the endothelial isoform of NO synthase.

Antispermatogenic Effect

Sixteen healthy chronic marijuana smokers were associated with a decline in sperm concentration and total sperm count during the fifth and sixth weeks after 4 weeks of high-dose smoking (8–20 cigarettes/day). The dried aerial part, taken by inhalation daily, decreases the quantity as well as quality of spermatozoa. Ethanol (80%) extract of the dried aerial parts, administered intragastrically to langurs at a dose of 14 mg/kg daily for 90 days, was equivocal. Ethanol (95%) extract of the dried aerial parts, administered intraperitoneally to mice at a dose of 2 mg/animal daily for 45 days, produced a complete arrest of spermatogenesis. The effect was reversible.

Antistress Activity

The leaf smoke, in combination with hashish smoke, administered to rats housed in a wire cage inside a larger cage with a cat, was equivocal. The rats' brains were dissected and measured for protein and catecholamine levels.

Anti-tumor Activity

Arachidonyl ethanolamide, in three cervical carcinoma (CxCa) cell lines at increasing doses with or without antagonists to receptors to arachidonyl ethanolamide, induced apoptosis of CxCa cell lines via aberrantly expressed vanilloid receptor-1. Arachidonyl ethanolamide-binding to the classical CB1 and CB2 cannabinoid receptors mediated a protective effect. A strong expression of the three forms of arachidonyl ethanolamide receptors was observed in ex vivo CxCa biopsies. Three cannabis constituents, CBD, Δ-8-THC, and cannabinol displayed anti-proliferative activity in several human cancer cell lines in vitro. They were oxidized to their respective paraquinones 2,4, and 6. Quinone 2 significantly reduced cancer growth of HT-29 cancer in nude mice. Δ-9-THC binds and activates membrane receptors of the 7-transmembrane domain, G protein-coupled superfamily. Several putative endocannabinoids have been identified, including anandamide (AEA), 2-arachidonyl glycerol, and noladin ether. Synthesis of numerous cannabinomimetics has expanded the repertoire of cannabinoid receptor ligands with the pharmacodynamic properties of agonists, antagonists, and inverse agonists. These ligands have proven to be powerful tools both for the molecular characterization of cannabinoid receptors and the delineation of their intrinsic signaling pathways. Much of the understanding of the signaling mechanisms activated by cannabinoids has been derived from studies of receptors expressed by tumor cells. Cannabinoids and their derivatives exerted palliative effects in cancer patients by preventing nausea, vomiting, and pain and by stimulating appetite. These compounds have been shown to inhibit the growth of tumor cells in culture and animal models by modulating key cell-signaling pathways. Cannabinoids are usually well tolerated, and do not produce the generalized toxic effects of conventional chemotherapies.

Anxiolytic Activity

AEA, a primary endogenous ligand of the brain cannabinoid receptors, is released in selected regions of the brain and is deactivated through a two-step process consisting of transport into cells followed by intracellular hydrolysis. Pharmacological blockade of the enzyme fatty acid amide hydrolase (FAAH), which is responsible for intracellular AEA degradation, produced anxiolytic-like effects in rats without causing the wide spectrum of behavioral responses typical of direct-acting cannabinoid agonists. These findings suggest that AEA contributes to the regulation of emotion and anxiety, and that FAAH might be the target for a novel class of anxiolytic drugs.

Attention Deficit Hyperactivity Disorder

Attention defict hyperactivity disorder has been considered a mental and behavioral disorder of childhood and adolescence. It is being increasingly recognized in adults, who may have psychiatric comorbidity with secondary depression, or a tendency to drug and alcohol abuse. A 32-year-old woman known for years as suffering from borderline personality disorder and drug dependence (including cannabis, LSD, and ecstasy) and alcohol abuse that did not respond to treatment was reported. Only when correctly diagnosed as attention defict hyperactivity disorder and appropriately treated with the psychotropic stimulant methylphenidate (Ritalin®), was there significant improvement. She succeeded academically, which had not been possible previously, her craving for drugs diminished, and a drug-free state was reached.

Auditory Function

Eight male subjects (aged 22–30 years) who had previously used cannabis were investigated. They performed air conduction pure tone audiometry in both ears over 0.5–8 kHz. A simple test of frequency selectivity by detecting a 4-kHz tone under two masking noise conditions was also carried out in one ear. Three test sessions at weekly intervals were carried out, at the start of which they ingested a capsule containing either placebo, 7.5, or 15 mg of THC. These were administered in a randomized cross-over, double-blind manner. Auditory testing was carried out 2 hours after ingestion. Blood samples were also obtained at this time point and assayed for Δ-9-THC and 11-hydroxy-THC levels. No significant changes in threshold or frequency resolution were seen with the dosages employed in this study.

Behavioral Effect

A four-page, self-completed questionnaire was designed to determine the drugs used (licit, illicit, and doping substances) along with beliefs about doping and the psychosociological factors associated with their consumption. The questionnaire was distributed to high school students enrolled in a school sports association in eastern France. The completed forms were received from 1459 athletes: 4% stated that they had used doping agents at least once in their life (their main source of supply being peers and health professionals). Thirty-four percent of the sample smoked some tobacco, 66% used alcohol, 19% used cannabis, 4% took ecstasy, 10% took tranquillizers, 9% used hypnotics, 4% used creatine, and 41% used vitamins against fatigue. Beliefs about doping did not differ among doping agent users and nonusers, except for the associated health risks, which were minimized by users. Users of doping agents stated that the quality of the relations that they maintained with their parents was sharply degraded, and they reported that they were susceptible to influence and difficult to live with. More often than nondoping-agent users, these adolescents were neither happy, nor healthy, although paradoxically, they seemed less anxious and were more self-confident. Maternal exposure to Δ-9-THC in rats resulted in alteration in the pattern of ontogeny of spontaneous locomotor and exploratory behavior in the offspring. Adult animals exposed during gestational and lactational periods exhibited persistent alterations in the behavioral response to novelty, social interactions, sexual orientation, and sexual behavior. They also showed a lack of habituation and reactivity to different illumination conditions. Adult offspring of both sexes also displayed a characteristic increase in spontaneous and water-induced grooming behavior. Some of the effects were dependent on the sex of the animals being studied, and the dose of cannabinoid administered to the mother during gestational and lactational periods. Maternal exposure to low doses of THC sensitized the adult offspring of both sexes to the reinforcing effects of morphine, as measured in a conditioned place preference paradigm.

β-Endorphin interaction

Δ-9-THC administered to rats produced large increases in extracellular levels of β-endorphin in the ventral tegmental area and lesser increases in the shell of the nucleus accumbens (Nac). In rats that had learned to discriminate injections of THC from injections of vehicle, the opioid agonist morphine did not produce THC-like discriminative effects, but markedly increased discrimination of THC. The opioid antagonist naloxone reduced the discriminative effects of THC. Bilateral microinjections of β-endorphin directly into the ventral tegmental area, but not into the shell of the Nac, markedly increased the discriminative effects of ineffective threshold doses of THC, but had no effect when given alone. The increase was blocked by naloxone.

Binocular Depth Inversion Reduction

A study to assess whether the binocular depth inversion illusion (BDII) could detect subtle cognitive impairment owing to regular cannabis use was conducted. Ten regular cannabis users and 10 healthy controls from the same community sources, matched for age, sex, and premorbid intelligence quotient (IQ) were evaluated. The subjects were also compared on measures of executive functioning, memory, and personality. Regular cannabis users were found to have significantly higher BDII scores for inverted images. This was not to the result of a problem in the primary processing of visual information, as there was no significant difference between the groups for depth perception of normal images. There was no relationship between BDII scores for inverted images and time since the last dose, suggesting that the measured impairment of BDII more closely reflected chronic than acute effects of regular cannabis use. There were no significant differences between the groups for other neuropsychological measures of memory or executive function. A positive relationship was found between psychoticism as defined by the revised Eysenck Personality Questionnaire and cannabis, tobacco, and alcohol use. Cannabis users also used significantly larger amounts of alcohol. No relationship was found between BDII scores and drug use other than cannabis or psychoticism. Nabilone, a psychoactive synthetic 9-*trans*-ketocannabinoid, CBD, and a combined oral application of both substances on binocular depth inversion and behavioral states were investigated in nine healthy male volunteers. A significant impairment of binocular depth perception was found when nabilone was administered, but combined application with CBD revealed reduced effects on binocular depth inversion.

Birth-weight Effect

A total of 32,483 cannabis-using women giving birth to live-born infants were investigated. The largest reduction in mean birth-weight for any cannabis use during pregnancy was 48 g (95% CI, 83–14 g), with considerable heterogeneity among the five studies. Mean birth-weight was increased by 62 g (95% CI, 8-g reduction – 132-g increase; p heterogeneity, 0.59) among infrequent users (≤ weekly), whereas cannabis use at least four times per week had a 131-g reduction in mean birth- weight (95% CI 52–209-g reduction; p heterogeneity, 0.25). From the five studies of low birth-weight, the pooled odds ratio for any use was 1.09 (95% CI 0.94–1.27; p heterogeneity, 0.19). In a cohort study consisted of a multiethnic population of 7470 pregnant women. Information on the use of drugs was obtained from personal interviews at entry to the study and assays of serum obtained during pregnancy. Pregnancy outcome data (low birth-weight [<2500 g], pre-term birth [<37 weeks gestation], and abruptio placentae) were obtained with a standardized study protocol. A total of 2.3% of the women used cocaine and 11% used cannabis during pregnancy. Cannabis use was not associated with low birth-weight (1.1, 0.9–1.5), pre-term delivery (adjusted odds ratio [OR] 1.1, CI 0.8–1.3), or abruptio placentae (1.3, 0.6–2.8).

Bladder Dysfunction

Two whole-plant extracts of *Cannabis sativa* were administered to patients with advanced multiple sclerosis (MS) and refractory troublesome lower urinary tract symptoms. The patients took the extracts containing Δ-9-THC and CBD (2.5 mg of each per spray) for 8 weeks followed by THC-only (2.5 mg THC per spray) for a further eight weeks, and then into a long-term extension. Assessments included urinary frequency and volume charts, incontinence pad weights, cystometry, and visual analog scales for secondary troublesome symptoms. Twenty-one patients were recruited and data from 15 were evaluated. Urinary urgency, the number and volume of incontinence episodes, frequency, and nocturia all decreased significantly following treatment ($p < 0.05$, Wilcoxon's signed rank test). Daily total voided, catheterized and urinary incontinence pad weights also decreased significantly for both extracts. Patient self-assessment of pain, spasticity, and quality of sleep improved significantly ($p < 0.05$, Wilcoxon's signed rank test) with pain improvement continuing up to a median of 35 weeks. There were few troublesome side effects, suggesting that cannabis-based medicinal extracts are a safe and effective treatment for urinary and other problems in patients with advanced MS.

Blood Pressure Stress Reactivity Effect

Data from an ascorbic acid (AA) trial (Cetebe 3 g/day for 14 days, $n = 108$) were compared by substance use level regarding systolic blood pressure (SBP) stress reactivity to the anticipation and actual experience phases of a standardized psychological stressor (10 minutes of public speaking and arithmetic). Self-reported never users of cannabis, persons not currently smoking tobacco, and persons consuming three or more caffeine beverages daily all exhibited AA SBP stress reactivity protection to the actual stressor, but not during the anticipation phase. Self-reported ever cannabis users, current tobacco smokers, and persons consuming less than three caffeine beverages daily exhibited the AA SBP protection during the anticipation phase, but only the lower caffeine consumption group exhibited AA protection during both phases. Covariates (neuroticism, extraversion and depression scores, age, sex, body mass index) were not significant.

Blood-borne Sexually Transmitted Infections

Substance use, including alcohol and illicit drugs, increases the risk for the acquisition and transmission of sexually transmitted infection

(STI). The prevalence of blood-borne STI including human immunodeficiency virus (HIV), human T-cell lymphotrophic virus type 1, hepatitis B virus, and syphilis in residents of a detoxification and rehabilitation unit in Jamaica were investigated. The demographic characteristics and the results of laboratory investigations for STI in 301 substance abusers presented during a 5-year period were reviewed. The laboratory results were compared with those of 131 blood donors. The substances used by participants were alcohol, cannabis, and cocaine. None of the clients was an intravenous drug user. Female substance abusers were at higher risk for STI.

The prevalence of STI in substance abusers did not differ significantly from that in blood donors (12% vs 10%). The prevalence of syphilis in substance abusers was significantly higher than that in blood donors (6% vs 3%, $p < 0.05$). The prevalence of syphilis was dramatically increased in female substance abusers and female blood donors (30%, $p < 0.001$ and 13%, $p < 0.05$, respectively). An excess of human T-cell lymphotrophic virus type 1 was also observed in female compared with male substance abusers. Unemployment was identified also as a risk factor for sexually transmitted disease in substance abusers.

Brain Aging Effect

The impact of duration of education, cannabis addiction and smoking on cognition and brain aging was studied in 211 healthy Egyptian volunteers with mean age of 46.4 ± 3.6 years (range, 20–76 years). The subjects were classified into two groups: Gr I ($n = 174$; mean age, 49.9 ± 3.8 years; range, 20–76 years), nonaddicts, smokers, and nonsmokers, educated and noneducated, and Gr II cannabis addicts ($n = 37$; mean age, 43.6 ± 2.6 years; range, 20–72 years) all smokers, educated and noneducated. Outcome measures included the Paced Auditory Serial Addition test for testing attention and the Trailmaking test A and Trailmaking test B (TMb) for testing psychomotor performance. Age correlated positively with score of TMb in the nonaddict group and in the addict group (Trailmaking test A and TMb). Years of education correlated negatively with scores of TMb in the nonaddict group (Gr I) but not the addict group (Gr II). Cannabis addicts (Gr II) had significantly poorer attention than nonaddict normal volunteers (Gr I). It was determined that impairment of psychomotor performance is age related whether in normal nonaddicts or in cannabis addicts. A decline in attention was detected in cannabis addicts and has been considered a feature of pathological aging.

Brain Cannabinoid Receptor

In humans, psychoactive cannabinoids produce euphoria, enhancement of sensory perception, tachycardia, antinociception, difficulties in concentration, and impairment of memory. The cognitive deficiencies persist after withdrawal. The toxicity of cannabis has been underestimated for a long time, since recent findings revealed that Δ-9-THC-induced cell death with shrinkage of neurons and DNA fragmentation in the hippocampus. The acute effects of cannabinoids, as well as the development of tolerance, are mediated by G protein-coupled cannabinoid receptors. The CB1 receptor and its splice variant, CB1A, are found predominantly in the brain with highest densities in the hippocampus, cerebellum, and striatum. The CB2 receptor is found predominantly in the spleen and in hemopoietic cells and has only 44% overall nucleotide sequence identity with the CB1 receptor. The existence of this receptor provided the molecular basis for the immunosuppressive actions of cannabis. The CB1 receptor mediates inhibition of adenylate cyclase, inhibition of N- and P/Q-type calcium channels, stimulation of potassium channels, and activation of mitogen-activated protein kinase. The CB2 receptor mediates inhibition of adenylate cyclase and activation of mitogen-activated protein kinase. The discovery of endogenous cannabinoid receptor ligands, AEA (*N*-arachidonyl-ethanolamine), and 2-arachidonylglycerol made the notion of a central cannabinoid neuromodulatory system plausible. AEA is released from neurons on depolarization through a mechanism that requires calcium-dependent cleavage from a phospholipid precursor in neuronal membranes.

The release of AEA is followed by rapid uptake into the plasma and hydrolysis by fatty-acid amidohydrolase. The psychoactive cannabinoids increase the activity of dopaminergic neurons in the ventral tegmental area–mesolimbic pathway. Because these dopaminergic circuits are known to play a pivotal role in mediating the reinforcing (rewarding) effects of the most drugs of abuse, the enhanced dopaminergic drive elicited by the cannabinoids is thought to underlie the reinforcing and abuse properties of cannabis. Thus, cannabinoids share a final common neuronal action with other major drugs of abuse such as morphine, ethanol, and nicotine in producing facilitation of the mesolimbic dopamine system. Hippocampal slices from humans, guinea pigs, rats, and mice, and cerebellar, cerebro-cortical, and hypothalamic slices from guinea pigs were incubated with [^{3}H] noradrenaline and then superfused. Tritium overflow was evoked either electrically (0.3

or 1 Hz) or by introduction of Ca^{2+} ions (1.3 μM) into Ca^{2+}-free, K^+-rich medium (25 μM) containing 1 μM of tetrodotoxin. The cyclic adenosone monophosphate (cAMP) accumulation stimulated by 10 μM of forskolin was determined in guinea pig hippocampal membranes. The following drugs were used: the cannabinoid receptor-agonists (-)-*cis*-3-[2-hydroxy-4-(1,1- dimethylheptyl) phenyl]-*trans*-4-(3-hydroxypropyl)cyclo-hexanol (CP-55,940) and R(+) -[2,3 -dihydro-5-methyl-3 - [(morpholinyl) methyl]pyrrolo[1,2, 3-de]-1 ,4-benzoxazinyl] -(1 -naphthalenyl)methanone (WIN 55,212-2 [WIN]), the inactive *S*(-)-enantiomer of the latter (WIN 55,212-3) and the CB1 receptor antagonist *N*-piperidino-5-(4-chlorophenyl)-1-(2,4-dichlorophenyl) -4-methyl-3-pyrazole-carboxamide (SR 141716). The electrically evoked tritium overflow from guinea pig hippocampal slices was reduced by WIN (peak inhibitory concentration 30%, 6.5) but not affected by WIN 55,212-3 up to 10 m*M*.

The concentration–response curve of WIN was shifted to the right by SR 141716 (0.032-μM) (apparent pA2 8.2), which by itself did not affect the evoked overflow. WIN (1 μM) also inhibited the Ca^{2+}-evoked tritium overflow in guinea pig hippocampal slices and the electrically evoked overflow in guinea pig cerebellar, cerebro-cortical, and hypothalamic slices, as well as in human hippocampal slices, but not in rat and mouse hippocampal slices. SR 141716 (0.32 μM) markedly attenuated the WIN-induced inhibition in guinea pig and human brain slices. SR 141716 (0.32 μM) by itself increased the electrically evoked tritium overflow in guinea pig hippocampal slices, but failed to do so in slices from the other brain regions of the guinea pig and in human hippocampal slices, but failed to do so in slices from the other brain regions of the guinea pig and in human hippocampal slices. The cAMP accumulation stimulated by forskolin was reduced by CP-55,940 and WIN.

The concentration-response curve of CP-55,940 was shifted to the right by SR 141716 (0.1 μM; apparent pA2 8.3), that by itself did not affect cAMP accumulation. In conclusion, cannabinoid receptors of the CB1 subtype occur in the human hippocampus, where they may contribute to the psychotropic effects of cannabis, and in the guinea pig hippocampus, cerebellum, cerebral cortex, and hypothalamus. The CB1 receptor in the guinea pig hippocampus is located presynaptically, was activated by endogenous cannabinoids, and may be negatively coupled to adenylyl cyclase. The acute administration of AEA or THC in rats increased the maximum binding capacity (B_{max}) of cannabinoid receptors in the cerebellum and, particularly, in the hippocampus. This effect was also observed after 5 days of a daily exposure to AEA or THC. The increase in the B_{max} after the acute treatment seemed to be caused by changes in the receptor affinity (high K_d). The increase after the chronic exposure may be attributed to an increase in the density of receptors. The [^{3}H]CP-55,940 binding to cannabinoid receptors in the striatum, the limbic forebrain, the mesencephalon, and the medial basal hypothalamus was not altered after the acute exposure to AEA or THC. The chronic exposure to THC significantly decreased the B_{max} of these receptors in the striatum and nonsignificantly in the mesencephalon. This effect was not elicited after the chronic exposure to AEA and was not accompanied by changes in the K_d.

Cannabinoid Hyperemesis

Nineteen patients were identified with chronic cannabis abuse and a cyclical vomiting illness. Follow-up was provided with serial urine drug screen analysis and regular clinical consultation to chart the clinical course. Of the 19 patients, five refused consent and were lost to follow-up, and five were excluded based on cofounders. In all cases, chronic cannabis abuse predated the onset of the cyclical vomiting illness. Cessation of cannabis abuse led to cessation of the cyclical vomiting illness in seven cases. Three cases did not abstain and continued to have recurrent episodes of vomiting. Three cases rechallenged themselves after a period of abstinence and suffered a return to illness. Two of these cases abstained again and became, and remain, well. The third case did not and remains ill. A novel finding was that 9 of the 10 patients, including the previously published case, displayed an abnormal washing behavior during episodes of active illness.

Cannabinoid-induced Fos Expression

Cannabinoid CB1 receptor agonist CP-55,940 in Lewis and Wistar rats was investigated. A moderate (50 μg/kg) and a high (250 μg/kg) dose level were used. The 250-μg/kg dose caused locomotor suppression, hypothermia, and catalepsy in both strains, but with a significantly greater effect in Wistar rats. The 50-μg/kg dose provoked moderate hypothermia and locomotor suppression but in Wistar rats only. CP-55,940 caused significant Fos immunoreactivity in 24 out of 33 brain regions examined. The most dense expression was seen in the paraventricular nucleus of the hypothalamus, the islands of Callej a, the lateral septum (ventral), the central nucleus of the amygdala, the bed nucleus of the stria terminalis (lateral division), and the ventrolateral periaqueductal gray. Despite having a similar distribution of CP-55,940-induced Fos expression, Lewis rats showed less overall Fos expression than Wistar rats in nearly every brain region counted. This held equally true for anxiety-related brain structures (e.g., central nucleus of the amygdala, periaqueductal gray, and the paraventricular nucleus of the hypothalamus) and reward-related sites (Nac and pedunculopontine tegmental nucleus). In a further experiment, Wistar rats and Lewis rats did not differ in the amount of Fos immunoreactivity produced by cocaine (15 mg/kg). These results indicate that Lewis rats are less sensitive to the behavioral, physiological and neural effects of cannabinoids.

Cannabis withdrawal Effect

A 35-year-old male was cognitively assessed prior to cessation of 18 years of daily cannabis use and monitored for several weeks post-cessation. Brain event-related potential measures of selective attention reflecting a difficulty in filtering out complex irrelevant information showed no indication of improvement over 6 weeks of abstinence. When tested in the acutely intoxicated state prior to cessation of use, a dramatic normalization of the event-related potential signature was observed. A treatment program based on supportive–expressive psychotherapy was administered and depression, anxiety, and general psychological health were monitored over the course of withdrawal from cannabis.

Cannabis–amphetamine Interaction

Cannabinoid–amphetamine interactions were studied as follows:

1. 30 minutes after acute injection of (-)-Δ-9-THC (0.1 or 6.4 mg/kg, intraperitoneally).
2. 30 minutes after the last injection of 14-daily treatment with (-)-Δ-9-THC (0.1 or 6.4 mg/kg).
3. 24 hours after the last injection of 14- daily treatment with (-)-Δ-9-THC (6.4 mg/kg).

Acute cannabinoid exposure antagonized the amphetamine-induced dose-dependent increase in locomotion, exploration, and the decrease in inactivity. Chronic treatment with (-)-Δ-9-THC resulted in tolerance to this antagonistic effect on locomotion and inactivity but not on exploration, and potentiated amphetamine-induced stereotypes. Lastly, 24 hours of withdrawal after 14 days of cannabinoid treatment resulted in sensitization to the effects of D-amphetamine on locomotion, exploration, and stereotypes.

Cannabis-induced Coma

Two cases of cannabis-induced coma were reported following accidental ingestion of cannabis cookies. The possibility of cannabis ingestion should be considered in cases of unexplained coma in a previously healthy young child if signs of conjunctival hyperemia, pupillary dilatation, and tachycardia were present and other causes, such as central nervous system infection or trauma were unlikely.

Cannabis-related Arteritis

A 19-year-old man who presented with plantar claudication associated with necrosis in a toe underwent diagnostic arteriography and surgery for popliteal artery entrapment type III was studied. Surgical clearance resolved the popliteal artery entrapment but left the clinical symptoms unchanged. Closer questioning disclosed a history of cannabis consumption and intravenous vasodilatory therapy was started. After the 21-day course of vasodilator agents, the pain disappeared and the toe necrosis regressed. The patient stopped taking cannabis and had no signs of recurrence. A 24-year-old woman who was a heavy cannabis smoker with progressive Raynauld's phenomenon and digital necrosis, was investigated. Systemic sclerosis and other connective tissue disorders, as well as arteriosclerosis and arterial emboli were excluded with appropriate laboratory examinations. Arteriography revealed multiple forearm, palmar and digital occlusions with corkscrew-shaped vessels. Based on the characteristic arteriography and clinical findings, the diagnosis of cannabis arteritis was retained. With careful necrectomy, conservative wound dressings and secondary prostacyclin therapy a complete healing of digital necrosis was observed. There was no recurrence during the 6-month follow-up. Young men were presented with distal arteriopathy of the lower limbs in three cases, and of the left upper limb in the remaining patient. Symptoms occurred progressively, distal pulses had disappeared, and distal necrosis was constant. Three patients suffered from Raynauld's phenomenon, none of them presented with venous thrombosis. Radiological evaluation revealed distal abnormalities in all cases, and proximal arterial thrombosis in one case. The four patients were cannabis smokers for at least four years. With cannabis interruption and symptomatic treatment, lesions improved for three patients. For one of them, recurrence of arteriopathy occurred when he resumed smoking cannabis. For the fourth who never stopped cannabis, an amputation was necessary. Ten male moderate tobacco smokers and regular cannabis users with a median age of 23.7 years, developed subacute distal ischemia of the lower or upper limbs, leading to necrosis in the toes and/or fingers and sometimes to distal limb gangrene. Two of the patients also presented with venous thrombosis and three patients were suffering from a recent Raynauld's phenomenon. Biological test results did not show evidence of the classical vascular risk factors for thrombosis. Arteriographic evaluation in all of the cases revealed distal abnormalities in the arteries of feet, legs, forearms, and hands resembling those of Buerger's disease. A collateral circulation sometimes with opacification of the vasa nervorum was noted. In some cases, arterial proximal atherosclerotic lesions and venous thrombosis were observed. Despite treatment with ilomedine and heparin in all cases, five amputations were necessary in four patients. The vasoconstrictor effect of cannabis on the vascular system has been known for a long time. It has been shown that Δ-8-THC and Δ-9-THC may induce peripheral vasoconstrictor activity. Cannabis arteritis resembles Buerger's disease, but patients were moderate tobacco smokers and regular cannabis users.

Cannabis-related Flashback

A young man who offended a friend without any objective reason was reported. The report of the forensic psychiatrist demonstrated that the offense was committed under the influence of a cannabis flashback. The last time the offender had consumed cannabis was 2 weeks before the acts. A plasmatic detection was realized and showed a level of 6 ng/mL, 30 minutes after the beginning of the flashback.

Capgras Syndrome

A report describes an apparently greater incidence of Capgras syndrome among the Maori population compared with the European population. Five cases of Capgras syndrome were identified in the eastern catchment area where 19% of the population identified as Maori, 75% as European, and 6% as other or nonspecified. All of the cases occurred in Maori patients. No cases were identified in the western catchment area where 12% of the population identified as Maori, 87% as European, and 1% as other or nonspecified. Four of five cases were females. Two cases had a history of cannabis use. Three cases had exhibited dangerous behavior towards family members.

Carcinogenic Activity

The dried leaf, administered intraperitoneally to rats of both sexes at a dose of 7 mg/kg/week, was active. The animals were irradiated with y radiation between 40 and 50 days of age and observed for 78 weeks. There was a greater incidence of tumors in animals given marijuana extract and γ radiation than either marihuana or γ radiation alone.

Cardiorespiratory Effect

Fifty stable patients (25 males, 25 females) with methadone maintenance treatment (MMT) programs were investigated. Forty-six MMT patients were current tobacco smokers, 19 were current cannabis users, and none were currently using opioids other than prescribed methadone. Abnormalities of respiratory function were defined as those results outside the 95% confidence interval of reference values for normal subjects adjusted for age, weight, height, and sex. Thirty-one (62%) MMT patients had reduced carbon monoxide transfer factor; 17 (34%) had elevated single breath alveolar volume, and 43 (86%) had a reduced carbon monoxide transfer factor–alveolar volume ratio. Six patients (12%) had reduced forced expiratory volume in 1 second (FEV1); one (2%) had reduced forced vital capacity (FVC); and nine (18%) had an obstructive ventilatory defect. Ten (20%) patients had arterial CO_2 pressure higher than 45 mmHg and 14 (28%) had alveolar to arterial oxygen gradient higher than 15 mmHg. Chest X-ray, echocardiography, and electrocardiogram showed no significant abnormalities. The potent cannabinoid receptor agonists WIN55,212-2 (0.05, 0.5, or 5 pmol/50 nL) and HU-210 (0.5 pmol/50 nL) or the CB1 receptor antagonist/inverse agonist AM28 1 (1 pmol/100 nL) were microinjected into the rostral ventrolateral medulla oblongata (RVLM) of urethane-anesthetized, immobilized and mechanically ventilated male Sprague-Dawley rats ($n = 22$). Changes in splanchnic nerve activity, phrenic nerve activity, mean arterial pressure, and heart rate in response to cannabinoid administration were recorded. The CB1 receptor gene was expressed throughout the ventrolateral medulla oblongata. Unilateral microinjection of WIN 55,212-2 into the RVLM evoked short-latency, dose-dependent increases in splanchnic nerve activity (0.5 pmol; 175 ± 8%, $n = 5$) and mean arterial pressure (0.5 pmol; 26 ± 3%, $n = 8$), and abolished phrenic nerve activity (0.5 pmol; duration of apnea: 5.4 ± 0.4 seconds, $n = 8$), with little change in heart rate ($p < 0.005$). HU-210, structurally related to Δ-9-THC, evoked similar effects when microinj ected into the RVLM ($n = 4$). Prior micro-injection of AM281 produced agonist-like effects, and significantly attenuated the response to subsequent injection of WIN (0.5 pmol, $n = 4$).

Cardiovascular Effects

The leaf, smoked by adults of both sexes, at a dose of 600 mg/person (1–1.5% THC), produced no adverse effects on blood pressure, electrocardiogram, and the heart. Cannabis and Δ-9-THC increase heart rate, slightly increase supine blood pressure, and on occasion produced marked orthostatic hypotension. Cardiovascular effects in animals are different, with bradycardia and hypotension the most typical responses. Cardiac output increases, and peripheral vascular resistance and maximum exercise performance decrease. Tolerance to most of the initial cardiovascular effects appears rapidly. With repeated exposure, supine blood pressure decreases slightly, orthostatic hypotension disappears, blood volume increases, heart rate slows, and circulatory responses to exercise and Valsalva maneuver are diminished, consistent with centrally mediated, reduced sympathetic, and enhanced parasympathetic activity. Receptor-mediated and probably nonneuronal sites of action account for cannabinoid effects. The endocannabinoid system appears important in the modulation of many vascular functions. Cannabis' cardiovascular effects are not associated with serious health problems for most young, healthy users, although occasional myocardial infarction, stroke, and other adverse cardiovascular events are reported.

CB1 Cannabinoid Receptor in Human Placenta

CB1 (G protein-coupled) receptor and FAAH expression in human term placenta were investigated by immunohistochemistry. CB1 receptor was found in all layers of the membrane, with particularly strong expression in the amniotic epithelium and reticular cells and cells of the maternal decidua layer. Moderate expression was observed in the chorionic cytotrophoblasts. The expression of FAAH was highest in the amniotic epithelial cells, chorionic cytotrophoblast, and maternal decidua layer. The results suggest that the human placenta is a likely target for cannabinoid action and metabolism. This is consistent with a placental site of action of endocannabinoids and cannabis being responsible, at least in part, for the poor outcomes associated with cannabis consumption and pathology in the endocannabinoid system during pregnancy.

Central Nervous System Depressant Activity

Fluidextract of the aerial parts, administered intraperitoneally to rats at a dose of 25 mg/kg, was active. The fluidextract, administered orally to dogs, produced ataxia. The leaf, smoked by human adults, produced a decrease in psychomotor performance.

Central Nervous System Effect

Δ-9-THC activates the two G protein-coupled receptors CB1 and CB2. The endogenous ligands of these receptors were identified as lipid metabolites of arachidonic acid, named endocannabinoids. The two most studied endocannabinoids are AEA and 2-arachidonyl-glycerol. The CB1 receptor is massively expressed throughout the central nervous system, whereas CB2 expression seems restricted to immune cells. Following endocannabinoid binding, CB1 receptors modulate second messenger cascades (inhibition of adenylate cyclase, activation of mitogen-activated protein kinases and of focal-adhesion kinases), as well as ionic conductances (inhibition of voltage-dependent calcium channels, activation of several potassium channels). Endocannabinoids transiently silenced synapses by decreasing neurotransmitter release. They play major roles in various forms of synaptic plasticity because of their ability to behave as retrograde messengers and activate noncannabinoid receptors (such as vanilloid receptor type-1). Mice strain with a disrupted *CB1* gene (CB1 knockout mice) appeared healthy and fertile, but they had a significantly increased mortality rate. They also displayed reduced locomotor activity, increased ring catalepsy, and hypoalgesia in hotplate and formalin tests. Δ-9-THC-induced ring catalepsy, hypomobility, and hypothermia were completely absent in CB1 mutant mice. In contrast, Δ-9-THC-induced analgesia in the tail-flick test and other behavioral (licking of the abdomen) and physiological (diarrhea) responses after Δ-9-THC administration were found. Results indicate that most, but not all, central nervous system effects of Δ-9-THC are mediated by the CB1 receptor.

Central Nervous System Stimulant Activity

The resin, ingested by a 4-year-old girl, showed signs of stupor alternating with brief intervals of excitation and foolish laughing with atactic movements. Her temperature, blood pressure, pulse, hemoglobin, leukocytes, serum electrolytes, and serum urea were normal. Respiratory rate was 12 beats per minute. Blood sugar elevated. Recovery was complete within 24 hours with no treatment.

Cerebellar clock-altering effect

Twelve volunteers who smoked cannabis recreationally about once weekly, and 12 volunteers who smoked daily for a number of years performed a self-paced counting task during positron emission tomography imaging, before and after smoking cannabis and placebo cigarettes. Smoking cannabis increased regional cerebral blood flow in the ventral forebrain and cerebellar cortex in both groups, but resulted in significantly less frontal lobe activation in chronic users. Counting rate increased after smoking cannabis in both groups, as did a behavioral measure of self-paced tapping, and both increases correlated with regional cerebral blood flow in the cerebellum. Results indicate that smoking cannabis appears to accelerate a cerebellar clock-altering self-paced behaviors.

Clinical Endocannabinoid Deficiency

Clinical endocannabinoid deficiency, and the prospect that it could underlie the pathophysiology of migraine, fibromyalgia, irritable bowel syndrome, and other functional conditions alleviated by clinical cannabis were studied. Migraine has numerous relationships to endocannabinoid function. AEA potentiated 5 -hydroxytrayptamine (HT1A) and inhibited 5-HT2A receptors supporting therapeutic efficacy in acute and preventive migraine treatment. Cannabinoids also demonstrated dopamine-blocking and anti-inflammatory effects. AEA is tonically active in the periaqueductal gray matter, a migraine generator. THC modulated glutamatergic neurotransmission via *N*-methyl-D-aspartic acid-receptors. Fibromyalgia is now conceived as a central sensitization state with secondary hyperalgesia. Cannabinoids have similarly demonstrated the ability to block spinal, peripheral and gastrointestinal mechanisms that promote pain in headache, fibromyalgia, irritable bowel syndrome and related disorders.

Cognitive Functioning

Cognitive performance was examined in 145 adolescents aged 13–16 years for whom prenatal exposure to cannabis and cigarettes had been ascertained. The subjects were from a low-risk, predominantly middle-class sample participating in an ongoing, longitudinal study. The assessment battery included tests of general intelligence, achievement, memory, and aspects of executive functioning. Consistent with results obtained at earlier ages, the strongest relationship between prenatal maternal cigarette smoking and cognitive variables was seen with overall intelligence and aspects of auditory functioning, whereas prenatal exposure to marijuana was negatively associated with tasks that required visual memory, analysis, and integration. A multisite, retrospective, cross-sectional, neuropsychological study was conducted among 102 near-daily cannabis users (51 long-term users: mean, 23.9 years of use; 51 shorter-term users: mean, 10.2 years of use), compared with 33 nonuser controls. Measures from nine standard neuropsychological tests that assessed attention, memory, and executive functioning were administered prior to entry into a treatment program following a median 17-hour abstinence. Long-term cannabis users performed significantly less well than shorter-term users and controls on tests of memory and attention. On the Rey Auditory Verbal Learning Test, long-term users recalled significantly fewer words than either shorter-term users ($p = 0.001$) or controls ($p = 0.005$). There was no difference between shorter-term users and controls. Long-term users showed impaired learning ($p = 0.007$), retention ($p = 0.003$), and retrieval ($p = 0.002$) compared with controls. Both user groups performed poorly on a time estimation task ($p < 0.001$ vs controls). Performance measures often correlated significantly with the duration of cannabis use, being worse with increasing years of use, but were unrelated to withdrawal symptoms and persisted after controlling for recent cannabis use and other drug use. A patient with a history of traumatic brain injury along with current mood disorder and cannabis use was reported. The impact of cannabis use appeared to have a detrimental effect on his mood. Treatment of the mood disorder resulted in larger cognitive gains. Sixty healthy volunteers (a negative urine drug-screening test was prerequisite) were investigated. On the first day, baseline data were obtained from a physical examination and a psychological test battery for the investigation of visual and verbal memory and cognitive-perceptual performance. On the second day, subjects received a regular cigarette or one containing 290 mg/kg body weight of THC. Physical and psychological assessments were performed immediately (15 minutes) after subjects smoked their cigarettes. Twenty- four hours later, physical and psychological examinations were repeated. Results suggest that perceptual motor speed and accuracy, two very important parameters of driving ability, seem to be impaired immediately after cannabis consumption. The analyses included 1318 participants under age 65 years who completed the Mini-Mental State Examination (MMSE) during three study waves in 1981, 1982, and 1993–1996. Individual MMSE score differences between waves two and three were calculated for each study participant. After 12 years, study participants' scores declined a mean of 1.20 points on the MMSE (standard deviation, 1.90), with 66% having scores that declined by at least one point. Significant numbers of scores declined by three points or more (15% of participants in the 18–29-year-old age group). There were no significant differences in cognitive decline between heavy users, light users, and nonusers of cannabis. There were also no male–female differences in cognitive decline in relation to cannabis use. From 250 individuals consuming cannabis regularly, 99 healthy, free of any other past or present drug abuse, or history of neuropsychiatric disease cannabis users were selected. After an interview, physical examination, analysis of routine laboratory parameters, plasma/urine analyses for drugs, and Minnesota Multiphasic Personality Inventory testing, users and respective controls were subjected to a computer-assisted attention test battery comprising visual scanning, alertness, divided attention, flexibility, and working memory. Of the potential predictors of test performance within the user group, including present age, age of onset of cannabis use, degree of acute intoxication (THC + THC–OH plasma levels), and cumulative toxicity (estimated total life dose), an early age of onset turned out to be the only predictor, predicting impaired reaction times exclusively in visual scanning. Early-onset users (onset before age 16; $n = 48$) showed a significant impairment in reaction times in this function, whereas late-onset users (onset after age 16; $n = 51$) did not differ from controls ($n = 49$). Male volunteers ($n = 5$) with histories of moderate alcohol and cannabis use were administered three doses of alcohol (0.25, 0.5, or 1 g/kg), three doses of cannabis (4.8, or 16 puffs of 3.55% Δ-9-THC), and placebo in random order under double blind conditions in seven separate sessions. Blood alcohol concentration (10–90 mg/dL) and THC levels (63–188 ng/mL) indicated that active drug was delivered to subjects dose dependently. Alcohol and cannabis produced dose-related changes in subjective measures of drug effect. Ratings of perceived impairment were identical for the high doses of alcohol and cannabis. Both drugs produced comparable impairment in digit–symbol substitution and word recall tests, but had no effect in time perception and reaction time tests. Alcohol, but not cannabis, slightly impaired performance in a number recognition test.

Comorbid Dysthymia and Substance Disorder

A total of 642 patients were assessed. Thirty-nine had substance-related disorder and dysthymia (SRD-dysthymia) and 308 had SRD only. Data on past use were collected by a research associate using a questionnaire. The patients with SRD-dysthymia and SRD did not differ with regard to use of alcohol, tobacco, and benzodiazepines. The patients with SRD-dysthymia started caffeine use at an earlier age, had shorter "use careers" of cocaine, amphetamines, and opiates, and had fewer days of cocaine and cannabis use in the last year. They also had a lower rate of cannabis abuse/dependence. The results indicated that patients with dysthymia and SRD have exposure to most substances of abuse that was comparable to patients with SRD only. They selectively use certain substances less often than patients with SRD only. A course and severity of SRD among 642 patients with comorbid major depressive disorder (MDD) was analyzed by means of both retrospective and concurrent data. Data on course included lifetime use, age at first use, years of use, use in the last year, periods of abstinence, and current diagnosis. Data on severity included two measures of SRD-associated problems, substance abuse vs dependence, self-help activities, and number of substances being abused. SRD-MDD patients tended to manifest lower levels of cannabis, opiate, and cocaine use, and more SRD-only patients were abusing three or more substances. Men with SRD-MDD demonstrated longer mean durations of abstinence compared with men with SRD-only, whereas SRD-MDD women demonstrated shorter mean durations of abstinence, compared with women with SRD-only. MDD-SRD patients showed slightly less substance abuse, but SRD severity was comparable with SRD-only patients.

Covariation among Risk Behaviors

A sample of 913 sexually active high school students completed a self-administered questionnaire that required mainly "yes" or "no" answers to questions involving participation in a range of risk behaviors. Contraceptive nonuse was not significantly associated with use of cigarettes, alcohol, or inhalants; perpetration or being a victim of violence; exposure to risk of physical injury; and suicidality. For males only, there was a significant inverse association between contraceptive nonuse and use of cannabis in the previous month. This was not the case for lifetime cannabis use for either gender.

Cytochrome P450 and 2C6 Expression

Hashish (cannabis) and heroin effect on the expression of cytochrome P450 2E1 (CYP 2E1) and cytochrome P450 2C6 (CYP 2C6) was measured after single (24 hours) and repeated-dose treatments (four consecutive days). The expression of CYP 2E1 was slightly induced after single-dose treatments and markedly induced after repeated-dose treatments of mice with hashish (10 mg/kg body weight). It is believed that *N*-nitrosamines are activated principally by CYP 2E1 and the activity of *N*-nitrosodimethylamine was found to be increased after single- and repeated-dose treatments of mice with hashish by 23 and 41%, respectively. Hashish treatments of mice increased the total hepatic content of CYP by 112 and 206%, respectively; aryl hydrocarbon hydroxylase activity by 110 and 165%, respectively; nicotinamide adenine dinucleotide phosphate–cytochrome c reductase activity by 21 and 98%, respectively, and glutathione level by 81 and 173%, respectively. The level of free radicals was potentially decreased after single- or repeated-dose treatments with either hashish or heroin.

Cytotoxic Effect

THC, in leukemic cell lines (CEM, HEL-92, and HL60) and in peripheral blood mononuclear cells, 6 hours after exposure induced apoptosis, even at one times the IC_{50}. THC did not appear to act synergistically with cytotoxic agents, such as cisplatin. THC-induced cell death was preceded by significant changes in the expression of genes involved in the mitogen-activated protein kinase signal transduction pathways. Both apoptosis and gene expression changes were altered independent of p53 and the cannabinoid 1 and 2 receptors (CB1-R and CB2-R).

Depressant Activity

Heavy cannabis use and depression are associated and evidence from longitudinal studies suggests that heavy cannabis use may increase depressive symptoms among some users. Participants (n = 1920) were reassessed as part of a follow-up study. The analysis focused on two cohorts: those who reported no depressive symptoms at baseline (n = 849) and those with no diagnosis of cannabis abuse at baseline (n = 1,837). Symptoms of depression, cannabis abuse, and other psychiatric disorders were assessed with the Diagnostic Interview Schedule. In participants with no baseline depressive symptoms, those with a diagnosis of cannabis abuse at baseline were four times more likely than those with no cannabis abuse diagnosis to have depressive symptoms at the follow-up assessment, after adjusting for age, gender, antisocial symptoms, and other baseline covariates. These participants were more likely to have experienced suicidal ideation and anhedonia during the follow-up period. Among the participants who had no diagnosis of cannabis abuse at baseline, depressive symptoms at baseline failed to significantly predict cannabis abuse at the follow-up assessment. The relationship between depressive symptoms and polydrug use (alcohol, cannabis, and cocaine) among blacks in a high-risk community was studied. A street sample (n = 570) from four high-risk communities was collected through personal interviews. Interviewers asked respondents about their drug use behavior during the past 30 days and their depressive symptoms during the past week. Odds ratios and logistic regressions, adjusted for age and sex, were used to assess the relationship between depressive symptoms and drug and polydrug use (drug use involving cocaine). Results showed that depressive symptoms are significantly associated with polydrug use. Depressive symptoms were not associated with alcohol use or with the combination of alcohol and cannabis use.

Diabetic Ketoacidosis

One hundred fifty-eight young adults, aged 16–30 years, with type 1 diabetes, attending an urban diabetes clinic, were sent an anonymous confidential postal questionnaire to determine the prevalence of street drug use. Eighty-five completed responses were received. Twenty-nine percent of respondents admitted to using street drugs. Of those, 68% habitually took street drugs more than once a month. Seventy-two percent of users were unaware of the adverse effects on diabetes. Results indicated that the street drug usage in young adults with type 1 diabetes is common and may contribute to poor glycemic control and serious complications of diabetes.

Digital Necrosis

An 18-year-old woman, with a history of severe anorexia nervosa of 5 years' duration, who acknowledged regular use of tobacco and cannabis, was hospitalized for necrosis of the left index and thumb that had occurred shortly after left radial artery puncture for blood gas analysis. Acrocyanosis of the four limbs had been present since the onset of anorexia nervosa. Arteriography of the upper limbs showed major spasm of the left radial and cubital arteries and thromboses in the left inter-digital arteries of the left index and thumb. The distal portions of the arteries were then on the left and on the right. The necrotic lesions healed after intravenous administration of ilomedine and interruption of tobacco and cannabis. Acrocyanosis of the four limbs persisted.

Discriminative Stimulus Effect

Rhesus monkeys, trained to discriminate Δ-9-THC from vehicle in a two-lever drug discrimination procedure, were tested with a variety of psychoactive drugs, including cannabinoids or drugs from other classes. The results indicated that Δ-9-THC discrimination showed pharmacological specificity, in that none of the noncannabinoid drugs fully substituted for Δ-9-THC. The classical cannabinoids, Δ-9-THC and Δ-8-THC, and the novel cannabinoids, WIN and 1-butyl-2-methyl-3-(1-naphthoyl)indole, produced full dose-dependent substitution for Δ-9-THC in all monkeys. A heptyl indole derivative failed to substitute for Δ-9-THC, but it also did not displace [^{3}H] CP-55,940 from its binding site.

Dopamine Metabolism

The effect of repeated administrations of THC or WIN, a synthetic cannabinoid receptor agonist, on dopamine turnover in the prefrontal cortex, striatum, and Nac in rats, was investigated. THC or WIN (twice daily for seven or 14 days) caused a persistent and selective reduction in medial prefrontal cortical dopamine turnover. No significant alterations of dopamine metabolism were observed in the Nac or striatum. These dopaminergic deficits in the prefrontal cortex were observed after a drug-free period of up to 14 days. The cognitive dysfunction produced by heavy, long-term cannabis use may be subserved, in part, by drug-induced alterations in frontal cortical dopamine turnover. Two weeks' administration of THC to rats, reduced dopamine transmission in the medial prefrontal cortex, whereas dopamine metabolism in striatal regions was unaffected.

Dopamine Release

A 38-year-old drug-free schizophrenic patient took part in a single photon emission computerized tomographic study of the brain, and smoked cannabis secretively during a pause in the course of an imaging session. Cannabis had an immediate calming effect, followed by a worsening of psychotic symptoms a few hours later. A comparison of the two sets of images, obtained before and immediately after smoking cannabis, indicated a 20% decrease in the striatal dopamine D2 receptor-binding ratio, suggestive of increased synaptic dopaminergic activity.

Dopamine Transmission Modulation

The endogenous cannabinoid system is a new signaling system composed by the central (CB1) and the peripheral (CB2) receptors, and several lipid transmitters including AEA and 2-arachidonylglycerol. Cannabinoid CB1 receptors are present in dopamine projecting brain areas. In primates and certain rat strains it is also located in dopamine cells of the A8, A9, and A10 mesencephalic cell groups, as well as in hypothalamic dopaminergic neurons controlling prolactin secretion. CB1 receptors co-localize with dopamine D1/D2 receptors in dopamine projecting fields. Manipulation of dopaminergic transmission is able to alter the synthesis and release of AEA, as well as the expression of CB1 receptors. CB1 receptors can switch their transduction mechanism to oppose to the ongoing dopamine signaling. Acute blockade of CB1 receptor potentiates the facilitatory role of dopamine D2 receptor agonists on movement. CB1 stimulation results in sensitization to the motor effects of indirect dopaminergic agonists.

Dyskinetic Activity

A 4-week dose escalation study was performed to assess the safety and tolerability of cannabis in six patients with Parkinson's disease (PD) with levodopa (L-DOPA)-induced dyskinesia. Then a randomized, placebo-controlled crossover study was performed, in which 19 patients with PD were randomized to receive oral cannabis extract followed by placebo or vice versa. Each treatment phase lasted for 4 weeks with an intervening 2-week washout phase. The primary outcome measure was a change in Unified Parkinson's Disease Rating Scale (UPDRS) (items 32 to 34) dyskinesia score. Secondary outcome measures included the Rush scale, Bain scale, tablet arm drawing task, and total UPDRS score following a levodopa challenge, as well as patient-completed measures of a dyskinesia activities of daily living scale, the PDQ-39, on–off diaries, and a range of category rating scales. Seventeen patients completed the study. Cannabis was well tolerated and had no pro- or antiparkinsonian action. There was no evidence for a treatment effect on L-DOPA-induced dyskinesia as assessed by the UPDRS, or any of the secondary outcome measures. An anonymous questionnaire sent to all patients attending the Prague Movement Disorder Center revealed that 25% of 339 respondents had taken cannabis and 45.9% of these described some form of benefit. 2,4,5-Trihydroxyphenethylamine (6-hydroxydopamine)-lesioned rats were treated with the enantiomers of the synthetic cannabinoid 7-hydroxy-Δ6-THC 1,1-dimethylheptyl. Treatment with its (-)- (3R, 4R) enantiomer (code name HU-210), a potent cannabinoid receptor type 1 agonist, reduced the rotations induced by L-DOPA/carbidopa or apomorphine by 34 and 44%, respectively. Treatment with the (+)-(3S, 4S) enantiomer (code name HU-211), an *N*-methyl-D-aspartate antagonist, and the psychotropically inactive cannabis constituent: CBD and its primary metabolite, 7-hydroxy-cannabinol, did not show any reduction of rotational behavior. The results indicate that activation of the CB1 stimulates the dopaminergic system ipsilaterally to the lesion, and may have implications in the treatment of PD.

Dystonic Activity

The neural mechanisms underlying dystonia involve abnormalities within the basal ganglia—in particular, overactivity of the lateral globus pallidus. Cannabinoid receptors are located presynaptically on γ-aminobutyric acid receptor (GABA) terminals within the globus pallidus internus, where their activation reduces GABA reuptake. Cannabinoid receptor stimulation may thus reduce overactivity of the globus pallidus, and thereby reduce dystonia. A double-blind, randomized, placebo-controlled, crossover study using the synthetic cannabinoid receptor agonist nabilone in patients with generalized and segmental primary dystonia showed no significant reduction in dystonia following treatment with nabilone.

Endocrine Effect

Animal models have demonstrated that cannabinoid administration acutely altered multiple hormonal systems, including the suppression of the gonadal steroids, growth hormone, prolactin, and thyroid hormone and the activation of the hypothalamic–pituitary–adrenal (HPA) axis. These effects were mediated by binding to the endogenous cannabinoid receptor in or near the hypothalamus. Despite these findings in animals, the effects in humans have been inconsistent, and discrepancies were likely owing in part to the development of tolerance. Intravenous administration of three cannabinoid agonists to nine castrated male calves under stress-free conditions provoked immediate increases of serum cortisol and respiration rate, and produced rapid hypoalgesia to cutaneous pain and thermal stimuli. AEA and methanandamide did not affect serum prolactin. Administration of WIN increased serum prolactin abruptly. None of the cannabinoid receptor agonists affected serum growth hormone.

Environmental Stress and Cannabinoids Interaction

Anxiety and panic are the most common adverse effects of cannabis intoxication. Data suggest that cannabinoid CB1 receptor modulation of amygdalar activity contributes to these phenomena. Using Fos as a marker, it was tested the hypothesis that environmental stress and CB1 cannabinoid receptor activity interact in the regulation of amygdalar activation in male mice. Both 30 minutes of restraint and CB1 receptor agonist treatment (Δ-9-THC [2.5 mg/kg]) or CP-55,940 (0.3 mg/kg); by intraperitoneal injection) produced barely detectable increases in Fos expression within the central amygdala (CeA). The combination of restraint and CB1 agonist administration produced robust Fos induction within the CeA, indicating a synergistic interaction between environmental stress and CB1 receptor activation. An inhibitor of endocannabinoid transport, AM404 (10 mg/kg), produced an additive interaction with restraint within the CeA. In contrast, FAAH inhibitor-treated mice (URB597, 1 mg/kg) and FAAH (–/–) mice did not exhibit any differences in amygdalar activation in response to restraint compared with control mice. In the basolateral amygdala and medial amygdala, restraint stress produced a low level of Fos induction, which was unaffected by cannabinoid treatment. The CB1 receptor antagonist SR141 716 dose-dependently increased Fos expression in the BLA and CeA.

Epileptic Effect

Δ-9-THC at a dose of 1 μM, significantly depressed evoked depolarizing postsynaptic potentials (PSPs) in rat olfactory cortex neurones. A standardized cannabis extract (SCE) and Δ-9-THC-free SCE significantly potentiated evoked PSPs (all results were fully reversed by the CB1 receptor antagonist SR141716A, 1 μM). The potentiation by Δ-9-THC-free SCE was greater than that produced by SCE. On comparing the effects of Δ-9-THC-free SCE on evoked PSPs and artificial PSPs (aPSPs; evoked electrotonically following brief intracellular current injection), PSPs were enhanced, whereas aPSPs were unaffected, suggesting that the effect was not resulting from changes in background input resistance. Similar recordings made using CB1 receptor-deficient knockout mice and wild-type littermate controls revealed cannabinoid or extract-induced changes in membrane resistance, cell excitability and synaptic transmission in wild-type mice that were similar to those seen in rat neurones, but no effect on these properties were seen in CB1 receptor-deficient knockout mice cells. Results indicated that the unknown extract constituent(s) effects over-rode the suppressive effects of Δ-9-THC on excitatory neurotransmitter release, which may explain some patients' preference for herbal cannabis rather than isolated Δ-9-THC (owing to attenuation of some of the central Δ-9-THC side effects) and possibly account for the rare incidence of seizures in some individuals taking cannabis recreationally. A SCE with pure Δ-9-THC, at matched concentrations of Δ-9-THC, and a Δ-9-THC-free extract (Δ-9-THC-free SCE) in in vitro rat brain slice model of epilepsy were examined. In the in vitro epilepsy model, in which sustained epileptiform seizures were induced by the muscarinic receptor agonist oxotremorine-M in immature rat piriform cortical brain slices, SCE was a more potent and again more rapidly-acting anticonvulsant than isolated Δ-9-THC. Δ-9-THC-free extract also exhibited anticonvulsant activity. CBD did not inhibit seizures, nor did it modulate the activity of Δ-9-THC in this model. These results demonstrated that not all of the therapeutic actions of cannabis herb might be a result of the Δ-9-THC content.

Estrogen Receptors Stimulating Effect

THC, CBD, and desacetyllevonantradol, in estrogen-induced MCF-7 breast cancer cells at concentrations of no more than 10 μM, produced no effect. THC failed to antagonize the response to estradiol under conditions in which the antiestrogen LY156758 (keoxifene; raloxifene) was effective. The phytoestrogen formononetin behaved as an estrogen at high concentrations, and this response was antagonized by LY156758. THC, desacetyllevonantradol, or CBD did not stimulate transcription of an *EREtkCAT* reporter gene transiently transfected into MCF-7 cells.

Estrous Cycle Disruption Effect

Ethanol (95%) extract of the dried aerial parts, administered intraperitoneally to gerbils at a dose of 2.5 mg/animal daily for 60 days, was active. Petroleum ether extract of the dried aerial, administered intraperitoneally to mice and rats at doses of 1 and 5 mg/animal, respectively, for 64 days, was active. Petroleum ether extract of the aerial parts, administered intraperitoneally to female rats, produced weak activity. Petroleum ether extract of the entire plant, administered by gastric intubation to female mice at doses of 75 mg/kg and 150 mg/kg, was active. A dose of 3 mg/kg produced weak activity. Petroleum ether extract of the resin, administered intraperitoneally to female rats at doses of 10 and 20 mg/kg, was active. Resin, administered orally to female rats at doses of 3, 15, and 75 mg/kg daily for 72 days, was active.

Familial Mediterranean Fever

A patient with familial Mediterranean fever was presented with chronic relapsing pain and inflammation of gastrointestinal origin. After determining a suitable analgesic dosage, a double-blind, placebo-controlled, crossover trial was conducted using 50 mg of Δ-9-THC daily in five doses in the active weeks and measuring effects on parameters of inflammation and pain. Although no anti-inflammatory effects of Δ-9-THC were detected during the trial, a highly significant reduction ($p < 0.001$) in additional analgesic requirements was achieved.

Food Intake Modulation

Cannabis sativa stimulates appetite, especially for sweet and palatable food. Cannabinoid action has proposed a central role of the cannabinoid system in obesity. Dronabinol, a commercially available form of a THC, has been used successfully for increasing appetite in patients with HIV wasting disease. Cannabinoid receptor antagonist may reduce obesity. To determine the prevalence of substance use in adolescents with eating disorders, the results of a data set of Ontario high school students were compared. One hundred and one female adolescents who met the *Diagnostic and Statistical Manual of Mental Disorders*, 4th edition's criteria for an eating disorder were followed up in a tertiary care pediatric treatment center. They were asked to participate in a cross-sectional study using a self-administered questionnaire assessing substance use and investigating reasons for use and nonuse; 95 agreed to participate and 77 completed the questionnaire (mean age, 15.2 years). The patients were divided into two groups: 63 with restrictive symptoms only, 17 with purging symptoms. The rates of drug use between subjects and their comparison groups were compared by *Z*-scores, with the level of significance set at 0.05. During the preceding year, restrictors used significantly less tobacco, alcohol, and cannabis than grade- and sex-matched comparison populations, and purgers used these substances at rates similar to those of comparison subjects. Other drugs seen frequently in the purgers included hallucinogens, tranquilizers, stimulants, LSD, phencyclidine, cocaine, and ecstasy. Both groups used caffeine and laxatives, but few used diet pills. Restrictors said they did not use substances because they were bad for their health, tasted unpleasant, were contrary to their beliefs, and were too expensive. Purgers generally used substances to relax, relieve anger, avoid eating, and "get away" from problems. Female adolescents with eating disorders who have restrictive symptoms use substances less frequently than the general adolescent population but do not abstain from their use. Those with purging symptoms use substances with a similar frequency to that found in the general adolescent population.

Gene Expression Effect

Cannabinoids can cross the placental barrier and be secreted in the maternal milk. Through this way, cannabinoids affect the ontogeny of various neurotransmitter systems leading to changes in different behavioral patterns. Dopamine and endogenous opioids are among the neurotransmitters that result more affected by perinatal cannabinoid exposure, which, when animals mature, produce changes in motor activity, drug-seeking behavior, nociception, and other processes. These disturbances are likely originated by the capability of cannabinoids to influence the expression of key genes for both neurotransmitters, in particular, the enzyme tyrosine hydroxylase and the opioid precursor proenkephalin. Cannabinoids seem to be able to influence the expression of genes encoding for neuroglia cell adhesion molecules, which supports a potential influence of cannabinoids on the processes of cell proliferation, neuronal migration or axonal elongation in which these proteins are involved. CB1 receptors, which represent the major targets for the action of cannabinoids, are abundantly expressed in certain brain regions, such as the subventricular areas, which have been involved in these processes during brain development. Cannabinoids might also be involved in the apoptotic death that occurs during brain development, possibly by influencing the expression of Bcl-2/Bax system. CB1 receptors are transiently expressed during brain development in different group of neurons which do not contain these receptors in the adult brain.

Glaucoma Effect

Nine patients with glaucoma unresponsive to treatment were treated with orally administered Δ-9-THC capsules or inhaled cannabis in addition to their existing therapeutic regimen. An initial decrease in intraocular pressure was observed in all patients, and the investigator's therapeutic goal was met in four of the nine patients. The decreases in intraocular pressure were not sustained, and the patients elected to discontinue treatment within 1–9 months for various reasons.

Gliomatous Effect

Gliomas, in particular glioblastoma multiform or grade IV astrocytoma, are the most frequent class of malignant primary brain tumors and one of the most aggressive forms of cancer. Cannabinoids and their derivatives slowed the growth of different types of tumors, including gliomas, in laboratory animals. Cannabinoids induced apoptosis of glioma cells in culture via sustained ceramide accumulation, extracellular signal-regulated kinase activation and Akt inhibition. Cannabinoid treatment inhibited angiogenesis of gliomas in vivo. Cannabinoids killed glioma cells selectively and could protect nontransformed glial cells from death.

Gynecomastic Effect

A retrospective analysis was carried out on 175 men over the age of 16 years who were presented with breast enlargement and/or "lumps" during a 7-year period to a single surgeon. The patients had complete biochemical assessment (liver function tests, γ-glutamyl transferase, prolactin, α-fetoprotein, and β-human chorionic gonadotropin), and mammography and/or ultrasound with fine-needle biopsy if indicated. Thirty-nine of the patients had bilateral true gynecomastia and 88 had unilateral gynecomastia (53% left). Carcinoma of the breast was diagnosed in eight, pseudo-gynecomastia in 18, 13 had physiological pubertal changes only, and 9 had other diagnoses. Adverse drug reactions were possibly implicated in the etiology of 47 patients, alcohol in seven patients, cannabis in one patient, testicular malignancy in four patients, and hepatocellular carcinoma in one patient. Five patients were found to have hyperprolactinemia. Twenty-four percent of patients were reassured without intervention; 18% failed to attend follow-up.

Hepatitis C Risk Factor

The study of a dually diagnosed population estimated the prevalence of hepatitis C virus (HCV) to be 29.7% or 16 times higher than that in the general population. A high correlation was found between the use of tobacco and HCV infection. This appears to be beyond the risk factor conveyed by intravenous drug use. Of the patients whose primary diagnoses were cocaine, opiate, amphetamine, or polysubstance dependence (drugs often used intravenously), 42% of the tobacco users were HCV-positive, whereas only 20% of the nontobacco using patients with similar primary diagnoses were HCV-positive. The association of tobacco use with HCV was found to be strong for females with alcohol, sedative/hypnotic, inhalant, or cannabis dependence, as none of the 17 nontobacco using female patients with these diagnoses were HCV-positive, whereas 14 of the 45 (31%) tobacco-using females with these diagnoses did test positive for HCV.

HIV Involvement

The prevalence, predictors, and patterns of cannabis use—specifically medicinal cannabis use among patients with HIV—were examined. Any cannabis use in the year prior to interview and self-defined medicinal use were evaluated. A cross-sectional multicenter survey and retrospective chart review were conducted to evaluate overall drug utilization in HIV, including cannabis use. HIV-positive adults were identified through the HIV Ontario Observational Database; 104 consenting patients were interviewed. Forty-three percent of the patients reported cannabis use, whereas 29% reported medicinal use. Reasons for use were similar by gender although a significantly higher number of women used cannabis for pain management. The most commonly reported reason for medicinal cannabis use was appetite stimulation/weight gain. Male gender and history of intravenous drug use were predictive of any cannabis use. Age, gender, HIV clinical status, antiretroviral use, and history of intravenous drug use were not significant predictors of medicinal cannabis use. Despite the frequency of medicinal use, minimal changes in the pattern of cannabis use on HIV diagnosis were reported with 80% of current medicinal users also indicating recreational consumption. HIV patients (n = 252) were recruited via consecutive sampling in public health care clinics. Structured interviews assessed patterns of recent cannabis use, including its perceived benefit for symptom relief. Associations between cannabis use and demographic and clinical variables were examined using univariate and multivariate regression analyses. Overall prevalence of smoked cannabis in the previous month was 23%. Reported benefits included relief of anxiety and/or depression (57%), improved appetite (53%), increased pleasure (33%), and relief of pain (28%). Recent use of cannabis was positively associated with severe nausea (OR = 4,

$p = 0.004$) and recent use of alcohol (OR = 7.5, $p < 0.001$) and negatively associated with being Latino (OR = 0.07, $p < 0.001$). No associations between cannabis use and pain symptoms were observed. No safety problems specific to HIV or protease inhibitors were found in a study in which volunteers stayed in a research hospital 24 hours a day and were randomly assigned to either smoke cannabis, take oral THC, or take an oral placebo. Cannabis and THC use was associated with weight gain.

Hyperglycemic Activity

Ethanol (95%) extract of the leaf, administered intravenously to rats at a dose of 300 mg/kg, produced an increase of 40 mg percentage 2 hours postinjection and a corresponding decrease in liver glycogen. Ethanol (95%) extract of the dried leaf, administered by gastric intubation to rabbits, produced an increase followed by a gradual decrease in blood sugar levels. The dried leaves, smoked by human adults, produced elevated glucose levels in two out of four subjects and no impairment of insulin release or changes in growth hormone levels.

Hypoglycemic Activity

Ethanol (95%) extract of the dried leaf, administered by gastric intubation to rabbits, produced an increase, followed by a gradual decrease, in blood sugar levels. Extract of the dried leaf, administered subcutaneously to rabbits at a dose of 0.5 mL/kg (approx 0.6 mg THC) for 9 weeks, further enhanced hypoglycemia induced by insulin. No hypoglycemic effect was seen in normal animals. Hot water extract of the resin, administered by gastric intubation to dogs at a dose of 20 g of air-dried resin/animal, produced weak activity. The dried leaf, smoked by adults at a dose of 2 g/person, was inactive.

Hypotensive Activity

Ethanol (50%) extract of the entire plant, administered intravenously to dogs at a dose of 50 mg/kg, was active. Ethanol (95%) and water extracts of the dried aerial parts, administered intravenously to cats, were inactive. The ethanol extract stimulated respiration, and the water extract had no effect.

Ilicit Drug in Plasmapheresis Donors

Seventy-five US plasma units from 10 different states in the United States and 75 German plasma units that had been analyzed principally for their protein composition were screened for drugs. Determinations were made, using automated immunoassays, of the presence of cannabis, cocaine, amphetamine, methamphetamine, MDMA, methyl enedioxyethylamphetamine (MDE), and opiates. Positive results were confirmed by gas chromatography–mass spectrometry. Eleven US plasma units were found to be positive for cocaine (14.6%), whereas all German samples were cocaine-negative ($p = 0.0007$). Fifteen US plasma units (20%) and one German unit (1.3%) were confirmed as positive for cannabis ($p = 0.0003$). Three out of 75 US plasma units were positive for both cannabis and cocaine. In none of the 150 samples were amphetamine, methamphetamine, MDMA, MDE, or opiates detected.

Immunomodulatory Effect

The smoking of cannabis showed a significant local immunosuppression of the bactericidal activity of human alveolar macrophages. In animal studies, cannabinoids were identified as potent modulators of cytokine production, causing a shift from T-helper-1 (Th1) to Th2 cytokines. In consequence, a compromised cellular immunity was observed in these animals, resulting in enhanced tumor growth and reduced immunity to viral infections. In vitro, immunosuppressive effects were shown in all immune cells, but only at high micro-molar cannabinoid concentrations not reached under normal clinical conditions. In conclusion, there was no evidence that cannabinoids induce a serious, relevant immunosuppression in humans, with the exception of cannabis smoking, which may affect local bronchoalveolar immunity. The immune function in 16 MS patients treated with oral cannabinoids was measured. A modest increase of tumor necrosis factor (TNF)-α in lipopolysaccharide-stimulated whole blood was found during cannabis plant-extract treatment ($p = 0.037$), with no change in other cytokines. In the subgroup of patients with high adverse event scores, an increase in plasma IL-12p40 was found ($p = 0.002$). The results indicate pro-inflammatory disease-modifying potential of cannabinoids in MS. THC and their metabolites inhibited production of IL-1 and γ-interferon, decreased a 33% of the lymphocytes activity and inhibited 66% of the lymphocytes adenylcyclase activity. The consumption of cannabis decreased immunological competence of macrophages, and alternated their essential role of trophicity of the central nervous system. Inhibiting actions of cannabinoids on the cyclo-oxygenase, promoted production of arachidonic acid degradation products. This compound mimics the action of histamine, induced a raise of the vascular permeability and bronchospasm, and contributed at delayed reaction of anaphylaxia.

Infant Mortality

For a period of 11 months, 2964 infants were enrolled and screened at birth for exposure to cocaine, opiate, or cannabinoid by meconium analysis. At birth, 44% of the infants tested positive for drugs, 30.5% positive for cocaine, 20.2% for opiate, and 11.4% for cannabinoids. Compared with the drug-negative group, a significantly higher percentage ($p < 0.05$) of the drug-positive infants had lower weight and smaller head circumference and length at birth and a higher percent of their mothers were single, multigravid, multiparous, and had little to no prenatal care. Within the first 2 years of life, 44 infants died: 26 were drug-negative (15.7 deaths per 1000 live births) and 18 were drug-positive (13.7 deaths per 1000 live births). The mortality rate among cocaine, opiate, or cannabinoid-positive infants were 17.7, 18.4, and 8.9 per 1000 live births, respectively. Among infants with birth-weight of 2500 g or less, infants who were positive for both cocaine and morphine had a higher mortality rate (OR = 5.9, CI = 1.4–24) than drug-negative infants. Eleven infants died from the sudden infant death syndrome (SIDS); 58% were positive for drugs, predominantly cocaine. The odds ratio for SIDS among drug-positive infants was 1.5 (CI = 0.46–5.01) and 1.9 (CI = 0.58–6.2) among cocaine-positive infants.

Infant Neurobehavioral Effect

The subjects and controls in this study were full- term infants of appropriate gestational age with no medical problems. At 1–2 days of age, 20 infants exposed to cocaine, alcohol, cannabis, and cigarettes, 17 infants exposed to alcohol and/or cannabis and cigarettes, and 20 drug-free infants were evaluated by using the Neonatal Intensive Care Unit Network Neurobehavioral Scale. Cocaine-exposed (CE) infants showed increased tone and motor activity, more jerky movements, startles, tremors, back arching, and signs of central nervous system and visual stress than unexposed infants. They also showed poorer visual and auditory following. There were no differences in how the examination was administered to CE and nonexposed infants. Reduced birth-weight and length were also observed in CE infants. Differences attributable to CE infants were related to muscle tone and motor performance, following during orientation, and signs of stress. CE infants were not more difficult to test, nor did they require an alteration in the examination. Both neurobehavioral patterns of excitability and lethargy were observed. The findings may have been a result of the synergistic effects of cocaine with alcohol and cannabis.

Inflammatory Effect

A case of a 17-year-old male regular cannabis user who developed a large swollen uvula and partial upper airway obstruction after smoking cannabis was evaluated. Symptoms resolved with the administration of corticosteroids and antihistamines. A healthy 17-year-old man who inhaled cannabis prior to general anesthesia is described. In the recovery room, after an uneventful general anesthetic, acute uvular edema resulted in postoperative airway obstruction and admission to the hospital. The uvular edema was treated successfully with dexamethasone.

Information-processing Effect

Information processes are thought to represent the basic building blocks of higher order cognitive processes. The inspection time task was used to investigate the effects of acute and subacute cannabis use on information processing in 22 heavy users compared with 22 nonusers. The findings indicated that users in the subacute state display significantly slowed information-processing speeds (longer inspection times) compared with controls. This deficit appeared to be normalized while users were in the acute state. These results may be explained as a withdrawal effect, but may also be owing to tolerance development because of long-term cannabis use.

Insecticidal Activity

Leaf extract, administered to larvae of *Chironomus samoensis*, produced paralysis leading to death. The extract brought a drastic change in the morphology of sensilla trichoidea, the general body cuticle, and a significant reduction in the concentration of magnesium and iron, whereas manganese showed only slight average increase. Because the sensilla trichoidea has nerve connections, it was assumed that the toxic principle of the leaf extract has affected the central nervous system.

Intestinal Motility Activity

Rat intestinal epithelia mounted in an Ussing chamber attached with voltage/current clamp were used for measuring changes of the short-circuit current across the epithelia. The intestinal epithelia were activated with current raised by serosal administration of forskolin 5 μM. Ethanol extracts of cannabis augmented the current additively when each was added after forskolin. In subsequent experiments, ouabain, and bumetanide were added prior to ethanol extract of cannabis to determine their effect on Na^+ and Cl^- movement. The results suggested that the extract may affect the Cl^- movement more directly than Na^+ movement in the intestinal epithelial cells.

Intraocular Pressure Reduction

Polysaccharide fraction of the dried entire plant, administered intravenously to rabbits at a dose of 1 μg/animal was active. Water extract of the dried aerial parts, administered intravenously to rabbits at a dose of 250 μg/animal, was active. A dose of 5 μg/animal was inactive on Rhesus monkeys and active on rabbits. A dose of 10 mg/animal, administered *per rectum* to Rhesus monkeys and rabbits, was inactive.

IQ Effect

Cannabis use for 70 individuals aged 17–20 years was determined through self-reporting and urinalysis. IQ scores were calculated by subtracting each person's IQ score at 9–12 years (before initiation of drug use) from his or her score at 17–20 years. The difference in IQ scores of current heavy users (at least five joints per week), current light users (less than five joints per week), former users (who had not smoked regularly for at least 3 months), and nonusers (who never smoked more than once per week and no smoking in the past 2 weeks) was compared. Current cannabis use was significantly correlated ($p < 0.05$) in a dose-related fashion with a decline in IQ over the ages studied. The comparison of the IQ difference scores showed an average decrease of 4.1 points in current heavy users ($p < 0.05$) compared with gains in IQ points for light current users (5.8), former users (3.5), and nonusers (2.6).

Lactate Inhibition

The dried leaf, smoked by adults at a dose of 2 g/person, decreased blood lactic acid.

Leutinizing Hormone-release Inhibition

The dried aerial part, smoked by menopausal women at a dose of 1 g/person, was inactive. When administered to normal and castrated male rats, at a dose of 75 mg/kg, was active.

Lower Limb Occlusive Arteriopathy

Seventy-three patients (60 males and 13 females less than 50 years of age) were divided into four groups: Buerger's disease (thromboangiitis obliterans [TAO]), atheromatous juvenile peripheral obstructive arterial diseases (POAD), autoimmune POAD, and arteriopathy of undetermined origin. The first symptoms occurred at 38 ± 8 years of age. Fourteen patients (20%) had TAO, 51 (70%) atheromatous POAD, 4 (5%) POAD with systemic or autoimmune disease, and 4 (5%) undetermined POAD. Age of onset was earlier in TAO (35 ± 8 vs 40 ± 8 years, $p = 0.046$), smoking was greater in the atheroma group (33 ± 16 vs 24 ± 14 pack/years, $p = 0.033$). Fifty-three patients with POAD had dyslipidemia and 26% had hypertension. Regular cannabis intake was more frequent in the TAO group (21% vs 8%). At the time of medical care, Fontaine's stage was more frequently stage II in atheroma patients (57% vs 14%) and stage IV in TAO patients (86% vs 35%). TAO was diagnosed in 43% cannabis users and in 19% nonusers. Results indicated that the main etiology of juvenile POAD is atheroma, followed by TAO. Cannabis users accounted for at least 10% of these patients. They were characterized by lower tobacco intake, more distal lesions, more frequent involvement of the upper limbs. They presented more frequently as TAO. A case of a 30-year-old woman who smoked cannabis and developed intermittent claudication of the lower limbs was reported. Results indicated that cannabis could be involved not only in the pathogenesis of juvenile obstructive arteriopathy, but also in the development of atheromatous lesions.

Lung Function

A group of over 900 young adults derived from a birth cohort of 1037 subjects were studied at age 18, 21, and 26 years. Cannabis and tobacco smoking were documented at each age using a standardized interview. Lung function, as measured by the FEV1–vital capacity (VC) ratio, was obtained by simple spirometry. A fixed effects regression model was used to analyze the data and to account for confounding factors. When the sample was stratified for cumulative use, there was evidence of a linear relationship between cannabis use and FEV1–VC ($p < 0.05$). In the absence of adjusting for other variables, increasing cannabis use over time was associated with a decline in FEV1–VC with time; the mean FEV1–VC among subjects using cannabis on 900 or more occasions was 7.2, 2.6 and 5% less than nonusers at ages 18, 21, and 26, respectively. After controlling for potential confounding factors (age, tobacco smoking, and weight) the negative effect of cumulative cannabis use on mean FEV1–VC was only marginally significant ($p < 0.09$). Age ($p < 0.001$), cigarette smoking ($p < 0.05$), and weight ($p < 0.001$) were all significant predictors of FEV1–VC. Cannabis use and daily cigarette smoking acted additively to influence FEV1–VC. Results indicated that longitudinal observations over 8 years in young adults revealed a dose-dependent relationship between cumulative cannabis consumption and decline in FEV1–VC. When confounders were accounted for the effect was reduced and was only marginally significant, but given the limited time frame over which observations were made, the trend suggests that continued cannabis smoking has the potential to result in clinically important impairment of lung function.

Memory Impairment

The effects of combined exposure to ethanol and Δ-9-THC in a memory task was investigated in rats. Ethanol, voluntarily ingested in alcohol-preferring rats, and THC, given by intraperitoneal injection, had a synergic action to impair object recognition when a 15-minute interval was adopted between the sample phase and the choice phase of the test. Ingestion of ethanol, or 2 or 5 mg/kg of THC were not able to modify object recognition in these experimental conditions. When voluntary ethanol ingestion was combined with administration of these doses of THC, object recognition was markedly impaired. THC impaired object recognition only at the dose of 10 mg/kg, when its administration was not combined with that of ethanol. The selective cannabinoid CB1 receptor antagonist SR 141716A (*N*-(piperidin-1-yl)-5-(4-chlorophenyl)-1(2,4-dichloro-phenyl)-4-methyl-1 H-pyrazole carboxamide HCl) at the dose of 1 mg/kg reversed the amnesic effect of 10 mg/kg of THC. This indicated that the effect is mediated by the receptor subtype. The synergism of ethanol and THC was not detected when an intertrial interval of 1 minute was adopted.

Memory Improvement

Extract from fructus cannabis (EFC), administered intragastrically to mice with drug-induced dysmnesia at doses of 0.2, 0.4, and 0.8 g/kg, for 7 days, prolonged the latency and decreased the number of errors in the step-down test, and enhanced the spatial resolution of amnesic mice in water maze test. EFC at the dose of 0.2 g/kg overcame amnesia of three stages of memory process. EFC activated calcineurin activity at a concentration range of 0.01–100 g/L. The maximal value of EFC on calcineurin activity (35% ± 5 %) appeared at a concentration of 10 g/LCS320. EFC with activation of calcineurin, extracted from Chinese traditional medicine, was used to determine the effects on memory and immunity in mice. In the stepdown-type passive avoidance test, the plant extract (0.2 g/kg) significantly improved amnesia induced by drugs, and greatly enhanced the ability of cell-mediated type hypersensitivity and nonspecific immune responses in normal mice.

Mitochondrial Function Disruption

Δ-9-THC in the pulmonary transformed cell line A549 produced a rapid and extensive depletion of cellular energy stores. Adenosine 5'-triphosphatase levels declined dose dependently with an IC_{50} of 7.5 μg/mL of THC after 24 hours of exposure. Cell death was observed only at concentrations greater than 10 μg/mL. Studies using JC-1, a fluorescent probe for mitochondrial membrane potential, revealed diminished mitochondrial function at THC concentrations as low as 0.5 μg/mL. At concentrations of 2.5 and 10 μg/mL of THC, a decrease in mitochondrial membrane potential was observed 1 hour after THC exposure. Mitochondrial function remained diminished for at least 30 hours after THC exposure. Flow cytometry studies on cells exposed to particulate smoke extracts indicated that JC-1 red fluorescence was fivefold lower in cells exposed to cannabis smoke extract compared with tobacco smoke-exposed cells. Comparison with a variety of mitochondrial inhibitors demonstrated that THC produced effects similar to that of carbonyl cyanide *p*-trifluoromethoxyphenylhydrazone, suggesting uncoupling of electron transport. Loss of red JC-1 fluorescence by THC was suppressed by cyclosporin A, suggesting mediation by the mitochondrial permeability transition pore. This disruption of mitochondrial function was sustained for at least 24 hours after removal of THC by extensive washing.

Mitogenic Effect

The resin was inactive on the human and rat white blood cells.

Molluscicidal Activity

Ethanol (95%) and water extracts of the dried flowering tops, at a concentration of 1000 ppm, produced weak activity on *Biomphalaria straminea* and *Biomphalaria glabrata*. Water saturated with essential oil of the aerial parts, at a concentration of 1:2, produced weak activity on *Biomphalaria glabrata*.

Motor Function

Nine cannabis smokers and 16 controls were studied to determine the attentional areas related to motor function, and primary and supplementary motor cortices. Echo planar images and high-resolution molecular resonance images were acquired. The challenge paradigm included left and right finger sequencing. Group differences in cerebral activation were examined for Brodmann areas (BA) 4, 6, 24, and 32 using region of interests analyses in statistical parametric mapping. Cannabis users, tested within 4–36 hours of discontinuation, exhibited significantly less activation than controls in BA 24 and 32 bilaterally during right- and left-sided sequencing and for BA 6 in all tasks except for left-sided sequencing in the left hemisphere. There were no statistically significant differences for BA 4. None of these regional activations correlated with urinary cannabis concentration and verbal IQ for smokers. The results suggested that recently abstinent chronic cannabis smokers produce reduced activation in motor cortical areas in response to finger sequencing compared with controls.

Multiple Sclerosis

One hundred fifty-seven drug-naïve, first-episode schizophrenic patients were examined. A significantly elevated brain-derived neurotrophic factor (BDNF) serum concentrations in patients with chronic cannabis abuse ($n = 35, p < 0.001$) or multiple substance abuse ($n = 20, p < 0.001$) prior to disease onset were found. Drug-naive schizophrenic patients without cannabis consumption showed similar results to normal controls and cannabis controls without schizophrenia. Elevated BDNF serum levels were not related to schizophrenia and/or substance abuse itself but may reflect a cannabis-related idiosyncratic damage of the schizophrenic brain. Disease onset was 5.2 years earlier in the cannabis-consuming group ($p = 0.0111$). A cannabis-based medicinal extract (CBME) was administered to 160 patients with multiple sclerosis experiencing significant problems from at least one of the following: spasticity, spasms, bladder problems, tremor, or pain. The interventions were oromucosal sprays of matched placebo, or whole plant CBME containing equal amounts of Δ-9-THC and CBD at a dose of 2.5–120 mg of each daily, in divided doses. The primary outcome measure was a Visual Analogue Scale (VAS) score for each patient's most troublesome symptom. Additional measures included VAS scores of other symptoms, and measures of disability, cognition, mood, sleep and fatigue. Following CBME the primary symptom score reduced from mean 74.36 (11.1) to 48.89 (22.0) following CBME and from 74.31 (12.5) to 54.79 (26.3) following placebo. Spasticity VAS scores were significantly reduced by CBME (Sativex) in comparison with placebo ($p = 0.001$). There were no significant adverse effects on cognition or mood and intoxication was generally mild. A SCE with pure Δ-9-THC, at matched concentrations of Δ-9-THC, and a Δ-9-THC-free extract (Δ-9-THC-free SCE) in a mouse model of MS, were examined. Although SCE inhibited spasticity in the mouse model of MS to a comparable level, it caused a more rapid onset of muscle relaxation and a reduction in the time to maximum effect compared with Δ-9-THC alone. The Δ-9-THC-free extract or CBD caused no inhibition of spasticity. In an experimental allergic encephalomyelitis (EAE), an animal model of MS, it was demonstrated that the cannabinoid system is neuroprotective during EAE. Mice, deficient in the cannabinoid receptor CB1, tolerated inflammatory and excitotoxic insults poorly, and developed substantial neurodegeneration following immune attack in EAE. Exogenous CB1 agonists can provide significant neuroprotection from the consequences of inflammatory central nervous system disease in an experimental allergic uveitis model.

Mutagenic Activity

Petroleum ether extract of the aerial parts, in the ration of *Drosophila* at concentrations of 0.5, 1, and 5% of the diet, was active. Petroleum ether extract of the dried leaf, administered by gastric intubation to male mice at a dose of 50 mg/kg, was active. Water and methanol extracts of the seed, on agar plate at a concentration of 100 mg/mL, were inactive on *Bacillus subtilis* H-17 (Rec+) and *Salmonella typhimurium* TA100 and TA98. Metabolic activation had no effect on the results.

Myocardial Infarction

A young man who suffered a myocardial infarction after taking Viagra in combination with cannabis was investigated. Viagra is metabolized predominantly by the CYP450 3A4 hepatic microsomal isoenzyme. Cannabis is a known inhibitor of CYP450 3A4 isoenzyme. The effect of the Viagra was thus potentiated by the effect of cannabis.

Natural-killer Cells Effect

Leukemia susceptible BALB/c and resistant C57BL/6 mice were infected with Friend leukemia virus complex and its helper component Rowson-Parr virus. At different time points, their natural-killer cells were separated from spleens and treated with 0–10 μg/mL of THC, subsequently mixed with Yac-1 target cells for 4 and 18 hours. The natural-killer cell activity in both mouse strains infected by either virus complex or helper virus weakened on days 2–4 postinfection, normalized by day 8 and enhanced on days 11–14. Natural-killer cell activity on the effect of low concentration (1–2.5 μg/mL) of THC slightly increased in BALB/c, was unaffected in C57BL/6, especially in the 18 hour assays. In the combined effects of cannabis and retrovirus, damages by cannabis dominated over those of retroviruses. Inhibition or reactive enhancement of natural-killer cell activity on the effect of viruses were similar to those of infected but cannabis-free counterparts, but on the level of uninfected cells treated with cannabis. The effects of cannabis and retrovirus were additive resulting in anergy of natural-killer cells.

Neonatal Abstinence Syndrome

The relationship of maternal drug abuse to symptoms, the effectiveness of pharmacological agents in controlling symptoms, and the length of in-patient stay were investigated in infants with neonatal abstinence syndrome. Pharmacological treatment was oral morphine sulphate (0.2 mg four to six times hourly), phenobarbitone (3–7 mg/kg/day), or combination of the two were administered to infants with a serial Finnegan score greater than 8. The average maternal age was 24.6 years, (18–34 years). Drug use volunteered by the mothers was methadone alone in 6 cases, methadone and benzodiazepines in 14, methadone and heroin and benzodiazepines in 7, methadone and heroin in 10, heroin alone in 2, and other multiple drug use including oral morphine sulphate, dothiepin, and cannabis in 4. Average gestational age was 40.3 (35–42 weeks). The average birth-weight was 2.81 kg (1.89–3.91 kg). Time-to-onset of withdrawal symptoms was 2.8 (1–13) days. The duration of pharmacological treatment (oral morphine sulphate and/or phenobarbitone) was 21.8 (1–62) days. The total hospital stay for the 43 infants was 1011 days.

Neuroendocrine Abnormalities

Prolactin response to D-fenfluramine was assessed in abstinent ecstasy (MDMA) users with concomitant use of cannabis only (13 males, 11 females) and in two control groups: healthy nonusers (13 females) and exclusive cannabis users (seven males). Prolactin response to D-fenfluramine was slightly blunted in female ecstasy users. Both male user samples exhibited a weak prolactin response to D-fenfluramine, but this was weaker in the group of cannabis users. Baseline prolactin and prolactin response to D-fenfluramine were associated with the extent of previous cannabis use. The results indicated that the endocrinological abnormalities of ecstasy users might be closely related to their coincident cannabis use.

Neurogenic Symptoms Alleviation

Whole-plant extracts of Δ-9-THC, CBD, 1:1 CBD: THC, or placebo were self-administered by sublingual spray to 24 patients with MS (n = 18), spinal cord injury (n = 4), brachial plexus damage (n = 1), and limb amputation owing to neurofibromatosis (n = 1), at doses determined by titration against symptom relief or unwanted effects within the range of 2.5–120 mg/24 hours for 2 weeks. The patients recorded symptoms, well-being, and intoxication scores on a daily basis using visual analog scales. At the end of each two-week period an observer rated severity and frequency of symptoms on numerical rating scales, administered standard measures of disability (Barthel Index), mood, cognition, and recorded adverse events. Pain relief associated with both THC and CBD was significantly superior to placebo. Impaired bladder control, muscle spasms, and spasticity were improved by cannabis medicinal extract (CME) in some patients with these symptoms. Three patients had transient hypotension and intoxication with rapid initial dosing of THC-containing CME. The results indicated that cannabis could improve neurogenic symptoms unresponsive to standard treatments. Unwanted effects were predictable and generally well tolerated.

Neuropathic Pain Relief

Forty-eight patients with at least one avulsed root and baseline pain score of four or more on an 11-point ordinate scale participated in a randomized, double-blind, placebo-controlled, three-period crossover study. The patients had intractable symptoms regardless of current analgesic therapy. They entered a baseline period of 2 weeks, followed by three, 2-week treatment periods; during each period they received one of three oromucosal spray preparations. These were placebo and two whole plant extracts of *C. sativa* L.: GW-1000-02 (Sativex), containing Δ-9-THC: CBD in an approx 1:1 ratio and GW-2000-02, containing primarily THC. The primary outcome measure was the mean pain severity score during the last 7 days of treatment. Secondary outcome measures included pain related quality of life assessments. The primary outcome measure failed to fall by the two points defined in our hypothesis. Both this measure and measures of sleep showed statistically significant improvements. The study medications were well tolerated with the majority of adverse events, including intoxication type mild to moderate in severity and resolving spontaneous reactions.

Neuroprotective Effect

The effect of cannabidiol on β-amyloid peptide-induced toxicity in cultured rat pheocromocytoma PC12 cells was investigated. Following exposure of cells to β-amyloid peptide (1 μg/mL), a marked reduction in cell survival was observed. This effect was associated with increased reactive oxygen species production and lipid peroxidation, and caspase 3 (a key enzyme in the apoptosis cell-signalling cascade) appearance, DNA fragmentation, and increased intracellular calcium. Treatment of the cells with CBD (10^{-7}–10^{-4} mol) prior to β-amyloid peptide exposure, significantly elevated cell survival, whereas it decreased reactive oxygen species production, lipid peroxidation, caspase 3 levels, DNA fragmentation, and intracellular calcium. CBD and other cannabinoids were examined as neuroprotectants in rat cortical neuron cultures exposed to toxic levels of glutamate. The psychotropic cannabinoid receptor agonist Δ-9-THC and cannabidiol, reduced *N*-

methyl-D-aspartate, α-amino-3 -hydroxy-5 -methyl-4-isoxazole propionic acid and kainate receptor mediated neurotoxicities. Neuroprotection was not affected by cannabinoid receptor antagonist, indicating a (cannabinoid) receptor-independent mechanism of action. CBD demonstrated a reduction in hydroperoxide toxicity in neurons. In this trial of the abilities of various antioxidants to prevent glutamate toxicity, cannabidiol was superior to both α-tocopherol and ascorbate in protective capacity.

Neuropsychological Effect

Cerebral blood flow was measured in 12 long-term cannabis users shortly after cessation of cannabis use (mean 1.6 days). The findings showed significantly lower mean hemispheric blood flow values and significantly lower frontal values in the cannabis subjects compared with normal controls. The results indicated that the functional level of the frontal lobes was affected by long-term cannabis use.

Neurotransmission inhibition

The BLA or the medial prefrontal cortex (PFC) stimulation in urethane-anesthetized rats induced generation of action potentials in the Nac neurons. This excitatory effect was strongly inhibited by the synthetic cannabinoid agonists WIN (0.062–0.25 mg/kg, iv [intravenously]) and HU-210 (0.125–0.25 mg/kg, iv), or Δ-9-THC (1 mg/kg, iv). D1 or D2 dopamine receptor antagonists (SCH23390 0.5–1 mg/kg, sulpiride 5–10 mg/kg, iv) or the opioid antagonist naloxone (1 mg/kg, iv) were not able to reverse the action of cannabinoids. The selective CB1 receptor antagonist/reverse agonist SR141716A (0.5 mg/kg, iv) fully suppressed the action of cannabinoid agonists, whereas *per se* had no significant effect.

Nicotine and Δ-9-THC Interaction

Δ-9-THC administration to mice significantly decreased the incidence of several nicotine withdrawal signs precipitated by mecamylamine or naloxone, such as wet-dog-shakes, paw tremor, and scratches. In both experimental conditions, the global withdrawal score was significantly attenuated by Δ-9-THC administration. The effect of Δ-9-THC was not to the result possible adaptive changes induced by chronic nicotine on CB1 cannabinoid receptors. The density and functional activity of these receptors were not modified by chronic nicotine administration in the different brain structures investigated. The consequences of Δ-9-THC administration on *c-Fos* expression in several brain structures after chronic nicotine administration and withdrawal were examined. *c-Fos* was decreased in the caudate putamen and the dentate gyrus after mecamylamine precipitated nicotine withdrawal. Δ-9-THC administration did not modify *c-Fos* expression under these experimental conditions. Δ-9-THC also reversed conditioned place aversion associated to naloxone precipitated nicotine withdrawal. The results indicated that Δ-9-THC administration attenuated somatic signs of nicotine withdrawal and this effect was not associated with compensatory changes on CB1 cannabinoid receptors during chronic nicotine administration. Δ-9-THC also ameliorated the aversive motivational consequences of nicotine withdrawal.

Night Vision Improvement

In a double-blind study, graduated THC administration at doses of 0–20 mg (as Marinol) on measures of dark adaptometry and scotopic sensitivity was evaluated. Field studies of night vision were performed among Jamaican and Moroccan fishermen, and mountain dwellers with the LKC Technologies Scotopic Sensitivity Tester-1. Improvements in night vision measures were noted after THC or cannabis. The effect was dose-dependent and cannabinoid-mediated at the retinal level.

Nocturnal Sleep Effect

Eight healthy volunteers (four males, four females; aged 21–34 years) were taking placebo, 15 mg Δ-9-THC, 5 mg THC combined with 5 mg CBD, and 15 mg THC combined with 15 mg CBD. These were formulated in 50:50 ethanol to propylene glycol and administered using an oromucosal spray during a 30-minute period from 10 PM. Electroencephalogram was recorded during the sleep period (11 PM to 7 AM). Performance, sleep latency, and subjective assessments of sleepiness and mood were measured from 8:30 AM (10 hours after drug administration). There were no effects of 15 mg THC on nocturnal sleep. With the concomitant administration of the drugs (5 mg THC and 5 mg CBD to 15 mg THC and 15 mg CBD), there was a decrease in stage 3 sleep, and with the higher dose combination, wakefulness was increased. The next day, with a 15-mg THC dose, memory was impaired, sleep latency was reduced, and the subjects reported increased sleepiness and changes in mood. With the lower dose combination, reaction time was faster on the digit recall task, and with the higher dose combination, subjects reported increased sleepiness and changes in mood. Fifteen milligrams of THC appeared to be sedative, and 15 mg CBD appeared to have alerting properties as it increased waking activity during sleep and counteracted the residual sedative activity of the 15 mg THC.

Occipital Stroke

A right occipital ischemic stroke occurred in a 37-year-old Albanese man with a previously uneventful medical history, 15 minutes after smoking a cigarette with approximately 250 mg of cannabis. Clinical manifestations of the stroke were left-sided hemiparesis, hemihypesthesia and blurred vision, which vanished spontaneously and almost completely after 3 days.

The patient has been smoking cannabis regularly from the age of 27, with a frequency of two to three cigarettes/cannabis per week during the 6 months that preceded his stroke. Except for cigarette smoking and slight dyslipidemia, classical risk factors for stroke/embolism were absent. The family history for cerebrovascular events, blood pressure, clotting tests, examinations for thrombophilia, vasculitis, extracranial and intracranial arteries, and cardiac investigations were normal or respectively negative; the stroke was attributed to the chronic cannabis consumption.

Oral Cancer

A study of 116 patients aged 45 years and younger, diagnosed with squamous cell carcinoma of the mouth was conducted. Two hundred and seven controls who had never had cancer, matched for age, sex, and area of residence, were recruited. The self-completed questionnaire contained items about exposure to the following risk factors: tobacco products, cannabis, alcohol, and diet. Conditional logistic analyses were conducted adjusting for social class, ethnicity, tobacco, and alcohol habits. All tests for statistical significance were two-sided. The majority of oral cancer patients reported exposure to the major risk factors of tobacco and alcohol even at the younger age. The estimated risks associated with tobacco or alcohol were low among both males and females. Only smoking for 21 years or more produced significantly elevated odds ratios (OR = 2.1; 95% CI: 1.1–4). Exposure associated with other major risk factors did not produce significant risks in this sample. Long-term consumption of fresh fruits and vegetables in the diet appeared to be protective for both males and females.

Oral Cytological Effect

The effects of cannabis, methaqualone, or tobacco smoking on the epithelial cells in 16 patients were evaluated. The site samples included the buccal mucosa (left and right sides), the posterior dorsum of the tongue, and the anterior floor of the mouth. There was a significant prevalence of bacterial cells in the smears and a greater number of degenerate and atypical squamous cells in cannabis users compared with controls. Epithelial cells in smears taken from cannabis users and tobacco-smoking controls showed koilocytic changes.

Pancreatic effect

A 29-year-old man presented with acute pancreatitis after a period of heavy cannabis smoking. Other causes of the disease were ruled out. The pancreatitis resolved itself after the cannabis was stopped and this was confirmed by urinary cannabinoid metabolite monitoring in the community. There were no previous reports of acute pancreatitis associated with cannabis use in the general population. Drugs of all types are related to the etiology of pancreatitis in approximately 1.4–2% of cases.

Panic Disorder

Sixty-six panic disorder patients were included in a study. All of whom met the DSM-IV diagnosis of panic disorder (n = 45) or panic disorder with agoraphobia ([PDA]; n = 21). Twenty-four patients experienced their first panic attack within 48 hours of cannabis use and then went on to develop panic disorder. All the patients were treated with paroxetine (gradually increased up to 40 mg/day). The two groups responded equally well to paroxetine treatment as measured at the 8 weeks and 12 months follow-up visits. There were no significant effects of age, sex, and duration of illness as covariates with response rates between the two groups. In addition, panic disorder or panic disorder with agoraphobia diagnosis did not affect the treatment response in either group. There were no significant differences in weight gain, sexual side effects, or relapse rates between patients according to gender or comorbid diagnosis.

Paroxysmal Atrial Fibrillation

A healthy young subject was observed for paroxysmal atrial fibrillation following cannabis intoxication. The abuse of this substance was the most possible and identifiable risk factor.

Place Conditioning Effect

THC was administered to female rats at doses of 1, 5, or 20 mg/kg) during gestation and lactation. Maternal exposure to low doses of THC (1 and 5 mg/kg), relevant for human consumption, produced an increased response to the reinforcing effects of a moderate dose of morphine (350 μg/kg), as measured in the place-preference conditioning paradigm (CPP) in the adult male offspring. These animals also displayed an enhanced exploratory behavior in the defensive withdrawal test. Only females born from mothers exposed to THC at a dose of 1 mg/kg exhibited a small increment in the place conditioning induced by morphine. The possible implication of the HPA was analyzed by monitoring plasma levels of adrenocorticotropic hormone (ACTH) and corticosterone in basal and moderate-stress conditions (after the end of the CPP test). Female offspring perinatally exposed to THC (1 or 5 mg/kg) displayed high basal levels of corticosterone and a blunted adrenal response to the HPA-activating effects of the CPP test. Male offspring born from mothers exposed to THC (1 or 5 mg/kg) displayed the opposite pattern: normal to low basal levels of corticosterone, and a sharp adrenal response to the CPP challenge. THC administration to rats at a low dose (1.5 mg/kg) resulted in failing to develop place conditioning, and developing a place aversion at a high dose (15 mg/kg). Administration of the cannabinoid antagonist SR141716A induced a CPP at both a low (0.5 mg/kg) and a high (5 mg/kg) dose.

Plant Germination Effect

Methyl chloride extract of the dried seed produced weak activity on *Amaranthus spinosus* (25.8%) inhibition. Methyl chloride extract of the dried leaves produced 17.5% inhibition of *Amaranthus spinosus*.

Plasma Norepinephrine Concentration

Forty-six newborn infants participated in a prospective study of the neonatal and long-term effects of prenatal cocaine exposure. Based on maternal self-report, maternal urine screening, and infant meconium analysis, 24 infants were classified as CE and 22 as unexposed. Between 24 and 72 hours postpartum, plasma samples for norepinephrine (NE), epinephrine, dopamine, and dihydroxyphenylalanine analysis were obtained. The Neonatal Behavioral Assessment Scale was administered at 1–3 days of age and at 2 weeks of age by examiners masked to the drug exposure status of the newborns. The CE newborns had increased plasma NE concentrations when compared with the unexposed infants (geometric mean, 923 pg/mL vs 667 pg/mL). There were no significant differences in plasma epinephrine, dopamine, or dihydroxyphenylalanine concentrations. Analysis for the effect of potential confounding variables revealed that maternal cannabis use was also associated with increased plasma NE, although birth-weight, gender, and maternal use of alcohol or cigarettes were not. Geometric mean plasma NE was 1164 pg/ mL in those infants with *in utero* exposure to both cocaine and cannabis compared to 812 pg/mL in those exposed to only cocaine and 667.0 pg/mL in those exposed to neither. Among the CE infants, plasma NE concentration correlated with an increased score for the depressed cluster ($r = 0.53$) and a decreased score for the orientation cluster (r

= –0.43) of the Neonatal Behavioral Assessment Scale administered at 1–3 days of age. Adjusting for cannabis exposure had no effect on these relationships between plasma NE and the depressed and orientation clusters.

Pneumonic Effect

A case–control study was conducted in 7001 individuals. Odds ratios were calculated by conditional logistic regression with substance use and social factors as cofounders. Pneumonia was not associated with kava use. Crude odds ratios = 1.26 (0.74–2.14, p = 0.386) increased after controlling for confounders (OR = 1.98, 0.63–6.23, p = 0.237) but was not significant. Adjusted odds ratios for pneumonia cases involving kava and alcohol users was 1.19 (0.39–3.62, p = 0.756). Crude odds ratios for associations between pneumonia and cannabis use (OR = 2.27, 1.18–4.37, p = 0.014) and alcohol use (OR = 1.95, 1.07–3.53, p = 0.026) were statistically significant and approached significance for petrol sniffing (OR = 1.98, 0.99–3.95, p = 0.056).

Postural Syncope

Twenty-nine volunteers participated in a randomized, double-blind, placebo-controlled study. Cerebral blood velocity, pulse rate, blood pressure, skin perfusion on forehead and plasma Δ-9-THC levels were quantified during reclining and standing for 10 minutes before and after THC infusions and cannabis smoking. Both THC and cannabis induced postural dizziness, with 28% reporting severe symptoms. Intoxication and dizziness peaked immediately after drug. The severe dizziness group showed the most marked postural drop in cerebral blood velocity and blood pressure and showed a drop in pulse rate after an initial increase during standing. Postural dizziness was unrelated to plasma levels of THC and other indices.

Prenatal Exposure

Data collected from the National Household Survey on Drug Abuse, a nationally representative sample survey of 22,303 non-institutionalized women aged 18–44 years, of whom 1249 were pregnant, were analyzed. During the 2-year study period, 6.4% of the non-pregnant women of childbearing age and 2.8% of the pregnant women reported that they used illicit drugs. Of the women who used drugs, the relative proportion of women who abstained from illicit drugs after recognition of pregnancy increased from 28% during the first trimester of pregnancy to 93% by the third trimester. However, because of postpregnancy relapse, the net pregnancy-related reduction in illicit drug use at postpartum was only 24%. Cannabis accounted for three-fourths of illicit drug use, and cocaine accounted for one-tenth of illicit drug use. Of those who used illicit drugs, over half of pregnant and two-thirds of non-pregnant women used cigarettes and alcohol. Among the sociodemographic subgroups, pregnant and non-pregnant women who were young (18–30 years) or unmarried, and pregnant women with less than a high school education had the highest rates of illicit drug use. Over 12,000 women at 18–20 weeks of gestation were enrolled in an Avon Longitudinal Study of Pregnancy and Childhood. Five percent of the mothers reported smoking cannabis before and/or during pregnancy; they were younger, of lower parity, better educated, and more likely to use alcohol, cigarettes, coffee, tea, and hard drugs. Cannabis use during pregnancy was unrelated to risk of perinatal death or need for special care, but the babies of women who used cannabis at least once per week before and throughout pregnancy were 216 g lighter than those of nonusers, had significantly shorter birth lengths, and smaller head circumferences. After adjustment for confounding factors, the association between cannabis use and birth-weight failed to be statistically significant ($p = 0.056$) and was clearly nonlinear. The adjusted mean birth-weights for babies of women using cannabis at least once per week before and throughout pregnancy were 90 g lighter than the offspring of other women. No significant adjusted effects were seen for birth length and head circumference. In two hospitals, 12,885 pregnant women answered questionnaires regarding consumption of alcohol, tobacco, cannabis, and other drugs. The prevalence of cannabis use was 0.8%. Women using cannabis, but no other illicit drugs were each retrospectively matched with four randomly chosen pregnant women in the same period and the same age group and with same parity. Eighty-four cannabis users were included. These women were socio-economically disadvantaged and had a higher prevalence of present and past use of alcohol, tobacco, and other drugs. No significant difference in pregnancy, delivery, or puerperal outcome was found. Children of women using cannabis were 150 g lighter, 1.2 cm shorter, and had 0.2 cm smaller head circumference than the control infants. A 27-year-old woman who smoked a joint (cannabis) and 20 cigarettes (tobacco) daily up to the time of a positive pregnancy test at 7 weeks and 4 days, was evaluated. On day 20 of pregnancy, she had a LSD minitrip. The patient had a spontaneous term delivery. The baby boy weight was between the 5th and the 50th percentile, length between the 50th and the 90th percentile, normal umbilical arterial and venous pH values, and Apgar scores of 7/9/10. There were no visible abnormalities, and behavior was normal. Eight hundred seven consecutive positive-pregnancy test urine samples were screened for a range of drugs, including cotinine as an indicator of maternal smoking habits. A positive test for cannabinoids was found in 117 (14.5%) of the samples. Smaller numbers of samples were positive for other drugs: opiates (11), benzodiazepines (4), cocaine (3), and one each for amphetamines and methadone. Polydrug use was detected in nine individuals. Only two samples tested positive for ethanol. The proportion with a urine cotinine level indicative of active smoking was 34.3%. The outcome of the pregnancy was traced for 288 of the subjects. Cannabis use was associated with a lower gestational age at delivery ($p < 0.005$), an increased risk of prematurity ($p < 0.02$), and reduction in birth-weight ($p < 0.002$). Maternal smoking was associated with a reduction in infant birth-weight ($p < 0.05$). This was less pronounced than the effect of other substance misuse. A sample of low-income women attending a prenatal clinic was assessed. The majority of the women decreased their use of cannabis during pregnancy. The assessments of child behavior problems included the Child Behavior Checklist, Teacher's Report Form, and the Swanson, Noland, and Pelham checklist. Multiple and logistic regressions were employed to analyze the relations between cannabis use and behavior problems of the children at age 10, while controlling for the effects of other extraneous variables. Prenatal cannabis use was significantly related to increased hyperactivity, impulsivity, and inattention symptoms as measured by the Swanson, Noland, and Pelham, increased delinquency as measured by the Child Behavior Checklist, and increased delinquency and externalizing problems as measured by the Teacher's Report Form. The pathway between prenatal cannabis exposure and delinquency was mediated by the effects of cannabis exposure on inattention symptoms. Attention and impulsivity of prenatally substance-exposed 6-year-olds were assessed as part of a longitudinal study. Most of the women were light-to-moderate users of alcohol and cannabis who decreased their use after the first trimester of pregnancy. Tobacco was used by a majority of women and did not change during pregnancy. The women, recruited from a prenatal clinic, were of low socioeconomic status. Attention and impulsivity were assessed using a Continuous Performance Task. Second and third trimester of tobacco exposure and first trimester of cocaine use predicted increased omission errors. Second trimester cannabis use predicted more commission errors and fewer omission errors. There were no significant effects of prenatal alcohol exposure. Lower Stanford-Binet Intelligence Scale composite scores, male gender, and an adult male in the household predicted more errors of commission. Lower Standford-Binet Intelligence Scale composite scores, younger child age, maternal work/school status, and higher maternal hostility scores predicted more omission errors. The neurophysiological effects of prenatal cannabis exposure on response inhibition were assessed in thirty-one participants aged 18–22. Ottawa Prenatal Prospective Study performed a blocked design Go/No-Go task while neural activity was imaged with functional magnetic resonance imaging. The Ottawa Prenatal Prospective Study is a longitudinal study that provides a unique body of information collected from each participant over 20 years, including prenatal drug history, detailed cognitive/ behavioral performance from infancy to young adulthood, and current and past drug usage. The functional magnetic resonance imaging results showed that with increased prenatal cannabis exposure, there was a significant increase in neural activity in bilateral PFC and right premotor cortex during response inhibition. There was also an attenuation of activity in left cerebellum with increased prenatal exposure to cannabis when challenging the response inhibition neural circuitry. Prenatally exposed offspring had significantly more commission errors than non-exposed participants, but all participants were able to perform the task with more than 85% accuracy. The findings were observed when controlling for present cannabis use and prenatal exposure to nicotine, alcohol, and caffeine, and suggest that prenatal cannabis exposure was related to changes in neural activity during response inhibition that last into young adulthood. The effects of prenatal cannabis and alcohol exposure on school achievement at 10 years of age were examined. Women were interviewed about their substance use at the end of each trimester of pregnancy, at 8 and 18 months, and at 3, 6, 10, 14, and 16 years.

The women were of lower socioeconomic status, high school-educated, and light-to-moderate users of cannabis and alcohol. At the 10-year follow-up, the effects of prenatal exposure to cannabis or alcohol on the academic performance of 606 children were assessed. Exposure to one or more cannabis joints per day during the first trimester predicted deficits in Wide Range Achievement Test- Revised reading and spelling scores and a lower rating on the teachers' evaluations of the children's performance. This relation was mediated by the effects of first-trimester cannabis exposure on the children's depression and anxiety symptoms. Second-trimester cannabis use was significantly associated with reading comprehension and under-achievement. Exposure to alcohol during the first and second trimesters of pregnancy predicted poorer teachers' ratings of overall school performance. Second-trimester binge drinking predicted lower reading scores. There was no interaction between prenatal cannabis and alcohol exposure. Each was an independent predictor of academic performance. Pregnant rats were treated daily with Δ-9-THC from the fifth day of gestation up to the day before birth (GD21). Then rats were sacrificed and their pups removed for analysis of the neural adhesion molecule L1-mRNA levels in different brain structures. The levels of L1 transcripts were significantly increased in the fimbria, stria terminalis, stria medullaris, corpus callosum, and in gray-matter structures (septum nuclei and the habenula). It remained unchanged in most of the gray-matter structures analyzed (cerebral cortex, BAL nucleus, hippocampus, thalamic and hypothalamic nuclei, basal ganglia, and sub-ventricular zones) and also in a few white-matter structures (fornix and fasciculus retroflexus). The increase in L1-mRNA levels reached statistical significance only in Δ-9-THC-exposed males but not females, where only trends or no effects were detected.

The results supported evidence on a sexual dimorphism, with greater effects in male fetuses, for the action of cannabinoids in the developing brain. Fetal cannabis exposure has no consistent effect on outcome. Prenatal cocaine exposure has not been shown to have any detrimental effect on cognition, except as mediated through cocaine effects on head size. Although fetal cocaine exposure has been linked to numerous abnormalities in arousal, attention, and neurological and neurophysiological function, most such effects appear to be self-limited and restricted to early infancy and childhood. Opiate exposure elicits a well-described withdrawal syndrome affecting the central nervous, autonomic, and gastrointestinal systems, which is most severe among methadone-exposed infants. Executive functioning in cocaine/polydrug (cannabis, alcohol, and tobacco)-exposed infants was assessed in a single session, occurring between 9.5 and 12.5 months of age. In an A-not-B task, infants searched—after performance-adjusted delays—for an object hidden in a new location. The CE infants did not differ from non-CE controls recruited from the same at-risk population. Comparison of heavier-CE (n = 9) with the combined group of lighter-CE (n = 10) and non-CE (n = 32) infants revealed significant differences on A-not-B performance, as well as on global tests of mental and motor development. Covariates investigated included socioeconomic status, marital status, race, maternal age, years of education, weeks of gestation, and birth-weight, as well as severity of prenatal cannabis, alcohol, and tobacco exposure. The relationship of heavier-CE status to motor development was mediated by length of gestation, and the relationship of heavier-CE status to mental development was confounded with maternal gestational use of cigarettes.

The relationship of heavier-CE status to A-not-B performance remained significant after controlling for potentially confounded variables and mediators, but was not statistically significant after controlling for the variance associated with global mental development. Weight, height, and head circumference were examined in children from birth to early adolescence for whom prenatal exposure to cannabis and cigarettes had been ascertained. The subjects were from a low-risk, predominantly middle-class sample participating in an ongoing longitudinal study. The negative association between growth measures at birth and prenatal cigarette exposure was overcome, sooner in males than females, within the first few years, and by the age of 6 years, the children of heavy smokers were heavier than control subjects. Pre- and postnatal environmental tobacco smoke did not have a negative effect on the growth parameters; however, the choice of bottle-feeding or shorter duration of breastfeeding by women who smoked during pregnancy appeared to play an important positive role in the catch-up observed among the infants of smokers. Prenatal exposure to cannabis was not significantly related to any growth measures at birth, although a smaller head circumference observed at all ages reached statistical significance among the early adolescents born to heavy cannabis users.

Prolactin Inhibition

The dried leaves, smoked by healthy female volunteers at a dose of 1 g/person, produced a decrease in plasma prolactin levels during the luteal phase of the menstrual cycle but not during the follicular phase. The results were significant at $p < 0.01$ level.

Propiospinal Myoclonus

A 25-year-old woman with clusters of myoclonus induced by a single exposure to inhaled cannabis was evaluated. Investigations excluded a structural abnormality of the spine. Multichannel surface electromyogram with parallel frontal electroencephalogram recording confirmed the diagnosis of propriospinal myoclonus.

Psoriatic Effect

Hot water extract of the dried seed, taken orally by 108 human adults with psoriasis at variable dosage level, was active. After 3–4 weeks of treatment, there was significant improvement. The extract was taken in combination with *Rehmannia glutinosa* (rhizome), *Salvia miltiorrhiza* (root), *Scrophularia ningpoensis* (root), *Isatis tinctoria* (branch and leaf), *Sophora subprostrata* (root), *Dictamnus dasycarpus* (rootbark), *Polygonum bistorta* (rhizome), and *Forsythia suspensa* (fruit).

Psychosocial Morbidity Association

Cannabis dependence is a prevalent comorbid substance use disorder among patients early in the course of a schizophrenia-spectrum disorder. Among 29 eligible patients, 18 participated in the study. First-episode patients with comorbid cannabis dependence ($n = 8$) reported significantly greater childhood physical and sexual abuse compared with those without comorbid cannabis dependence ($n = 10$). The result indicated the preliminary evidence of an association between childhood maltreatment and cannabis dependence among this especially vulnerable population. Childhood physical and sexual abuse may be a risk factor for the initiation of cannabis dependence and other substance use disorders in the early course of schizophrenia.

Psychotic Effect

Thirty five hundred representatives 19 years of age were examined in a cohort study. The subjects completed a 40-item Community Assessment of Psychic Experiences, measuring subclinical positive (paranoia, hallucinations, grandiosity, first- rank symptoms) and negative psychosis dimensions, depression, and drug use. Use of cannabis was associated positively with both positive and negative dimensions of psychosis, independent of each other and of depression. An association between cannabis and depression disappeared after adjustment for the negative psychosis dimensions. First use of cannabis younger than age 16 years was associated with a much stronger effect than first use after age 15 years, independent of lifetime frequency of use. The association between cannabis and psychosis was not influenced by the distress associated with the experiences, indicating that self-medication may be an unlikely explanation for the entire association between cannabis and psychosis. Cross-sectional epidemiological studies indicated that individuals with psychosis use cannabis more often than other individuals in the general population. It has long been considered that this association was explained by the self-medication hypothesis, postulating that cannabis is used to self-medicate psychotic symptoms. This hypothesis has been recently challenged. Several prospective studies carried out in population-based samples, showed that cannabis exposure was associated with an increased risk of psychosis. A dose–response relationship was found between cannabis exposure and risk of psychosis, and this association was independent from potential confounding factors, such as exposure to other drugs and preexistence of psychotic symptoms.

The brain mechanisms underlying the association have to be elucidated; they may implicate deregulation of cannabinoid and dopaminergic systems. Cannabis exposure may be a risk factor for psychotic disorders by interacting with a preexisting vulnerability for these disorders.

Refractory Neuropathic Pain

Seven patients (three women and four men) aged 60 ± 14 years suffering from chronic refractory neuropathic pain, received oral THC titrated to the maximum dose of 25 mg/day (mean dose: 15 ± 6 mg) during an average of 55.4 days (range: 13–128). Various components of pain (continuous, paroxysmal, and brush-induced allodynia) were assessed using visual analog scale scores. Health-related quality of life was evaluated using the Brief Pain Inventory, and the Hospital Anxiety and Depression scale was used to measure depression and anxiety. THC did not induce significant effect on the various pain, health-related quality of life and anxiety and depression scores. Numerous side effects (notably sedation and asthenia) were observed in five out of seven patients, requiring premature discontinuation of the drug in three patients.

Reproductive Effect

Cannabis use during pregnancy in developed nations is estimated to be approx 10%. Recent evidence suggests that the endogenous cannabinoid system, now consisting of two receptors and multiple endocannabinoid ligands, may also play an important role in the maintenance and regulation of early pregnancy and fertility. Drugs of abuse, like alcohol, opiates, cocaine, and cannabis, are used by many young people for their presumed aphrodisiac properties. The opioids inhibit the hypothalamus–pituitary–gonads axis (HPG), and increase the prolactin levels, which interferes with the male and female sexual response. Cannabis, at high doses, could inhibit the HPG axis and reduce fertility. Cannabis initially increases libido and potency, but chronic use causes sexual inversion. Long-term use of cannabis has been found to cause physiological changes that can alter individual reproductive potential. The effects of cannabis depend on the dose and can include death from depression of the respiratory system. Cannabis is absorbed rapidly and eliminated very slowly. Δ-9-THC is highly liposoluble and fixes to the serum proteins, passing to the lungs and liver for metabolization and to the kidneys and liver for excretion. As with estrogens, there is an enterohepatic circuit for reabsorption and elimination. Ninety percent is eliminated in the feces, 65% within 48 hours. Because of the enterohepatic circuit and liposolubility, elimination requires 1 week for completion. The other important biotransformation of the active principle is hydroxylation. The hydroxylated derivatives are responsible for the psychoactivity of cannabis. Cannabis affects both neuroendocrine function and the germ cells. Studies on experimental animals have indicated that THC can cause a decline in the pituitary hormones, follicle stimulating hormone, luteinizing hormone, and prolactin, and in the steroids progesterone, estrogen, and androgens. Human studies have shown that chronic users have decreased levels of serum testosterone. Because steroidogenesis can be restimulated with human chorionic gonadotropin, it appears that THC does not directly affect steroid production by the corpus luteum, but that its action is mediated by the hypothalamus. Because of its potent antigonadotropic action, THC is under study as an anovulatory agent. The same animal studies have shown that ovulation returns to normal 6 months after termination of use.

High rates of anovulation and luteal insufficiency have been observed in women smoking cannabis at least three times weekly. THC accumulates in the milk. Animal studies have shown that THC depresses the enzymes necessary for lactation and causes a diminution in the volume of the mammary glands. Significant amounts of the drug have been detected in both mothers' milk and the blood of newborns. Animal studies indicate that THC crosses the placenta, achieving concentrations in the fetus as high as those in the mother. Animal studies also demonstrated increasing frequency of abortions, intrauterine death, and declines in fetal weight. The effects were probably caused by an alteration in placental function. A human study likewise showed that cannabis use during pregnancy was significantly related to poor fetal development, low birth-weight, diminished size, and decreased cephalic circumference. Congenital malformations have been observed in experimental animals exposed to THC. Declines in sperm volume and count and abnormal sperm motility have been observed in chronic cannabis users. In vitro studies show that THC produces a marked degeneration of human sperm. Among sexually experienced girls, 39% (n = 123) reported using oral contraceptive pills(OCPs), 5.4% (n = 17) used Depo-Provera (medroxyprogesterone acetate) or Norplant (levonorgestrel), and 55.6% (n = 175) used no hormonal method. Logistic regression analysis revealed that the factors most significantly associated with the use of hormonal methods were older age (OR = 1.19; 95% CI, 1.07–1.33), not using a condom at last intercourse (OR = 0.55; CI, 0.34–0.90), and having had a well visit within 1 year (OR = 2.11; CI, 1.12–3.70). OCP users were less likely than Depo-Provera or

Norplant users to have used alcohol ($p = 0.041$), cigarettes ($p = 0.002$), or cannabis ($p = 0.018$) in the past 30 days. OCP users were less likely than nonusers of hormonal methods to have smoked cigarettes ($p = 0.034$) or cannabis ($p = 0.052$). The school-based clinic had a greater proportion of subjects using long-acting progestins ($p < 0.001$).

Respiratory Effect

Smoking a "joint" of cannabis resulted in exposure to significantly greater amounts of combusted material than with a tobacco cigarette. The histopathological effects of cannabis-smoke exposure included changes consistent with acute and chronic bronchitis. Cellular dysplasia has also been observed, suggesting that, like tobacco smoke, cannabis exposure has the potential to cause malignancy. Symptoms of cough and early morning sputum production are common (20–25%) even in young individuals who smoke cannabis alone. Almost all studies indicated that the effects of cannabis and tobacco smoking are addictive and independent. A small group of current male cannabis processors with a mean age of 43 years was studied. Questionnaire data, lung function, serial FEV1 and blood were collected from all workers. Seven workers (64%) complained of at least one respiratory symptom (one with byssinosis). The mean percentage predicted FEV1 was 91.5, FVC 97.7, peak expiratory flow 92.1, and forced expiratory flow between 25 and 75% of FVC 79.5. Serial FEV1 measurements in the two workers with work-related respiratory symptoms revealed a mean change in FEV1 on the first working day of -12.9%. This contrasted with +6.25% on the last working day. Respective values for the two workers without work-related symptoms were -1.4 and +3.2%. Nine hundred forty-three young adults from a birth cohort of 1037 were studied at age 21 years.

Standardized respiratory symptom questionnaires were administered. Spirometry and methacholine challenge tests were undertaken. Cannabis dependence was determined using DSM-III-R criteria. Descriptive analyses and comparisons between cannabis-dependent, tobacco- smoking, and nonsmoking groups were undertaken. Adjusted odds ratios for respiratory symptoms, lung function, and airway hyperresponsiveness (PC20) were measured. Ninety-one subjects (9.7%) were cannabis-dependent and 264 (28.1%) were current tobacco smokers. After controlling for tobacco use, respiratory symptoms associated with cannabis dependence included wheezing apart from colds, exercise-induced shortness of breath, nocturnal wakening with chest tightness, and early morning sputum production. These were increased by 61,

65, 72 (all $p < 0.05$), and 144% ($p < 0.01$) respectively, compared with non-tobacco smokers. The frequency of respiratory symptoms in cannabis-dependent subjects was similar to tobacco smokers of 1–10 cigarettes per day. The proportion of cannabis-dependent study members with an FEV1/FVC ratio of less than 80% was 36% compared with 20% for nonsmokers ($p = 0.04$). These outcomes occurred independently of coexisting bronchial asthma.

Reversal of Cannabinoid Addiction

Δ-9-THC was administered orally to mice at a dose of 10 mg/kg twice daily for 6 days to make them dependent on cannabinoids. Other groups of mice were administered orally with a Δ-9-THC and benzoflavone from *Passiflora incarnata* at doses of 10 or 20 mg/kg twice daily for 6 days. Mice receiving the Δ-9-THC and *Passiflora incarnata* extract developed significantly less dependence, worse locomotor activity, and less of typical withdrawal effects like paw tremors and headshakes, compared with mice receiving Δ-9-THC alone. Administration of SR-141716A, a selective cannabinoid-receptor antagonist (10 mg/kg, orally), to all groups on the seventh day resulted in an artificial withdrawal. Administration of 20 mg/kg of the *Passiflora incarnata* benzoflavone moiety to mice showing symptoms of withdrawal owing to administration of SR-141716A produced a marked attenuation of withdrawal effects.

Schizophrenic Effect

The nerve growth factor (NGF) serum levels of 109 consecutive drug-naïve schizophrenic patients were measured and compared with those of healthy controls. The results were correlated with the long-term intake of cannabis and other drugs. Mean (± standard deviation) NGF serum levels of 61 control persons (33.1 ± 31 pg/mL) and 76 schizophrenics who did not consume illegal drugs (26.3 ± 19.5 pg/mL) did not differ significantly. Schizophrenic patients with regular cannabis intake (> 0.5 g per day on average for at least 2 years) had significantly raised NGF serum levels of 412.9 ± 288.4 pg/mL ($n = 21$) compared with controls and schizophrenic patients not consuming cannabis ($p < 0.001$). In schizophrenic patients who abused not only cannabis, but also additional substances, NGF concentrations were as high as 2336.2 ± 1711.4 pg/mL ($n = 12$). On average, heavy cannabis consumers suffered their first episode of schizophrenia 3.5 years ($n = 21$) earlier than schizophrenic patients who abstained from cannabis. These results indicate that cannabis is a possible risk factor for the development of schizophrenia. This might be reflected in the raised NGF-serum concentrations when both schizophrenia and long-term cannabis abuse prevail.

Schizotypy Correlation

Two hundred eleven healthy adults who used cannabis showed higher scores on schizotypy, borderline, and psychoticism scales than never-users. Multivariate analysis, covarying lie scale scores, age, and educational level indicated that high schizotypal traits best discriminated subjects who had used cannabis from never-users, whether or not they reported having used other recreational drugs. The results indicated that cannabis use was related to a personality dimension of psychosis-proneness in healthy people.

Sedative and Stimulant Effects

A double-blind, placebo-controlled study assessed subjective effects of smoking cannabis with either a long or short breath-holding duration. During eight test sessions, 55 male volunteers made repeated ratings of subjective "high," sedation, and stimulation, as well as rating their perceptions of motivation and performance on cognitive tests. The long, relative to the short, breath-holding duration increased "high" ratings after smoking cannabis, but not placebo. Cannabis smoking increased sedation and a perception of worsened test performance, and decreased motivation with respect to test performance. Paradoxical subjective effects were observed in those subjects reporting some stimulation, as well as sedation after smoking cannabis, particularly with the long breath-holding duration. Breath-holding duration did not produce any subjective effects that were independent of the drug treatment (i.e., occurred equally after smoking of cannabis and placebo).

Sexual Receptivity

The effects of THC on sexual behavior in female rats and its influence on steroid hormone receptors and neurotransmitters in the facilitation of sexual receptivity was examined. Results revealed that the facilitatory effect of THC was inhibited by antagonists to both progesterone and dopamine D(1) receptors. To test further the idea that progesterone receptors (PR) and/or dopamine receptors (D[1]R) in the hypothalamus were required for THC-facilitated sexual behavior in rodents, antisense, and sense oligonucleotides to PR and D(1)R were administered intra-cerebroventricularly into the third cerebral ventricle of ovariectomized, estradiol benzoate-primed rats. Progesterone- and THC-facilitated sexual behavior was inhibited in animals treated with antisense oligonucleotides to PR or to D(1)R. Antagonists to cannabinoid receptor-1 subtype (CB1), but not to cannabinoid receptor-

2 subtype (CB2) inhibited progesterone- and dopamine-facilitated sexual receptivity in female rats. Adult female and male rats that had been perinatally exposed to hashish extracts were investigated. Adult males perinatally exposed to hashish extracts exhibited marked changes in the behavioral patterns executed in the sociosexual approach behavior test; these changes did not exist in females. Control males first visited the incentive male and took longer to visit the incentive female, whereas hashish-exposed males followed the opposite pattern. Hashish-exposed males spent more time in the vicinity of the incentive female, whereas they decreased their frequency of visits to, and the time spent in, the male incentive area. This behavior was observed during the first third of the test, but became normalized and even inverted during the last two- thirds. In the social interaction test, the normal reduction in the time spent in active social interaction following the exposure to a neophobic situation (high light levels) in controls did not occur in hashish-exposed males, although these exhibited a response in the dark-light emergence test similar to that of their corresponding controls. No changes were seen in spontaneous locomotor activity in both tests. These behavioral alterations observed in hashish-exposed males were paralleled by a significant decrease in L-3,4-dihydroxyphenylacetic acid contents in the limbic forebrain; this suggests a decreased activity of mesolimbic dopaminergic neurons. No effects were seen in females.

Smooth Muscle Relaxant Activity

Ethanol (95%) and water extracts of the dried aerial parts, at a concentration of 1:1, produced weak activity on the rabbit duodenum. The ethanol extract was equivocal on the guinea pig ileum. Petroleum ether extract of the dried entire plant, administered intraperitoneally to rats at a dose of 0.89 mg/kg, was active vs corneo-palpebral reflex.

Spasticity Treatment

Standardized plant extract was administered orally to 57 MS patients with poorly controlled spasticity, at a dose of 2.5 mg of THC and 0.9 mg of CBD. Patients in group A started with a drug escalation phase from 15 to a maximum of 30 mg of THC by 5 mg per day if well tolerated, being on active medication for 14 days before starting placebo. Patients in group B started with placebo for 7 days, crossed to the active period (14 days), and closed with a three-day placebo period (active drug-dose escalation and placebo sham escalation as in group A). Measures used included daily self-report of spasm frequency and symptoms, Ashworth Scale, Rivermead Mobility Index, 10-meter timed walk, nine-hole peg test, paced auditory serial addition test, and the digit span test. There were no statistically significant differences associated with active treatment compared with placebo, but trends in favor of active treatment were seen for spasm frequency, mobility, and getting to sleep. In the 37 patients (per-protocol set) who received at least 90% of their prescribed dose, improvements in spasm frequency ($p = 0.013$) and mobility after excluding a patient who fell and stopped walking were seen ($p = 0.01$). Minor adverse events were slightly more frequent and severe during active treatment, and toxicity symptoms, which were generally mild, were more pronounced in the active phase. Six hundred thirty participants with stable MS and muscle spasticity were treated with oral cannabis extract ($n = 211$), Δ-9-THC ($n = 206$), or placebo ($n = 213$) for 15 weeks. Six hundred eleven of 630 patients were followed up for the primary end point. No treatment effect of cannabinoids on the primary outcome ($p = 0.40$) was noted. The estimated difference in mean reduction in total Ashworth score for participants taking cannabis extract compared with placebo was 0.32 (95% CI, –1.04 to 1.67), and for those taking Δ-9-THC vs placebo it was 0.94 (–0.44 to 2.31). There was an evidence of a treatment effect on patient-reported spasticity and pain ($p = 0.003$), with improvement in spasticity reported in 61% ($n = 121$, 95% CI, 54.6–68.2), 60% ($n = 108$, 52.5–66.8), and 46% ($n = 91$, 39–52.9) of participants on cannabis extract, Δ-9-THC, and placebo, respectively.

Spatial Working Memory Effect

Functional magnetic resonance imaging was used to examine brain activity in 12 long-term heavy cannabis users, 6–36 hours after last use, and in 10 control subjects while they performed a spatial working memory task. Regional brain activation was analyzed and compared using statistical parametric mapping techniques. Compared with controls, cannabis users exhibited increased activation of brain regions typically used for spatial working memory tasks (such as PFC and anterior cingulate). Users also recruited additional regions not typically used for spatial working memory (such as regions in the basal ganglia). The findings remained essentially unchanged when reanalyzed using the subjects ages as a covariate. Brain activation showed little or no significant correlation with subjects years of education, verbal IQ, lifetime episodes of cannabis use, or urinary cannabinoid levels at the time of scanning.

Spontaneous Pneumomediastinum

Spontaneous pneumomediastinum is defined as pneumomediastinum in the absence of an underlying lung disease. It is the second most common cause of chest pain in young, healthy individuals (<30 years) necessitating hospital visits. Inhalational drug use (cocaine and cannabis) has been associated with a significant number of cases, although cases with no apparent etiological or incriminating factors are well-recognized. A case of an 18-year-old high school student with spontaneous pneumomediastinum was evaluated.

Sudden Infant Death Syndrome

In a nationwide case–control study of 369 cases and 1558 controls, two-thirds of SIDS deaths occurred at night (between 10 PM and 7:30 AM). The odds ratio (95% CI) for prone sleep position was 3.86 (2.67–5.59) for deaths occurring at night, and 7.25 (4.52–11.63) for deaths occurring during the day; the difference was significant. The odds ratio for maternal smoking and SIDS deaths occurring at night was 2.28 (1.52–3.42), and for the day, 1.27 (0.79–2.03). If the mother was single, the odds ratio was 2.69 (1.29–3.99) for a nighttime death, and 1.25 (0.76–2.04) for a daytime death. Both interactions were significant. The interactions between time of death and bed sharing, not sleeping in a cot or bassinet, ethnicity, late timing of prenatal care, binge drinking, cannabis use, and illness in the baby were also significant. All were more strongly associated with SIDS occurring at night. In a nationwide case–control study, 393 cases and 1592 controls were analyzed. Adjusting for ethnicity and maternal tobacco use, the SIDS odds ratio for weekly maternal cannabis use since the infant's birth was 2.23 (95% CI = 1.39, 3.57) compared with nonusers, and the multivariate odds ratio was 1.55 (95% CI = 0.87, 2.75).

Suicidal Effect

Standardized interview assessments were conducted with 2311 youths aged 8–15 years who used drugs before age 16. Approximately 15 years after recruitment, 1695 persons (mean age = 21 years) were reassessed. One hundred fifty-five of them made suicide attempts (SA) and 218 had onset of depression-related suicide ideation (SI). The relative risk, from survival analysis and logistic regression models, to study early use of tobacco, alcohol, cannabis, and inhalants, with covariate adjustments for age, sex, race/ethnicity, and other pertinent covariates were examined. Early-onset of cannabis use and inhalant use for females, but not for males, signaled a modest excess risk of SA (cannabis-associated RR = 1.9; $p = 0.04$; inhalant-associated RR = 2.2; $p = 0.05$). Early-onset of cannabis use by females (but not for males) signaled excess risk for SI (RR = 2.9; $p = 0.006$). Early-onset alcohol and tobacco use were not associated with later risk of

SA or SI. Two hundred seventy-seven same-sex twin pairs (median age: 30 years) discordant for cannabis dependence and 311 pairs discordant for early-onset cannabis use (before age 17 years) were examined. Individuals who were cannabis-dependent had odds of SI and SA that were 2.5–2.9 times higher than those of their noncannabis-dependent co-twin. Cannabis dependence was associated with elevated risks of major depressive disorder (MDD) in dizygotic, but not in monozygotic twins. Twins who initiated cannabis use before age 17 years of age had elevated rates of subsequent SA (OR, 3.5, 95% CI, 1.4–8.6) but not of MDD or SI. Early MDD and SI were significantly associated with subsequent risks of cannabis dependence in discordant dizygotic pairs, but not in discordant monozygotic pairs. The results indicated that the comorbidity between cannabis dependence and MDD likely arises through shared genetic and environmental vulnerabilities predisposing to both outcomes. In contrast, associations between cannabis dependence and suicidal behaviors cannot be entirely explained by common predisposing genetic and/or shared environmental predisposition.

Synergic Cytotoxicity

THC, in A549 lung tumor cells culture at concentrations of less than 5 μg/mL, produced no cytotoxic effect. At higher levels it induced cell necrosis, with a lethal concentration $(LC)_{50}$of 16–18 μg/mL. Butylated hydroxyanisole ([BHA], a food additive)alone at concentrations of 10–200 μM, produced limited cell toxicity and significantly enhanced the necrotic death resulting from concurrent exposure to THC. In the presence of BHA at 200 μM, the LC_{50} for THC decreased to 10–12 μg/mL. Similar results were obtained with smoke extracts prepared from cannabis cigarettes, but not with extracts from tobacco or placebo cannabis cigarettes (containing no THC). Experiments were repeated in the presence of either diphenyleneiodonium or dicumarol as inhibitors of the redox cycling pathway. Neither of the compounds protected cells from the effects of combined THC and BHA, but rather enhanced necrotic cell death. Measurements of cellular ATP revealed that both THC and BHA reduced ATP levels in A549 cells, consistent with toxic effects on mitochondrial electron transport. The combination was synergistic in this respect, reducing ATP levels to less than 15% of the control. Exposure to cannabis smoke in conjunction with BHA may promote deleterious health effects in the lung.

Teratogenic Activity

Resin, administered orally to pregnant rabbits at a dose of 1 mL/kg, was active. Alcohol extract of the dried leaves, administered intragastrically to pregnant rats at a dose of 125 mg/kg from days 7 to 16 of gestation, was active. The fetuses showed several gross abnormalities, visceral anomalies, and skeletal malformations. Water extract of the dried leaf, administered intragastrically to pregnant rats at doses of 125, 200, 400, and 800 mg/kg, produced various types of malformations in the fetuses. Petroleum ether extract of the aerial parts, administered orally to rats and rabbits, was inactive.

Tourette Syndrome

Tourette syndrome (TS) is a complex inherited disorder of unknown etiology, characterized by multiple motor and vocal tics. Involvement of the central cannabinoid (CB1) system was suggested because of therapeutic effects of cannabis consumption and Δ-9-THC-treatment in TS patients. The central cannabinoid receptor (CNR1) gene encoding the *CNR1* was considered as a candidate gene for TS and systematically screened by single-strand conformation polymorphism analysis and sequencing. Compared with the published *CNR1* sequence, three single-base substitutions were identified: 1326T→A, 1359G→A, 1419 + 1G→C. The change at position 1359 is a common polymor-phism (1359 G/A) without allelic association with TS. 1326T→A was present in only one TS patient and is a silent mutation, which does not change codon 442 (valine). 1419 + 1G→C affects the first nucleotide immediately following the coding sequence. It was first detected in three of 40 TS patients and none of 81 healthy controls. This statistically significant association with TS ($p = 0.034$) could not be confirmed in two subsequent cohorts of 56 TS patients (one heterozygous for 1419 + 1G→C) and 55 controls, and 64 patients and 66 controls (one heterozygous for 1419 + 1G→C), respectively. Transcript analysis of lymphocyte RNA from five 1419 + 1G→C carriers revealed no systematic influence on the expression level of the mutated allele. In addition, segregation analysis of 1419 + 1G→C in affected families gave evidence that 1419 1G→C does not play a causal role in the etiology of TS. It was concluded that genetic variations of the *CNR1* gene are not a plausible explanation for the clinically observed relation between the cannabinoid system and TS. A single-dose, cross-over study in 12 patients, and a 6-week, randomized trial in 24 patients, demonstrated that Δ9-THC, the most psychoactive ingredient of cannabis, reduced tics in TS patients. No serious adverse effects occurred and no impairment on neuropsychological performance was observed. In the randomized, double-blind, placebo-controlled study, 24 patients with TS, according to DSM-III-R criteria, were treated over a 6-week period with up to 10 mg/day of THC. Tics were rated at six visits (visit 1, baseline; visits 2–4, during treatment period; visits 5–6, after withdrawal of medication) using the Tourette Syndrome Clinical Global Impressions scale (TS-CGI), the Shapiro Tourette-Syndrome Severity Scale (STSSS), the Yale Global Tic Severity Scale (YGTSS), the self-rated Tourette Syndrome Symptom List (TSSL), and a videotape-based rating scale. Seven patients dropped out of the study or had to be excluded, but only one because of side effects. Using the TS-CGI, STSSS, YGTSS, and video rating scale, there was a significant difference ($p < 0.05$) or a trend toward a significant difference ($p < 0.1$) between THC and placebo groups at visits 2, 3, and/or 4. Using the TSSL at 10 treatment days (between days 16 and 41) there was a significant difference ($p < 0.05$) between both groups. Analysis of variance also demonstrated a significant difference ($p = 0.037$). No serious adverse effects occurred. In the randomized, double-blind, placebo-controlled study, the effect of a treatment with up to 10 mg Δ-9-THC over a 6-week period on neuropsychological performance in 24 patients suffering from TS was investigated. During medication and immediately, as well as 5–6 weeks after, withdrawal of Δ-9-THC treatment, no detrimental effect was seen on learning curve, interference, recall and recognition of word lists, immediate visual memory span, and divided attention. A trend towards a significant immediate verbal memory span improvement during and after treatment was found.

A randomized double-blind placebo-controlled crossover single-dose trial of Δ-9-THC (5, 7.5, or 10 mg) in 12 adult TS patients was performed. Tic severity was assessed using the TSSL and examiner ratings (STSSS, YGTSS, TS-CGS). Using the TSSL, patients also rated the severity of associated behavioral disorders. Clinical changes were correlated to maxi-mum plasma levels of THC and its metabolites 11-OH-THC and 11-nor-Δ-9- tetrahydrocannabinol-9-carboxylic acid. Using the TSSL, there was a significant improvement of tics ($p = 0.015$) and obsessive-compulsive behavior ($p = 0.041$) after treatment with Δ-9-THC compared with placebo. Examiner ratings demonstrated a significant difference for the subscore "complex motor tics" ($p = 0.015$) and a trend towards a significant improvement for the subscores "motor tics" ($p = 0.065$), "simple motor tics" ($p = 0.093$), and "vocal tics" ($p = 0.093$). No serious adverse reactions occurred. Five patients experienced mild, transient side effects. There was a significant correlation between tic improvement and maximum 11-OH-THC plasma concentration.

Toxic Effect

Petroleum ether extract of the dried leaf, administered by gastric intubation to pregnant rats at a dose of 150 mg/kg, produced a reduction of food and water consumption and maternal weight gain. The weight of pups at birth was reduced by approx 10% of the litter size, and pup mortality at birth was not affected significantly. Water extract of the aerial parts, administered intravenously to male adults, was active. The resin, ingested by a 4-year-old girl, showed signs of stupor alternating with brief intervals of excitation and foolish laughing with atactic movements. Her temperature, blood pressure, pulse, hemoglobin, leukocytes, serum electrolytes, and serum urea were normal. Respiratory rate was 12 beats per minute. Blood sugar elevated. Recovery was complete within 24 hours with no treatment. Four patients suffered gastrointestinal disorders and psychological effects after eating salad prepared with hemp seed oil. The concentration of THC in the oil far exceeded the recommended tolerance dose. From January 1998 to January 2002, 213 incidences were recorded of dogs that developed clinical signs following oral exposure to cannabis, with 99% having neurological signs and 30% exhibiting gastrointestinal signs. The cannabis ingested ranged from 0.5 to 90 g. The lowest dose at which signs occurred was 84.7 mg/kg and the highest reported dose was 26.8 g/kg. Onset of signs ranged from 5 minutes to 96 hours, with most signs occurring within 1–3 hours after ingestion. The signs lasted from 30 minutes to 96 hours. Management consisted of decontamination, sedation (with diazepam as drug of choice), fluid therapy, thermoregulation, and general supportive care. All followed animals made full recoveries. The suspension prepared from the benzene washing solution of cannabis seeds, administered intravenously to mice at a dose of 3 mg/kg, produced hypothermia, catalepsy, pentobarbital-induced sleep prolongation, and suppression of locomotor activity. These pharmacological activities of benzene washing solution of cannabis seeds were significantly higher than those of Δ-9-THC (3 mg/kg, iv).

Trauma Injuries

An association between combat-related posttraumatic stress disorder (C-PTSD) and other mental disorders was studied in co-twin (male monozygotic twin pairs in the Vietnam Era Twin Registry). Logistic regression analyses demonstrated that combat exposure, adjusted for C-PTSD, was significantly associated with increased risk for alcohol and cannabis dependence and that C-PTSD mediated the association between combat exposure and both major depression and tobacco dependence. Sera from 111 patients with trauma injuries who presented during a 3-month period were screened for blood alcohol. Urine specimens were analyzed for metabolites of cannabis and cocaine. Sixty-two percent of patients were positive for at least one substance and 20% for two or more. Positivity rates were as follows: cannabis, 46%; alcohol, 32% (with 71% of these having blood alcohol levels >80 mg/ dL); and cocaine (6%). Substance usage was most prevalent in the third decade of life. The patients who yielded a positive result were significantly younger than those negatives. There was no significant difference in age or substance usage between the victims of interpersonal violence or road traffic accidents. In the group designated "other accidents," patients were significantly older and had a lower incidence of substance usage than the other two groups. Cannabis was the most prevalent substance in all groups. Fifty and 55% of victims of road accidents and interpersonal violence, respectively, were positive for cannabis compared with 43 and 27% for alcohol, respectively. There was no significant difference in hospital stay or injury severity score between substance users and nonusers.

Trigeminovascular System Effect

Arachidonylethanolamide is believed to be the endogenous ligand of the cannabinoid CB1 and CB2 receptors. Known behavioral effects of AEA are antinociception, catalepsy, hypothermia, and depression of motor activity, similar Δ-9-THC, the psychoactive constituent of cannabis. A role of the CB1 receptor in the trigeminovascular system, using intravital to study the effects of AEA against various vasodilator agents was examined. AEA inhibited dural blood vessel dilation brought about by electrical stimulation by 50%, calcitonin gene-related peptide (CGRP) by 30%, capsaicin by 45%, and NO by 40%. CGRP(8–37) attenuated NO-induced dilation by 50%. The AEA inhibition was reversed by the CB1 receptor antagonist AM251. AEA also reduced the blood pressure changes caused by CGRP injection, this effect was not reversed by AM251.

Tumor-promoting Effect

A 28-year-old man who abused alcohol, nicotine, and cannabis for several years was investigated. He suffered simultaneously from a squamous cell carcinoma of the hypopharynx with bilateral cervical metastases, an adenocarcinoma of the transverse colon and a primary hepatocellular carcinoma. There were occurrences of three separate malignant tumors with different histologies in the aerodigestive tract, which could be related to a chronic abuse of cannabis.

Turning Behavior

Cannabinoid agonists: WIN (1–100 ng/mouse), CP-55,940 (0.1– 50 ng/mouse), and AEA (0.5–50 ng/mouse), administered unilaterally into the mouse striatum, dose-dependently induced turning behavior. SR 141716A [*N*-(piperidin-1-yl)-5-(4-chlorophenyl)-1-(2,4-dichlorophenyl)-4-methyl-1H-pyrazole-3-carboxamide hydrochloride], the selective antagonist of CB1 receptor, antagonized the three cannabinoid receptor agonists-induced turning with similar effective dose$_{50}$ (0.13–0.15 mg/kg, intraperitoneally). Spiroperidol (a D2 receptor blocker), (+)-SCH 23390 (a D1 receptor blocker), or prior 6-hydroxydopamine lesions of the striatum blocked WIN- and CP-55,940-induced turning, thus suggesting the involvement of DA transmission in cannabinoid-induced turning.

Uterine Stimulant Effect

Ethanol (50%) extract of the entire plant was inactive on the rat uterus. Ethanol (95%) and water extracts of the dried aerial parts, at a concentration of 1:1, produced strong activity on the non-pregnant rat uterus. Water extract of the flowering tops produced strong activity on the rat uterus.

Ventricular Septal Defect

A Birth Defect Case–Control Study was used to identify 122 isolated simple ventricular septal defect (VSD) cases and 3029 control infants. Exposure data on alcohol, cigarette, and illicit drug use were obtained through standardized interviews with mothers and fathers. Associations between lifestyle factors and VSD were calculated using maternal self-reports; associations were also calculated using paternal proxy reports of the mother's exposures. Maternal self-report of heavy alcohol consumption and paternal proxy report of the mother's moderate alcohol consumption were associated with isolated simple VSD. A twofold increase in risk of isolated simple VSD was identified for maternal self- and paternal proxy-reported cannabis use. Risk of isolated simple VSD increased with regular (≥3 days per week) cannabis use for both maternal self- and paternal proxy report, although the association was significant only for maternal self-report.

Visuospatial Memory Effect

Twenty-five college students who were heavy cannabis smokers (who had smoked a median of 29 of the last 30 days) were compared with 30 light smokers (1 day in thc last 30 days). The subjects were tested after a supervised period of abstinence from cannabis and other drugs lasting at least 19 hours. Differences between the overall groups of heavy and light smokers did not reach statistical significance on the four subtests of attention administered. On examining data for the two sexes separately, marked and significant differences were found between heavy- and light-smoking women on the subtest examining visuospatial memory. On this test, subjects were required to examine a 6×6 "checkerboard" of squares in which certain squares were shaded. The shaded squares were then erased and the subject was required to indicate with the mouse which squares had formerly been shaded. Increasing numbers of shaded squares were presented at each trial. The heavy-smoking women remembered significantly fewer squares on this test, and they made significantly more errors than the light-smoking women. These differences persisted despite different methods of analysis and consideration for possible confounding variables.

Wilson's disease

A patient with generalized dystonia owing to Wilson's disease obtained mrked improvement in response to smoking cannabis.

Winiwarter-Buerger disease

Two young men aged 18 and 20 years with juvenile endarteritis were evaluated. Both developed acute distal ischemia of the lower or upper limbs with arteriographic evidence suggestive of Winiwarter-Buerger disease. Both smoked regularly but not excessively, and both used cannabis regularly. In one case, the therapeutic response to withdrawal of cannabis was good. In the second, use of cannabis continued and arterial disease persisted. The main clinical and radiographical features in this condition are the same as in Winiwarter-Buerger disease.

5

HORDEUM VULGARE

Hordeum vulgare is grass that may be either a winter or a spring annual of the Poaceae (Graminae) family. It forms a rosette type of growth in fall and winter, developing elongated stems and flower heads in early summer. Winter varieties form branched stems or tillers at the base, so several stems rise from a single plant. The stems of both winter and spring varieties may vary in length from 30 to 120 cm, depending on variety and growing conditions. Stems are round, hollow between nodes, and develop five to seven nodes below the head. At each node, a clasping leaf develops. In most varieties, the leaves are coated with a waxy chalk-like deposit. Shape and size of leaves vary with variety, growing conditions, and position on the plant. The spike contains the flowers and consists of spike- lets attached to the central stem or rachis. Stem intervals between spikelets are 2 mm or less in dense-headed varieties and up to 4–5 mm in lax or open-headed kinds. Three spikelets develop at each node on the rachis. *Hordeum vulgare* is six-row variety, where all three of the spikelets at each node develop a seed. Each spikelet has two linear to lanceolate glumes rising from near the base and flat and terminates in an awn. The glumes, minus the awn, are approximately half the length of the kernel in most varieties, but this varies from less than half to equal to the kernel in length. Glumes may be covered with hairs, weakly haired, or hairless. The awns on the glumes may be shorter than the gume, equal in length, or longer. The barley kernel consists of the caryopsis, or internal seed, the lemma, and palea. In most barley varieties, the lemma and palea adhere to the caryopsis and are a part of the grain following threshing. The lemmas in barley are usually awned. Awns vary in length from very short up to as much as 12 in. Edges of awns may be rough or "barbed" (bearded) or nearly smooth. Awnless varieties are also known. In six-row barley, awns are usually more developed on the central spikelets than on the lateral ones. The barley kernel is generally spindle shaped. In commercial varieties, the length ranges from 7 to 12 mm.

Origin and Distribution

Grains found in pits and pyramids in Egypt indicated that barley was cultivated there more than 5000 years ago. The most ancient glyph or pictograph found for barley is dated approx 3000 BC. References to barley and beer are found in the earliest Egyptian and Sumerian writings. The origin of barley is still not known. There are differing views among researchers regarding whether the original wild forms were indigenous to Eastern Asia, particularly Tibet, or to the Near East, Eastern Mediterranean area, or both. Varieties are constantly changing as new ones are developed and tested while others pass out of cultivation.

Traditional Uses

Afghanistan. Flowers are taken orally by females for contraception.

Argentina. Decoction of the dried fruit is taken orally for diarrhea and to treat respiratory and urinary tract infections.

China. Decoction of the dried fruit is taken orally for diabetes.

Egypt. Dried fruits are smoked as a treatment for schistosomiasis. The fruit is used intravaginally as a contraceptive before and after coitus. Fifty-three percent of 1200 puerperal women interviewed practiced this method, of whom 47% depended on indigenous method and/or prolonged lactation.

Guatemala. Hot water extract of the dried seed is taken orally for renal inflammation and kidney disease. Hot water extract of the dried seed is used externally for dermatitis, inflammations, erysipelas, and skin eruptions.

India. Powdered flowers of *Calotropis procera*, fruits of *Piper nigrum*, seed ash of *Hordeum vulgare*, and rose water are taken orally for cholera.

Iran. Flour is used as a food. A decoction of the dried seed is used externally as an emollient and applied on hemorrhoids and infected ulcers. A decoction of the dried seed is taken orally as a diuretic and antipyretic and used for hepatitis, diarrhea, scorbutism, nephritis, bladder inflammation, gout, enema, and its tonic effect. Decoction of the dried seed is applied to the nose to reduce internasal inflammation.

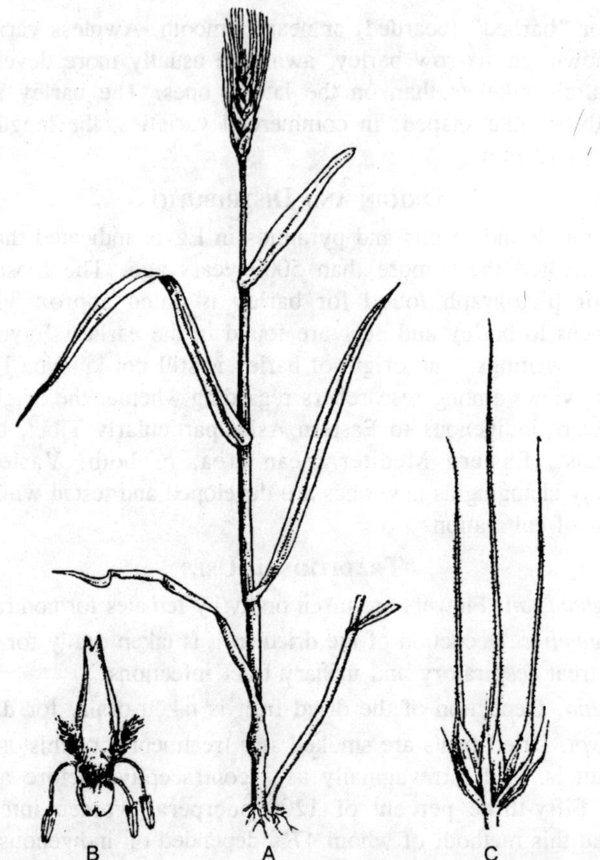

Fig. 5.1. Hordeum vulgare. A–Flowering plant, B–Flower, C–Spikelet.

Italy. Seeds are eaten as a urinary antiseptic. Compresses of boiled seeds are used to soothe rheumatic and joint pains. Infusion of the dried seed is used as a galactogogue.

Korea. Hot water extract of the dried entire plant is taken orally for beriberi, coughs, influenza, measles, syphilis, nephritis, jaundice, dysentery, and ancylostomiasis; for thrush in infants; and as a diuretic. Extract of the dried entire plant is used externally for prickly heat.

Peru. Hot water extract of dried fruits is used externally for measles and as an emollient and taken orally as a diuretic.

South Korea. Hot water extracts of the fruit and dried seeds are taken orally by pregnant women to induce abortion. Hot water extracts of the fruits taken orally by females as a contraceptive.

Turkey. Decoction of the fruit is taken orally for common colds.

United States. Infusion of the dried seed is taken orally for dysentery, diarrhea, and colic and for digestive and gastrointestinal disorders.

Medicinal Uses

Acquired Immune Deficiency Syndrome Therapeutic Effect

Hot water extract of the dried fruit, administered orally to patients with acquired immuodeficiency syndrome (AIDS) four to five doses/week for a year for the purpose of in clearing heat and detoxifying the blood, produced an improvement in the patient's health.

Allergenic Activity

Extract of the dried seed, administered externally to male adults at a concentration of 10%, was active. Allergens, administered by ingestion or inhalation to 40 children aged 3–6 months who suffered from diarrhea, vomiting, eczema, or weight loss after the introduction of the cereal in the diet and to 18 food-allergic adults and eight patients with Baker's asthma, produced strong effect in children. Protein Z(4), administered to four patients with beer allergy, provoked weak positive response to skin testing in two of the patients and was recognized by the four individual sera tested. Lipid transfer protein 1 showed reactivity with three of four individual sera and induced strong positive skin prick test responses in all four of the patients tested. Two cases of severe systemic reactions resulted from beer ingestion: one case of anaphylaxis requiring emergency care and one of generalized urticaria and angioedema were reported. Barley was recognized as the specific ingredient responsible for the observed allergic reaction. Beer and malt allergens were found in three patients with urticaria. Urticaria from beer was an immunoglobulin E (IgE)-mediated hypersensitivity reaction induced by a protein component of approx 10 kDa deriving from barley. A 50-year-old man, who developed bronchial asthma after exposure to barley flour, was confirmed by skin prick test and serum-specific IgE. Bronchial challenge test with every allergen showed no response, except for an immediate response to barley flour. The most relevant clinical feature was an immediate asthmatic response developed after oral provocation with either barley-made beer or barley flour itself that indicated IgE-mediated food-induced bronchial asthma. A 32-year-old storeman developed occupational asthma resulting from barley grain dust in the packaging of flour, barley, and peanuts. He developed immediate symptoms of sneezing, cough, and dyspnea on exposure to barley only. Bronchial provocation test to the barley confirmed the diagnosis.

Anti-atherogenic Activity

β-Glucan in barley cellulose, administered to Syrian golden F(1)B hamsters at doses of 2, 4, or 8 g/100 g in a semipurified hyper-cholesterolemic diet of 0.15 g/g cholesterol, 20 g/100 g of hydrogenated coconut oil and 15 g/100 g of cellulose, produced cholesterol-lowering effect. Compared with control hamsters, dose-dependent decreases that were similar in magnitude in plasma total and low-density lipoprotein cholesterol concentrations were observed in hamsters fed the β-glucan diet at weeks 3,6, and 9. Liver cholesterol concentrations were also reduced significantly in hamsters consuming 8 g/100 g β-glucan.

Antibacterial Activity

Decoction of the dried fruit, on agar plate, was inactive on *Pseudomonas aeruginosa*. Ethanol (95%) and water extracts of the dried fruit, on agar plate at a concentration of 50 μL/plate, were inactive on *Staphylococcus aureus*. Water extract of the dried fruit, on agar plate, at a concentration of 1 mg/mL, was inactive on *Salmonella typhi*. Hot water extract of the dried fruit, on agar plate at a concentration of 62.5 mg/mL, was inactive on *Escherichia coli* and *Staphylococcus aureus*. Tincture of the dried seed, on agar plate at a concentration of 30 μL/disc, was inactive on *Escherichia coli*, *Pseudomonas aeruginosa*, and *Staphylococcus aureus*. Extract of 10 g plant material in 100 mL ethanol was used.

Anticoagulation Activity

Serpin BSZx (an inhibitor of trypsin and chemotrypsin) inhibited thrombin, plasma kallikrein, factor VIIa/tissue factor, and factor Xa at heparin-independent association rates. Only factor Xa turned a significant fraction of BSZx over as substrate. Activated protein C and leukocyte elastase were slowly inhibited by BSZx, whereas factor XIIa, urokinase and tissue type plasminogen activator, plasmin and pancreas kallikrein, and elastase were not or only weakly affected. Trypsin from *Fusarium* was not inhibited, while interaction with subtilisin Carlsberg and Novo was rapid, but most BSZx was cleaved as a substrate.

Antidiabetic Activity

The plant, administered to diabetic rats, produced a decrease of blood glucose concentration, water consumption, and weight loss. No differences were found in healthy animals.

Antidiarrheal Activity

Extract of the germinating seeds, administered in the ration of male rats, was active vs cecocolectomy-induced diarrhea. Germinated barley and scutellum fraction of germinated barley, administered to rats, prevented diarrhea caused by cecocolectomy and increased the protein content and sucrose activity of small intestinal mucosa. The aleurone and scutellum fractions of barley grains before and after germination, administered to rats with diarrhea, were active. The addition of fractions of germinated barley and not barley collected before germination increased the fecal output and jejunal mucosal protein content. The effect of malted barley was similar to that of germinated barley food-stuff.

Antifungal Activity

Dried stem, on agar plate, was active on *Sphacelia segetum*. Hot water extract of the dried fruit, on agar plate at a concentration of 62.5 mg/mL, was inactive on *Aspergillus niger*. Water extract of the seed, on agar plate at a concentration of 5 mg/mL, was inactive on *Helicobacter pylori*. Protein fraction of the seed without seed coat, on agar plate at a concentration of 2 μg/disc, was active on *Neurospora crassa* and *Trichoderma* sp. Protein fraction of the seed without seed coat, on agar plate at a concentration of 2 μg/disc, was active on *Neurospora crassa*.

Antihepatotoxic Activity

Methanol extract of the dried fruit, administered by gastric intubation to rabbits at a dose of 0.5 g/kg, was active vs CCl_4-induced hepatotoxicity. A mixture of *Machilus* sp., *Alisma* sp., *Amomum xanthioides*, *Bulboschoenus maritimus*, *Artemisia iwaymogis*, *Atractylodes japonica*, *Crataegus cuneata*, *Hordeum vulgare*, *Citrus sinensis*, *Polyporus umbellatus*, *Agastache rugosa*, *Raphanus sativus*, *Poncirus trifoliatus*, *Curcuma zeodaria*, *Citrus aurantium*, *Saussurea lappa*, *Glycyrrhiza glabra*, and *Zingiber officinale* was used.

Antihypercholesterolemic Activity

Dried bran, administered in ration of male rats, was active. Methanol extract of the dried fruit, administered by gastric intubation to rabbits at a dose of 500 mg/kg, was active vs CCl_4-induced hepatotoxicity. A mixture of *Machilus* sp., *Alisma* sp., *Amomum xanthioides*, *Bulboschoenus maritimus*, *Artemisia iwaymogis*, *Atractylodes japonica*, *C. cuneata*, *Hordeum vulgare*, *Citrus sinensis*, *Polyporus umbellatus*, *Agastache rugosa*, *Raphanus sativus*, *Poncirus trifoliatus*,

Curcuma zeodaria, *Citrus aurantium*, *Saussurea lappa*, *Glycyrrhiza glabra*, and *Zingiber officinale* was used. Results were significant at $p < 0.01$ level. Gum, administered orally to male rats for 4 weeks, was active. Biological activity reported had been patented. Chromatographic fraction of the green leaf juice, administered to rats at a dose of 1% of diet, was active vs cholesterol-loaded animals. The results were significant at $p < 0.005$ level. Seeds, administered to 20 men with hypercholesterolemia aged 41 ± 5 years, resulted in significant fall in serum total cholesterol, LDL cholesterol, and phospholipids, and LDL and very low-density lipoprotein (VLDL). A dose of 50/50 w/w mix with rice, administered to seven women with mild hypercholesterolemia aged 56 ± 7 years twice daily for 2–4 weeks, produced a significant improvement of serum lipid profiles. In the normolipemic subjects, serum lipids were unaffected. Bran flour and oil extract, administered to 79 patients with hypercholesterolemia, aged 48.2 years at a dose of 3 g oil extract or 30 g flour for 30 days, significantly decreased total serum cholesterol. LDL cholesterol was decreased 6.5% with addition of bran flour and 9.2% with oil. High-density lipoprotein (HDL) cholesterol decreased significantly in the bran flour group but not in the oil group. Fiber (nonstarch polysaccharides), administered to 21 men with mild hypercholesterolemia aged 30–59 years for 4 weeks, produced a significant fall in plasma total cholesterol and LDL cholesterol. The triglyceride and glucose concentrations did not change significantly.

Antihyperglycemic Activity

Dried seeds, administered orally to six patients with noninsulin-dependent diabetes mellitus at a dose of 50 g/person, was active. A single dose resulted in a glycemic index of 53.4. Water extract of the dried fruit, administered intragastrically to rats at a dose of 150 mg/kg, produced weak activity on blood vs streptozotocin-induced hyperglycemia. Dried seeds, administered orally to eight adults with normal glucose tolerance at a dose of 50 g/person, were active. Flour, administered in the ration of male rats, was active vs streptozotocin-induced hyperglycemia. Barley gum, administered to 4-week-old male Sprague–Dawley rats as a 2% dietary supplement for 14 days, lowered serum cholesterol concentration and suppressed the elevation of serum and liver triglyceride concentrations. Thick and thin rolled oat made from raw or preheated kernel, administered to healthy subjects, produced high glucose, insulin, and metabolic responses. Barley flour naturally high in β-glucan and β-glucan-enriched flour, administered to 11 healthy men, resulted in a decrease of the insulin response. Plasma glucose and insulin concentrations increased significantly. Cholesterol concentration dropped below the fasting concentration 4 hours after the meal and was significantly lower than after low-fiber meal. The cholecystokinin remained elevated for a long time after the barley-containing meals. Boiled intact and milled kernels with different amylase-amylopectin ratios, administered to healthy subjects, produced lower metabolic responses and higher satiety scores when compared to white wheat bread. The boiled flours produced higher glucose and insulin responses than did the corresponding boiled kernels. The impact of amylase to amylopectin on the metabolic responses was marginal. The intact kernels, administered to healthy subjects at concentrations of 40 and 80% (SCB-40 and SCB-80), produced the glycemic and insulinemic indices 39 and 33 for SCB-80, compared to pumpernickel bread 69 and 61, respectively. The glycemic index for SCB-40 was 66.

Anti-inflammatory Activity

Germinated barley foodstuff, administered to mice with DSS-induced colitis, prevented disease activity and loss of body weight after induction of colitis. Serum interleukin (IL)-6 level, mucosal STAT3 expression, necrosis factor-κB activity, and mucosal damages were decreased and cecal butyrate content increased. The germinated barley foodstuff-fed mice had lower bile acid concentration than the control group. Green barley extract, in LPS-activated human monocytes cell line culture (THP-1), was active.

Antioxidant Activity

Water extract of the roasted seed, at a concentration of 1 mg/mL, produced strong activity vs a liposome model system. A concentration of 25 mg/mL was active vs 2,2-diphenyl-1-picryl-hydrazyl-hydrate-induced radical. A concentration of 5 mg/mL was inactive vs linoleic acid system. Ethanol (80%) extract of the freeze-dried leaf, at a concentration of 60 μg/mL, was active vs oxidation of ethyl linoleate by Fenton's reagent. Young leaf extract, administered orally to 36 patients with type 2 diabetes at a dose of 15 g daily for 4 weeks, enhanced the scavenging of oxygen free radicals, saved the LDL-vitamin E content, and inhibited LDL oxidation. Purified green barley extract, in human mononuclear culture of cells isolated from perithelial blood and synovial fluid of patients with rheumatoid arthritis, was active. Leaf essence, administered to atherosclerotic New Zealand White male rabbits at a dose of 1% of diet, produced a decrease of plasma total cholesterol, triacylglycerol, lucigenin-chemiluminescence, and luminal-chemiluminescence levels. The value of T_{50} of red blood cell hemolysis and the lag phase of LDL oxidation increased in barley-treated group compared with the control. Ninety percent of the intimal surface of the thoracic aorta was covered with atherosclerotic lesions in the control group, but only 60% of the surface was covered in the barley group. This inhibition was associated with a decrease in plasma lipids and an increase in antioxidative abilities.

Anti-tumor Activity

Commercial barley bran (13% dietary fiber) from the aleurone/subaleurone layer; outer-layer barley bran, including the germ (25.5% dietary fiber); and spent barley grain bran (product of the brewery including the hull) (47.7% dietary fiber) were administered to male Sprague–Dawley rats as a 5% dietary supplement for 7 months. Commercial barley bran was most effective in reducing tumor incidence and burden. Tumor burden and tumor mass index were reduced significantly by outer-layer barley bran and spent barley grain bran. Commercial barley bran and spent barley grain bran, administered to rats with 1,2-dimethylhydrazine-induced intestinal tumor, produced a higher incidence and burden of tumor. Fiber was administered to 4-week-old male Sprague–Dawley rats with dimethylhydrazine-induced tumors at a dose of 5% of diet. The insoluble fiber-rich fraction (spent barley grain) was significantly more effective at preventing induced tumors than the soluble commercial barley bran. The incidence of rats affected, tumor mass index, and plasma cholesterol concentration were reduced by spent barley grain. Outer-layer barley bran was moderately effective in cancer prevention. The crude and partially purified lunasin, in stably *ras*-transfected mouse fibroblast cell culture, suppressed colony formation induced with isopropylthiogalactoside. This fraction also inhibited histone acetylation in mouse fibroblast (NIH 3T3) and human breast (MCF-7) cells in the presence of the histone deacetylase inhibitor sodium butyrate.

Anti-ulcer Activity

Water extract of the green leaf juice, administered by gastric intubation to rats at a dose of 500 mg/kg, was active vs stress-induced (restraint) ulcers. The results were significant at $p < 0.001$ level. Water extract was active vs acetic acid-induced and aspirin-induced ulcers. The results were significant at $p < 0.01$ and $p < 0.005$ levels, respectively. Water extract was inactive vs pylorus ligation-induced ulcers. Extract of the dried seedling, administered orally to adults at a dose of 30 g/person, was active. Germinated barley foodstuff, administered orally to male Sprague–Dawley rats on day 6 after initiation of colitis, was active vs dextran sodium sulfate-induced colitis. Germinated barley foodstuff treatment reduced colonic inflammation with an increase in cecal butyrate levels. Fiber and protein factions of germinated barley foodstuff, administered to dextran sodium sulfate-induced colitis Sprague–Dawley rats, significantly attenuated the clinical signs of colitis and decreased serum α1-acid glycoprotein levels, with an increase in cecal butyrate production, whereas germinated barley foodstuff- protein did not. Germinated barley foodstuff with or without salazosulfapyridine, administered to rats after the onset of colitis, accelerated colonic epithelial repair and improved clinical signs. Germinated barley foodstuff, administered orally to patients with mild to moderate active ulcerative colitis at a dose of 30 g/person daily for 4 weeks, produced a significant clinical and endoscopic improvement independent of disease extent. The improvement was associated with an increase in stool butyrate concentrations and in luminal *Bifidobacterium* and *Eubacterium* levels. After the end of treatment, the patients had an exacerbation of the disease. Germinated barley foodstuff with or without *Clostridium butyricum*, administered to 3% dextran sodium sulfate-induced colitis in Sprague–Dawley rats for 8 days, prevented bloody diarrhea and mucosal damage and increased the fecal short-chain fatty acid levels. Germinated barley foodstuff, administered to Sprague–Dawley rats for 5 days, prevented bloody diarrhea and mucosal damage, elevated fecal acetic acid and *N*-butyric acid levels, and tended to increase the number of *Eubacteria* and *Bifidobacteria*. The number of *Enterobacteriaceae*, the total number of aerobes and *Bacteroidaceae*, were lowered by germinated barley foodstuff treatment. Germinated barley foodstuff, administered to HLA-B27 transgenic rats for 13 weeks, produced an increase of bacterial butyrate production and the decrease of cecal occult blood, colonic mucosal hyperplasia, colonic mucosal necrosis factor-κB-DNA binding activity, and the production of IL-8. Butyrate from germinated barley foodstuff, administered orally or intracecally to Sprague–Dawley rats, produced reduction of mucosal damage only by intrathecal administration. Bacterial butyrate production and reduction of mucosal damage depended on the dose of germinated barley foodstuff in the diet. Germinated barley foodstuff and scutellum fraction of germinated barley, administered to Sprague–Dawley rats with colitis induced by 3% dextran sodium sulfate, prevented bloody diarrhea and mucosal damage in colitis. The germinated samples did not produce a protective effect. Germinated barley foodstuff increased mucosal protein and RNA content in the colitis model. Germinated barley foodstuff, administered to 18 patients with mildly to moderately active ulcerative colitis at a dose of 20–30 g germinated barley foodstuff daily for 4 weeks, produced a significant decrease in clinical activity index scores compared to the control group. No side effects related to germinated barley foodstuff were observed. Germinated barley foodstuff therapy increased fecal concentrations of *Bifidobacterium* and *Eubacterium limosum*. Germinated barley foodstuff, administered to patients with mild to moderate active ulcerative colitis, irresponsible to or intolerant of standard treatment at a dose of 20–30 g germinated barley foodstuff daily for 4 weeks, resulted in a significant clinical and endoscopic improvement associated with an increase in stool butyrate concentrations.

Cardiovascular Activity

β-glucan, administered to 18 men with mild hypercholesterolemia with a mean body weight index of 27.4 ± 4.6 at a dose of 8.1–11.9 g β-glucan per day, produced no significant change in total, LDL and HDL cholesterol, triacylglycerol, fasting glucose, and postprandial glucose.

Cholesterol Biosynthesis Inhibition

The inhibitor I from oily nonpolar fraction of flour, administered to chicken at a dose of 2.5–20 ppm, produced a significant decrease in hepatic cholesterogenesis and serum total and LDL cholesterol and an increase in lipogenic activity.

Cholesterol-7-α-hydroxylase inhibition

Petroleum ether extract of the fresh fruit, administered to pigs at a concentration of 3.5 g/kg of diet for 29 days, produced 40% inhibition of the hepatic enzyme activity.

Cytotoxic Activity

Water extract of the dried fruit, in cell culture at a concentration of 500 μg/mL, produced weak activity on CA-mammary-microalveolar. Ethanol (50%) extract of the seed, in cell culture, was inactive on CA-9KB, ED_{50} greater than 20 μg/mL. Methanol extract of the dried seed, in cell culture, was inactive on SNU-1 human cells, IC_{50} greater than 0.3 mg/mL, and on SNU-C4 human cells, IC_{50} greater than 0.3 mg/mL. Protein fraction of the seed without seed coat, in cell culture at a concentration of 2 μg/disc, was active on CA-Ehrlich ascites. Methanol extract of the aerial parts, in cell culture at a concentration of 50 mg/mL, was equivocal on CA-9KB.

Gastrointestinal Activity

Water extract of the green leaf juice, administered by gastric intubation to rats at a dose of 500 mg/kg, was inactive vs pylorus ligation-induced ulcers. Fiber, administered orally to young male Wistar rats at a dose of 500 g extrudates/kg of diet for 6 weeks, produced a higher concentration of neutral sterols in the intestinal content of the barley-fed group than in the control group ($p < 0.005$) and affected indirectly the amount of formed secondary bile acids. Fiber, administered orally to young male Wistar rats at a dose of 50 g/100 g extrudates or mixtures for 6 weeks, produced greater food intake in the last 2 weeks and increased ceca and colon masses, cecal and colon contents, concentration of resistant starch in cecal, most of colon contents, and β-glucan level in the small intestine, cecum, and colon. The numbers of coliforms and *Bacteroides* were lower and those of *Lactobacillus* were higher than in the control group. The dose increased weight gain in the sixth week. Short-chain fatty acids were higher in the cecal, colon, and feces content of the test group. The proportion of secondary bile acids was lower and the amount of neutral sterol was higher in feces of fiber-treated animals. The concentrations of excreted bile acids increased up to 30% during the feeding period. Germinated barley food-stuff, administered to Sprague–Dawley rats fed on various diets with the same protein and dietary fiber levels, produced an increase fecal output compared with commercial water-soluble and insoluble dietary fibers.

The dietary fiber from germinated barley foodstuff increased the fecal output and mucosal protein content. The protein fraction of germinated barley foodstuff degraded to the peptide form did not increase the fecal output or mucosal protein content[HV152]. Germinated barley foodstuff from aleurone and scutellum fractions of germinated barley, administrated to healthy volunteers at a dose of 9 g daily for 14 days, significantly increased fecal butyrate content and fecal *Bifidobacterium* and *Eubacterium*. Ten anaerobic microorganisms selected from intestinal microflora were cultured in vitro in germinated barley foodstuff medium. After 3 days of incubation, seven strains (*Bifidobacterium breve*, *Bifidobacterium longum*, *Lactobacillus acidophilus*, *Lactobacillus casei* ssp. *casei*, *Bacteroides ovatus*, *Clostridium butyricum,* and *Eubacterium limosum*) lowered the medium pH producing short-chain fatty acid. Germinated barley foodstuff changed the intestinal microflora and increased probiotics such as *Bifidobacterium*. Butyrate was produced by the mutual action of

Eubacterium and *Bifidobacterium*. Germinated barley foodstuff from aleurone layer, scutellum, and germ, administered to 10 healthy volunteers at a dose of 30 g/day/person for 28 days, produced an increased fecal butyrate content, fecal weight, and water content. There were no significant changes in body weight and major abnormalities in hematologic and urinary analysis. Fiber, administered to nine patients with ileostomies at a dose of 35 g/day, increased the ileal excretion of starch. Flaked and finely milled barley, eaten by patients with ileostomies, showed that only 2 ± 1% of starch remained undigested after the consumption of finely milled barley and 17 ± 1% resisted digestion, partly as oligosaccharides but largely as intact unpitted starch granules bound by intact cell walls. The energy excretion from the stoma was three times higher after flaked that after milled barley. Nonstarch polysaccharide, starch, and fat made almost equal contributions to the higher energy excretion. Bran flour, administered to 44 volunteers at a dose of 30 g/day, decreased the transit time by 8.02 hours from baseline and increased daily fecal weight by 48.6. Groats were administered to volunteers at a dose of 1 g carbohydrate/kg body weight, three times or at three different doses of 0.75, 1, and 1.5 g carbohydrate/kg body weight. After consumption of 1 g carbohydrate/kg body weight, produced a mean mouth to cecum transit time of 8.4 ± 0.4 hour. After consumption of the high dose, a mean mouth-to-cecum transit time of 9.0 ± 0.5 hours was produced. Particle size did not significantly affect the mouth-to-caecum transit time. Germinated barley food-stuff, administered to Sprague–Dawley rats, prevented diarrhea and mucosal damages; increased mucosal protein, DNA, and RNA content; and depressed bacterial translocation and elevation of myeloperoxidase activity induced by methotrexate. β-glucan-rich barley fraction, administered to ileostomy subjects at a dose of 13.0 g β-glucan/day for 2 days, increased the cholesterol excretion higher than with the oat bran with β-glucanase and wheat flour diets. Bile acid excretion was 755 (133–1187) mg/day. Carbohydrates, administered to healthy subjects at a dose of 90 g for dinner in random order 1 week apart, significantly increased the breath hydrogen and improved glucose tolerance. No difference in the rates of glucose disappearance or gut glucose absorption was observed. Serum-free fatty acid concentrations were significantly reduced the morning after the barley meal. Germinated barley foodstuff, administered to Sprague–Dawley rats with constipation induced by loperamide, produced an increase of bowel movements, fecal water content, and concentration of short-chain fatty acids in cecal content, especially butyrate.

Glucose Tolerance Effect

Fiber, administered orally to type 2 diabetic Goto– Kakizaki male rats for 9 months, improved the area under the plasma glucose concentration time curves, lowered the fasting plasma glucose and glycosylated hemoglobin levels, and decreased plasma total cholesterol, triglycerides, and free fatty acid levels. Fiber, administered orally to 8- week-old male Goto–Kakizaki strain rats, at a dose of 1.79 g/day/rat for 3 months, improved glucose tolerance and lowered the plasma cholesterol and triglyceride levels. The fasting plasma glucose level was significantly lower in comparison to rice and corn starch-fed rats. High barley (high-fiber diet), administered to 10 women (20.4 ± 1.3 year-old, 19.2 ± 2 kg/m^2) for 4 weeks with a 1-month interval, resulted in lowering plasma total and LDL cholesterol concentrations and reduced plasma triacylglycerol concentration. The barley diet increased stool volume. There was no significant difference in glucose tolerance between diet regimens. Barley bread containing lactic acid and reference barley bread, administered in the morning to 10 healthy men and women, produced a significant lowering of the incremental glycemic area and of the glucose response at 95 minutes after ingestion of the bread with lactic acid. At 45 minutes after the meal, the insulin level was significantly lower after the lactic acid bread, compared with reference barley bread.

Glucose-6-phosphate Dehydrogenase Inhibition

Petroleum ether extract of the fresh fruit, administered to pigs at a concentration of 3.5 g/kg of diet for 29 days, was active on hepatic enzymes.

Glutamate–Oxaloacetate–Transaminase Inhibition

Methanol extract of the dried fruit, administered by gastric intubation to rabbits at a dose of 500 mg/kg, was active vs CCl_4-induced hepatotoxicity. A mixture of *Machilus* sp., *Alisma* sp., *Amomum xanthioides*, *Bulboschoenus maritimus*, *Artemisia iwaymogis*, *Atractylodes japonica*, *Crataegus cuneata*, *Hordeum vulgare*, *Citrus sinensis*, *Polyporus umbellatus*, *Agastache rugosa*, *Raphanus sativus*, *Poncirus trifoliatus*, *Curcuma zeodaria*, *Citrus aurantium*, *Saussurea lappa*, *Glycyrrhiza glabra*, and *Zingiber officinale* was used. Results were significant at $p < 0.01$ level.

Glutamate–Pyruvate–Transaminase Inhibition

Methanol extract of the dried fruit, administered by gastric intubation to rabbits at a dose of 500 mg/kg, was active vs carbon tetrachloride (CCl_4)-induced hepatotoxicity. A mixture of *Machilus* sp., *Alisma* sp., *Amomum xanthioides*, *Bulboschoenus maritimus*, *Artemisia iwaymogis*, *Atractylodes japonica*, *Crataegus cuneata*, *Hordeum vulgare*, *Citrus sinensis*, *Polyporus umbellatus*, *Agastache rugosa*, *Raphanus sativus*, *Poncirus trifoliatus*, *Curcuma zeodaria*, *Citrus aurantium*, *Saussurea lappa*, *Glycyrrhiza glabra*, and *Zingiber officinale* was used. Result was significant at $p < 0.01$ level.

Hypocholesterolemic Activity

Fixed oil of the bran, administered orally to adults of both sexes at a dose of 30 mg/day, was active. Flour bran, administered orally to adults at a dose of 3 g/day, was active. Dried bran, administered in the ration of male rats, was active. Petroleum ether extract of the fresh fruit, administered to pigs at a concentration of 3.5 g/kg of diet, produced a decrease of serum total cholesterol, LDL cholesterol, and HDL cholesterol after 29 days of feeding. Flour, administered orally to adults with hypercholesterolemia at a dose of 44 g/day, produced a decrease of total and LDL cholesterol levels.

Hypoglycemic Activity

Water extract of the fermented root, administered intravenously to rabbits, was active. The dried seed, administered orally to eight healthy volunteers at a dose of 50 g/person, was active. A single dose resulted in a glycemic index of 68.7 and an insulinemic index of 71.1.

Hypolipemic Activity

Fiber, administered orally to nine adults with ileostomies at a dose of 13 g/day, increased the excretion of cholesterol. Petroleum ether extract of the fresh fruit, administered to pigs at a concentration of 3.5 g/kg of diet, was inactive. Purified green barley extract, in human mononuclear culture of cells isolated from perithelial blood and synovial fluid of patients with rheumatoid arthritis, was active. Leaf essence, administered to atherosclerotic New Zealand White male rabbits at a dose of 1% of diet, produced a decrease of plasma total cholesterol, triacylglycerol, lucigenin–chemiluminescence, and luminal–chemiluminescence levels. The value of T_{50} of red blood cell hemolysis and the lag phase of LDL oxidation increased in barley-treated group compared with the control. Ninety percent of the intimal surface of the thoracic aorta was covered with atherosclerotic lesions in the control group, but only 60% of the surface was covered in the barley group. This inhibition was associated with a decrease in plasma lipids and an increase in antioxidative abilities.

Hypotriglyceridemic Activity

Fixed oil of the bran, administered orally to adults of both sexes at a dose of 30 mg/day, was active. Flour bran, administered orally to adults at a dose of 3 g/day, was inactive.

Laxative Effect

Powdered dried bran, administered orally to 44 adults at a dose of 30 g/person, was active on gastrointestinal motility. Transit time decreased by 8 hours, and fecal mass increased by 48.6 g/day.

Lipid Metabolism

Fiber, administered orally to male type 2 diabetic Goto– Kakizaki rats for 9 months, improved the area under the plasma glucose concentration time curves, lowered the fasting plasma glucose and glycosylated hemoglobin levels, and decreased plasma total cholesterol, triglycerides and free fatty acid levels.

Lipolytic Effect

Ethanol (95%) extract of the dried entire plant in combination with *Rhizoma zingiberis*, *Ligustrum chuanxiong*, *Lilium brownii*, *Nephelium longa*, and *Polygonum multiflorum*, administered in drinking water to C57BL/6J obese mice at a concentration of 5%, was active.

Lung Function

Exposure of six men to barley dust for 2 days decreased ventilatory capacity. Five volunteers not previously exposed to barley dust, when exposed to the dust for 2 hours, decreased the ventilatory capacity ranging from 200 mL to 800 mL, with recovery taking up to 72 hours. All of the subjects had decreases in flow at 50% vital capacity but little or no change in flow at 75% vital capacity. In three subjects, there was a drop in specific conductance that lasted for less than 24 hours. Sixty-ine of 80 dockworkers handling grains reported evening feverish episodes/symptoms not related to smoking or atopic status. No gross deficits in lung function were detected.

Malic Enzyme Inhibition

Petroleum ether extract of the fresh fruit, administered to pigs at a concentration of 3.5 g/kg of diet for 29 days, was active on hepatic enzymes.

Mineral Utilization

Germinated barley foodstuff, administered orally to 5-week-old Sprague–Dawley rats for 14 days, promoted the absorption of calcium (Ca) and magnesium (Mg) by the gastrointestinal tract. The absorption of iron and potassium was not attenuated and mineral absorption was not inhibited. Barley husk, administered to 5- and 9-week-old rats at different doses, produced a lowering of zinc (Zn) and Ca absorption already at dose 20 g dietary fiber/kg dry matter and had a small negative effect on potassium absorption. Phytate did not appear as a major factor affecting mineral absorption in barley husk. All of the diets containing barley husk had very low molar ratios (phytate:Zn was 4). Processed or unprocessed barley was administered to healthy subjects in two single meals containing porridge or breakfast (60 g) cereals for 2 months. Zn absorption from hydrothermally-treated barley porridge was significantly higher than from the control porridge; Ca absorption did not differ. Zn absorption from breakfast cereals of malted barley with phytase activity was significantly higher than from flakes of barley without phytase activity; Ca absorption was not significantly different. Standard barley and β-glucan-enriched barley dehulled grains was administered to 10 healthy hydrogen-producing adults at a dose of 35 g. The percentage of the ^{13}C dose oxidized was greater after standard barley than after enriched barley consumption. The area under the curve for H_2 was greater after enriched barley intake. There was no difference in CO_2 production. Hull fiber extract, in Caco-2 cell culture, produced no effect on the rate of transepithelial ^{45}Ca transport across Caco-2 cell monolayers and the uptake of ^{45}Ca into Caco-2 cells. A low-phytate barley-fiber concentrate was administered to young women at a dose of 15 g barley fiber (high-fiber, high-protein diet) and 15 g barley fiber (high-fiber, low-protein diet). The mean daily intake of the cations was 25.4 and 22.9 mmol Ca, 10.1 and 10 mmol Mg, 166.8 and 119.3 μmol Zn, and 186.2 and 154 μmol Fe, respectively. Mean balances were 1.9 and –0.8 mmol Ca, –0.2 and –0.5 mmol Mg, –4.6, and –18.4 imol Zn, respectively. The mean apparent iron absorption was 5.4 and –23.2 μmol.

Monocytic Differentiation

Prodelphinidin B-3, T1, T2, and T3 from bran polyphenol extract, in HL60 human myeloid leukemia cell culture, induced 26–40% nitro blue tetrazolium positive cells and 22–32% α-naphthyl-butyrate esterase-positive cells. Proanthocyanidins potentiated all-*trans*-retinoic acid-induced granulocytic and sodium butyrate-induced monocytic differentiation in HL60 cells.

Mutagenic Activity

Ethanol (70%) extract of the dried seed, on agar plate at a concentration of 50 mg/mL, was inactive on *Escherichia coli* PQ 37.

The water and chloroform extracts of the ethanol (70%) extract were inactive. Metabolic activation had no effect on the results.

Oxidative Effect

Ethanol (95%) extract of the dried entire plant, administered in drinking water to C57BL/6J obese mice at a concentration of 5%, increased glucose oxidation in epididymal fat pads. Extract of mixture of following plants: *Hordeum vulgare*, *Rhizoma zingiberis*, *Ligustrum chuanxiong*, *Lilium brownii*, *Nephelium longa*, and *Polygonum multiflorum* was used.

Pepsin Inhibition

Water extract of the green leaf juice, administered by gastric intubation to rats at a dose of 500 mg/kg, was inactive vs pylorus ligation-induced ulcers.

Phosphogluconate Dehydrogenase Inhibition

Petroleum ether extract of the fresh fruit, administered to pigs at a concentration of 3.5 g/kg of diet for 29 days, was active on hepatic enzymes.

Proteinemic Effect

Methanol extract of the dried fruit, administered by gastric intubation to rabbits at a dose of 500 mg/kg, produced an increase in serum albumin and protein content vs CCl_4-induced hepatotoxicity. A mixture of *Machilus* sp., *Alisma* sp., *Amomum xanthioides*, *Bulboschoenus maritimus*, *Artemisia iwaymogis*, *Atractylodes japonica*, *Crataegus cuneata*, *Hordeum vulgare*, *Citrus sinensis*, *Polyporus umbellatus*, *Agastache rugosa*, *Raphanus sativus*, *Poncirus trifoliatus*, *Curcuma zeodaria*, *Citrus aurantium*, *Saussurea lappa*, *Glycyrrhiza glabra*, and *Zingiber officinale* was used. Results were significant at $p < 0.01$ level.

Respiratory Effect

Barley ear inhaled by a 2.5-year-old child produced fever, dyspnea, right paracardiac infiltrate with pleural reaction on X-rays, and normal bronchoscopy after 8 days. On day 11, extensive right pneumothorax, and on day 20, right axillary inflammatory lesion were observed. On day 28, the ear of barley was expulsed and there was complete recovery. Barley spike, inhaled into the tracheobronchial tree of 18 children under the age of 5 years, produced coughing and choking in 14 of the children. The spikes were removed by laryngoscopy in 12 patients and by rigid bronchoscopy in two. Four patients with history of cough, dyspnea, fever, and serious respiratory diseases, such as pneumothorax, lobar pneumonia, and pleural empyema, required surgical intervention. All of the children made satisfactory recoveries. Dust extract of barley, in cell culture on nonsensitized guinea pig tracheal smooth muscle pretreated with drugs, produced constrictor effect that was significantly inhibited by atropine indicating an interaction of the extracts with parasympathetic nerves. Inhibition of contraction of other mediators was less effective and varied with the dust extract.

Toxic Effect

β-glucan-enriched soluble barley fiber, administered orally to Wistar rats at concentrations of 0.7, 3.5, and 7.0% β-glucan for 28 days, increased the number of circulating lymphocytes in males. The increase was not dose-dependent and was not observed in females. A dose-dependent increase in full and empty cecum weight was observed. There were no adverse effects on general condition and behavior, growth, feed and water consumption, feed conversion efficiency, red blood cell and clotting potential parameters, clinical chemistry values, and organ weight. Necropsy and histopathology findings revealed no treatment-related changes in any organ evaluated. β-glucan (64%) preparation (barley β-fiber), administered to CD-1 mice at concentrations of 1,5, or 10% of diet (0.7, 3.5, and 7% β-glucan) for 28 days, produced no adverse effect in hematological or clinical chemistry measurements, in organ weights and immunopathology in either sex after treatment or after the recovery period. Azidoalanine and the azide-treated extracts, in Chinese hamster and normal human skin fibroblast cell cultures, significantly increased the frequency of sister chromatid exchanges observed in both cultures. This increase was approximately twofold, as compared with the control.

Toxicity Assessment

Ethanol (50%) extract of the seed, administered intraperitoneally to mice, produced a maximum tolerated dose of 1 g/kg. Hexane extract of the green leaf juice, administered in ration of rats, produced lethal $dose_{50}$ greater than 10 g/kg.

Weight Gain Inhibition

Ethanol (95%) extract of the dried entire plant, administered in drinking water to C57BL/6J obese mice at a concentration of 5%, was active. The extract also contained *Rhizoma zingiberis*, *Ligustrum chuangxiong*, *Lilium brownii*, *Nephelium longa*, and *Polygonum multiflorum*.

6

COCOS NUCIFERA

Cocos nucifera is an unbranched monoecious plant of the *Palmae* family. It grows to 30 m tall, with a crown of 25–35 paripinnate leaves, producing 12–16 new leaves per year. There is a central bud, which if cut off, leads to the death of tree. The trunk is straight or gently curved, with marked foliar scars, 30–50 cm in diameter, rises from a thickened base, and increases in height at a decreasing rate with age. The leaves are horizontal or somewhat hanging, 4–8 m in length and divided. Segments of leaves are numerous, linear-lanceolate, 0.5–1 m long and tapering. In the axil of each leaf is a spathe enclosing a long, stout, straw, or an orange-colored spadix. The spadix is composed of up to 40 branches, each bearing up to 300 small, fragrant, male flowers, and a few female, 2 cm long, globose flowers. The male and female flowers are produced separately in the leaf axils, usually on a long stalk. Approximately one-third of the female flowers develop into four to eight ripe fruits in 12–13 months, per inflorescence. The fruit is ovoid, three-angled drupe, up to 30 cm long, usually with thick, fibrous mesocarp (husk) and a hard, green-brown endocarp (shell) enclosing one seed. The seeds consist of 10 to 20 mm-thick white, fleshy endosperm (meat), covered by thin brown testa, surrounding a cavity party filled with a watery, sweet fluid (coconut water or milk). The latter is found mainly in immature fruits.

ORIGIN AND DISTRIBUTION

Evidence has been found that the place of origin is "submerged land to the north west of New Guinea." The major coconut areas lie between 20°N and 20°S of the equator. Although it is found beyond this region, 27° N and 27° S, cultivation has not been successful and the palm does not fruit. Many varieties are found in Melanesian region. It is most widely cultivated in the tropics: India, Ceylon, Malaysia, Indonesia, Philippines, South Sea Islands in the Pacific, East Africa, and Central and South Americas, up to 800 m above sea level, on humus-rich and porous soil or pure sand in coastal regions.

TRADITIONAL USES

Admiralty islands. The young root or leaves of the coconut plant are chewed for diarrhea.

Cook islands. Water extract of grated endosperm and *Citrus aurantium* juice is used to soak affected part in fractures and sprains. Endosperm is taken orally for asthenia. Oil, mixed with crushed *Phyllanthus virgatus*, is rubbed around the ear for ear infections. Extract of different dried parts of the palm are taken orally for filariasis. A water solution of the crushed dried bark or husk and grated bark of *Hibiscus tiliaceus* is used externally to soak fractures and sprains. Crushed aerial root tips of *Ficus prolixa* are fried with coconut cream made of fresh endosperm, and the resulting oil is taken orally as a laxative in treating serious diseases.

Ecuador. Hot water extract of the plant is taken orally by females for sterility.

Fiji. Oil is used externally to prevent hair loss. The oil is warmed with crushed onion and garlic and applied aurally for earache. Water of the unripe fruit is taken orally for kidney problems.

Ghana. Coconut milk is taken orally for diarrhea.

Guatemala. Hot water extract of the dried fruit is taken orally as a febrifuge and sudorific and for renal inflammation and scrofula. Hot water extract of the dried fruit is applied externally on wounds, ulcers, bruises, sores, skin infections, mucosa, dermatitis, inflammations, abscesses, and furuncles.

Haiti. Decoction of the dried pericarp is taken orally for amenorrhea. Fresh essential oil is applied externally on burns.

India. Infusion of the inflorescence is taken orally every morning for 3 days, coinciding with the menstrual cycle for leukorrhea and problems associated with the menstrual cycle. A dose of 50 g daily of a mixture of *Cocos nucifera* fruit and *Ficus-benghalensis* latex is taken for 3 months to increase sexual potency in men. Fruit is taken orally as a remedy for tapeworms.

Indonesia. Coconut oil is applied externally to treat wounds and injuries by the ethnic group of Ngada. Shell is used as incense. Hot water extract of the root is taken orally for fever, bloody diarrhea, and dysentery. Milk is taken orally by adults for poisoning. Fruit milk is taken orally by females as a contraceptive. Fruit juice is believed to diminish libido or fertility. Seed oil with lemon juice and various tree roots is taken orally as an abortifacient. Fruit ointment is applied externally for swollen legs. Fresh flowers, chewed with *Borassus flbellifer*, are used for gonorrhea.

Jamaica. Hot water extract of the dried shell is taken orally for diabetes. Hot water extract of the root, with seven other plants, wine, and rum, is used as a tonic. Extract of different parts are taken orally for diabetes.

Marquesas islands. Fruit juice is mixed with *Cordia subcordata* and other plants and used for general menstrual disorders.

Mexico. A plaster made of fresh milk mixed with egg white is applied externally to prevent miscarriage.

Mozambique. Fruit is eaten by males as an aphrodisiac and used for relief of tumors.

Papua New Guinea. The fresh root of a young coconut is dug out and washed, then chewed and swallowed to relieve stomachache. Fruit milk of *Cocos nucifera* is taken orally together with leaves of *Cleroden* sp., *Pouzolzia microhylla*, or *Macaranga tanarius* by pregnant women as an abortifacient agent.

Peru. Hot water extract of the fresh fruit is taken orally for blennorrhagia and asthma and as a diuretic, tenifuge, and galactagogue.

Thailand. Hot water extract of the fresh fruit juice is taken orally as a cardiotonic and neurotonic.

Tonga. Infusion of fresh kernel of coconut and Euodia hortensis are taken orally to treat retention of blood clots in the uterus after childbirth (locally called "toka-ala"). A mixture of *Cocos nucifera*, *Glochidion concolor*, *Vigna marina*, *Morinda citrifolia*, *Euodia hortensis*, and *Premna taitensis* with lemon juice is taken orally by pregnant women to treat severe bleeding during early pregnancy. Infusion of fresh kernel, taken twice daily with "tongan oil" prepared from *Aleurites moluccana* is used for dysuria caused by "kahi," a locally described syndrome affecting the gastrointestinal and genitourinary systems.

Trinidad. Hot water extract of the root is taken orally for amenorrhea.

Vanuatu. Hot water extract of the fruit, drunk in large quantities when very hot, is said to induce abortion. Juice of the crushed root mixed with water is drank after delivery to restore strengh.

Fig. 6.1. Cocos nucifera. A–Plant a fruiting stage; B–Fruit (entire); C–Fruit (L.S.).

West Indies. Hot water extract of the root is taken orally for amenorrhea. Hot water extract of the mesocarp is taken orally by females for dysmenorrhea, amenorrhea, and menorrhagia.

Medicinal Uses

Acquired Immunodeficiency Syndrome Activity

Coconut oil and "monolaurin"—a coconut oil byproduct—were administered to 12 women and 3 men who were in the early stage of human immunodeficiency virus infection. Ten patients took different doses of monolaurin, and five patients took coconut oil. It was believed that the treatment would lead to higher CD4 counts and a lower viral load. The trial was abandoned because it received only lukewarm approval from the government.

Allergenic Activity

Dried fruit juice, administered intraperitoneally to guinea pigs at a dose of 10 mL/animal, was active. Dried fruit juice, administered intravenously to guinea pigs at a dose of 5 mL/animal, was active[CN036]. A patient with coconut anaphylaxis was confirmed by skin prick test. In vitro serum specific immunoglobulin E (IgE) was present. Two patients with allergy manifested by life-threatening systemic reaction after consumption of coconut were investigated. Sera IgE from both patients indicated reduced coconut allergens with molecular weight of 35 and 36.5 kDa. IgE from 1 patient also bound a 55-kDa antigen. Preabsorption of sera with nut extracts suppressed IgE binding to coconut proteins. Preabsorption of sera with coconut produced a disappearance of IgE binding to protein bands at 35 and 36 kDa on a reduced immunoblot of walnut protein extract in one patient and suppression of IgE binding to a protein at 36 kDa in another patient. Three cases of individuals allergic to cocamide diethanolamide were investigated. In two of the cases, multiple other cutaneous allergies were present. In both instances, cocamide diethanolamide was present in several personal care products used by the patients. In the third case, occupational exposure was suspected. Coconut diethanolamide, administered to six patients with occupational contact dermatitis caused by coconut diethanolamide, produced sensitization from a barrier cream in two patients, hand-washing liquid in three, and one had been exposed to a hand-washing liquid and to a metalworking fluid containing coconut diethanolamide. Leave-on products (hand-protection foams) produced sensitization more rapidly (2–3 months) than rinse-off products (5–7 years). There was no contact allergy to another coconut-oil-derived sensitizer (cocamidopropyl betaine). Coconut diethanolamide, administered to a dentist, produced an occupational allergic contact dermatitis for hand-washing liquids. Pollen extract, administered to 24 patients with allergy, asthma, and rhinitis produced positive skin prick test in all cases, and 19 of them were phadezym radioallergosorbent test-positive. Bronchial provocation test was positive in seven out of eight patients, and no late response or nonspecific reactions were observed. An 8-month-old baby fed from birth with maternal milk was investigated. The first milk induced a severe gastrointestinal disorder, which disappeared when second milk was used. The third milk caused a relapse. The allergen was coconut, which was physicochemically modified in the second milk. It was confirmed by positive reintroduction test, positive skin test, and positive specific IgE test. Pollen extract immunotherapy was studied in 96 allergic patients for 6–12 months.

The clinical status measured by the symptom-medication scores demonstrated that the patients had significant clinical improvement after pollen extract immunotherapy. Serological study indicated a significant reduction of specific IgE and elevation of specific IgG in posttherapeutic patients' sera. There was no correlation between symptom-mediation scores and changes in specific serum IgE or IgG levels. Proteins from fresh coconut and commercial extracts of coconut, administered to a patient with coconut anaphylaxis, produced a positive skin prick test. In vitro serum-specific IgE was present for coconut, hazelnut, Brazil nut, and cashew. Immunoblots demonstrated IgE binding to 35- and 50-kDa protein bands in the coconut and hazelnut extracts. Inhibition assays using coconut demonstrated complete inhibition of hazelnut specific IgE, but inhibition assays using hazelnut showed only partial inhibition of coconut-specific IgE. A total of 100 patients (59 females and 41 males, aged 10–59 years, mean age 27.9 years) with allergic rhinitis underwent a skin prick test with 30 aeroallergens. Coconut produced 12% of positive results. Oil, administered to 3-lactoglobulin-treated brown Norway rats at a dose of 10% of diet supplemented with 0.5% curcumin for 3 weeks, lowered the circulatory release of rat chymase II in response to antigen. The triethanolamine (TEA) salt of the condensation product of coconut fatty acids with a complex of polypeptides and amino acids derived from collagen (TEA-Coco-hydrolyzed protein), administered to a 21-year-old woman, produced a severe dermatitis of the face after using a proprietary skin cleanser. Patch testing revealed delayed hypersensitivity to TEA-Coco-hydrolyzed protein but not to other ingredients of the cleanser and positive results with other condensates of fatty acids and protein hydrolysates.

Aminopeptidase Activity Influence

Oil, administered to mice testis using acrylamides as substrates, produced no difference in soluble glutamyl-aminopeptidase angiotensin among the groups tested. Soluble aspartyl-AP and soluble pyroglutamyl-AP progressively decreased with the degree of saturation of the fatty acids used in the diet. Membrane-bound glutamyl-AP progressively increased with the degree of saturation of the fatty acids used in the diet. For membrane-bound aspartyl-AP activity, mice that were fed diets containing fish oil showed significantly higher levels than those fed sunflower oil, olive oil, and lard, but not those fed coconut oil.

Antibacterial Activity

Ethanol (50%) extracts of the leaf, on agar plate at concentrations above 25 μg/mL, were inactive on *Bacillus subtilis*, *Escherichia coli*,

Salmonella typhosa, and *Staphylococcus aureus*. Tincture of the dried fruit (10 g plant material in 100 mL ethanol), on agar plate at a concentration of 30 μL/disc, was inactive on *Pseudomonas aeruginosa* and *Staphylococcus aureus* and produced weak activity on *Escherichia coli*. Water extract of the husk fiber and fractions from adsorption chromatography were active on *Staphylococcus aureus*. Hydrogenated oil, administered orally to Balb/c mice at a dose of 20% by weight for 4 wk and then infected with *Listeria monocytogenes* or treated with *N*-acetyl-L-cysteine (25 mg/mL intraperitoneally), produced no effect on survival. *N*-acetyl-L-cysteine reduced the recovery of *Listeria monocytogenes* from the spleen of the mice fed coconut oil diet. There was a reduction in lymphocyte proliferation. An important increase in the production of reactive oxygen species was found after 12 h of the incubation with *Listeria monocytogenes*. Hydrogenated oil, administered to *Listeria monocytogenes*-infected mice, produced a significant increase in peritoneal cells of coconut oil-fed mice and a reduction of bacterial recovery from the spleen. Oil, administered to mice injected with a nonlethal dose of *Escherichia coli* at a dose of 20% by weight for 5 weeks, produced a decrease of peak plasma tumor necrosis factor (TNF)-α, interleukin (IL)-1β, and IL-6 concentrations. Peak plasma IL-10 concentrations were higher in the coconut-fed group than in those fed the other diets, coconut oil diminished production of proinflammatory cytokines in vivo.

Anticarcinogenic Effect

Coconut cake, administered to rat with 1,2-dimethylhydrazine-induced colon cancer at a dose of 25% of diet for 30 weeks, produced a significant decrease of the incidence and number of tumors, and β-glucuronidase and mucinase activities. Hydrogenated oil, administered to Wistar female rats at doses of 8% and 24%, with or without a 24 mg/day/rat phytosterol supplement, produced no significant influence. Colonic glands were found in area of lymphoid follicles in all the groups but were more frequently in rats on high-fat diet.

Antifungal Activity

Ethanol (50%) extract of the leaf, on agar plate at a concentration of 25 μg/mL or more, was inactive on *Microsporum canis* and *Trichophyton mentagrophytes*. Oil, on agar plate at a concentration of 05 mL/plate, was active on *Absidia corymbifera*, *Aspergillus flavus*, *Aspergillus niger*, and *Penicillium nigricans*. Ethanol (95%) extract of the dried shell, on agar plate at a concentration of 100 μg/mL, was active on *Microsporum audouini*, *Microsporum canis*, *Microsporum gypseum*, *Trichophyton mentagrophytes*, *Trichophyton rubrum*, *Trichophyton tonsurans*, *Trichophyton violaceum*, and *Epidermophyton floccosum*. All fractions of the husk fiber, from adsorption chromatography, were inactive on *Candida albicans*, *Fonsecaea pedrosoi*, and *Cryptococcus neoformans*.

Antihypercholesterolemic Activity

Oil was administered orally to male Wistar rats at doses of 12 or 24% of diet *ad libitum* for 4 weeks. Absorption of oleic acid in rats fed 24% oil was significantly greater than in controls during 0–8 hours but was not significantly different during 0–24 hours. There were no differences among groups in the distribution of cholesterol and oleic acid either in the lymph lipoproteins or in the lipid classes. Oil was administered to 61 healthy males and 22 females aged 20–34 years, group I received a coconut–palm–coconut dietary sequence; group II, coconut–corn–coconut; group III, coconut oil during all three 5-week dietary periods. Compared with entry-level values, coconut oil raised the serum total cholesterol concentration greater than 10% in all three groups. The entry level of the ratio of low- density lipoprotein (LDL) to high-density lipoprotein (HDL) was not altered by coconut oil. Oil was administered to 6- month-old ovariectomized rats with or without taurine for 28 days. Body mass gain, food intake, liver weight, and plasma apoliprotein (apo) A-I, apo B, LDL, and very low-density lipoprotein (VLDL) concentrations were not affected by the diet. Taurine lowered the plasma total cholesterol and increased liver total lipid and triglyceride in rats that were fed corn oil but not in those fed coconut oil. Taurine increased the 3 -hydroxy-3-methylglutaryl coenzyme A (HMG-CoA) reductase messenger (m)RNA level in the liver of coconut-fed rats, but not in those fed corn oil. Structured lipids (10%) synthesized from oil triglycerides, coconut oil, and coconut oil–safflower oil blends (1:0.7 w/w) were administered to rats for 60 days. The structured lipids lowered serum and liver cholesterol levels. Most of the decrease observed in serum was found in LDL fraction. Oil, administered orally to male Wistar rats at a dose of 11% w/w for 6 months, produced a significant increase of serum total cholesterol (twofold), serum triglycerides (92.6%), LDL cholesterol (92.3%), and body weight gain (2.8-fold).

Antilipidemic Activity

Triglycerides structured lipids from coconut oil, administered to rats at a dose of 10% of diet for 60 days, produced a 15% decrease in total cholesterol and a 23% decrease in LDL cholesterol levels in the serum compared to coconut oil-fed rats. Total and free cholesterol levels in the liver of structured lipid-fed rats were lowered by 31 and 36%, respectively. The triglycerides in the serum and liver were decreased by 14 and 30%, respectively.

Anti-nociceptive Activity

Aqueous extract of the husk fiber, administered orally to mice at doses of 200 or 400 mg/kg, produced an inhibition of the acetic acid-induced writhing response.

Antioxidant Activity

Aqueous extract of the husk fiber, in cell culture, was active vs 2,2-diphenyl-1 -picryl-hydrazyl-hydrate radicals. Juice, in cell culture, was active vs 1,1 -diphenyl-2-picrylhydrazyl, 2,2'-azino-*bis*(3-ethylbenzthiazoline-6-sulfonic acid) and superoxide radicals but promoted the production of hydroxyl radicals and increased lipid peroxidation. The activity was most significant for fresh samples and diminished significantly when heated or treated with acid, alkali, or dialysis. Maturity of coconut drastically decreased the scavenging activity. Juice protected hemoglobin from nitrite-induced oxidation when added before the autocatalytic stage of the oxidation. Acid, alkali, or heat-treated or dialyzed juice showed a decreased ability in protecting hemoglobin from oxidation.

Antiparasitic Activity

Polyphenolic-rich extract of the husk fiber, in cell culture at a concentration of 10 μg/mL, produced a reduction approx 44% of the association index between peritoneal mouse macrophages and *Leishmania amazonensis* promastigotes. It inhibited the growth of promastigote and amastigote developmental stages after 60 minutes. There was a concomitant increase of 182% in nitric oxide production by the infected macrophage in comparison to nontreated macrophages.

Antiproteinemic Effect

Hydrogenated oil was administered orally to rats fed a protein-deficient diet at a dose of 200 g casein and 50 g coconut oil or 20 g casein and 50 g coconut oil for 28 days. The treatments produced a low concentration of protein and triacylglycerol (in serum VLDL, LDL–HDL 1 and 2–3), of cholesterol (in LDL–HDL1), and of phospholipids (in VLDL) in the protein-deficient groups. Relative amounts of linoleic and arachidonic acids in phospholipids of VLDL and HDL 2–3 were also lowered in the 20 g casein and 50 g coconut oil group. Coconut fat, administered to rats, produced an increase in plasma esterase-1 activity. Oil, administered to male Wistar rats at doses of 20% casein and 5% coconut oil or 2% casein and 5% coconut oil for 28 days, produced low concentration of protein, triacylglycerol, and VLDL in the plasma of 2% casein and 5% coconut oil group. Malondialdehyde content of the 2% casein and 5% coconut oil group was significantly higher than of control group. The lowest level of malondialdehyde was observed in the coconut oil group.

Anti-yeast Activity

Tincture of the dried fruit, on agar plate at a concentration of 30 μL/disc, was inactive on *Candida albicans*. Extract of 10 g plant material in 100 mL ethanol was used. Ethanol (50%) extract of the leaf, at a concentration of 25 μg/mL or more, was inactive on *Candida albicans* and *Cryptococcus neoformans*. Seed oil, on agar plate at a concentration of 0.05 mL, was active on *Candida albicans*.

Apo B Synthesis

Oil, administered to 1-month-old calves fed on a conventional milk replacer containing coconut oil, produced a twofold lower concentrations of total (^{35}S) proteins, (^{35}S) albumin, and (^{35}S) apo B in liver cells than in beef tallow-fed calves. The total amount of proteins secreted (including albumin) was similar in both groups. The amount of VLDL-(^{35}S) apo secreted was twofold lower in coconut oil-fed group. Oil, administered to rabbits at a dose of 14% of the diet for 4 weeks, produced an increase in HDL-cholesterol (C) level from 170 to 250% over chow-fed control, with peak differences occurring at 1 week. Plasma apo A-I levels were also increased from 160 to 180%. After 4 weeks, there was no difference in plasma VLDL-C or LDL-C levels in the both groups. Hepatic level of apo A-I mRNA was increased in the coconut-fed groups. Treatment of cultured rabbit liver cells and sera with various saturated fatty acids did not alter apo A-I mRNA levels as observed in vivo.

Atherosclerotic Influence

Oil was administered orally to rabbits fed a semipurified, cholesterol-free atherogenic diet at a dose of 14% fat: coconut oil (CNO), corn oil (CO), palm kernel oil (PO), and cocoa butter (CB). Serum cholesterol levels (mg/dL) at 9 months were CO, 64; PO, 436; CB, 220; and CNO, 474. HDL cholesterol (%) was CO, 37; PO, 8.6; CB, 25; and CNO, 7. average artherosclerosis (arch + thoracic/2) was CO, 0.15; PO, 1.28; CB, 0.53; and CNO, 1.60. Oil, administered to British Halflop rabbits at a dose of 3% of the diet, produced an increase of cholesterol, triacylglycerol, apo B, apo C-III, and apo E as compared to chow-fed rabbits. The apo A level did not differ from the chow-fed rabbits. There was a decrease in the low fractional catabolic rate of LDL-C. Oil, administered with cholesterol (30 g/kg) to male golden Syrian hamsters at a dose of 150 g/kg of diet, developed lipid-rich lesions in the ascending aorta and aortic arch after 4 weeks. The lesions continued to progress throughout the next 8 weeks. Removal of cholesterol from the diet halted this progression. Oil, administered without supplemental cholesterol in the same dose for 16 weeks, doubled the size of lesions in the ascending aorta and decreased linearly the lesion size in the aortic arch.

Blood Pressure Effect

Fruit juice, administered intravenously by infusion to dogs at a dose of 3 mL/minute for 100 minutes, was active. Initial effect was a decrease in blood pressure. Oil, administered to male weanling rats at a dose of 10% of diet for 5 weeks, produced significantly higher blood pressure than other groups. Systolic blood pressure was found related to the dietary intakes of saturated and unsaturated fatty acids. Prenatal exposure of the rats to a maternal low-protein diet abolished the hypertensive effect of the coconut oil diet.

Butyryl Cholinesterase Activity

Oil was administered to rats at different doses with or without clofibrate for 15 days. The hypolipidemic action of clofibrate was not influenced by the amount of fat. Clofibrate did not affect lower cholesterol concentration in rats fed the low fat diet, but it counteracted the rise in liver cholesterol seen in rats fed the high-fat diet. The high-fat diet produced slightly higher levels of butyryl cholinesterase in the small intestine but markedly raised intestinal esterase-1 activity.

Carcinogenic Activity

Coconut oil acid diethanolamine condensate in ethanol was administered externally to 10 male and 10 female F344/N rats at doses of 25, 50, 100, 200, or 400 mg/kg body weight five times per week for 14 weeks. All of the rats survived the study. Final mean body weights and body weight gains of the 200 and 400 mg/kg males and females were significantly less than those of the controls. Clinical findings included irritation of the skin at the site of application in the 100, 200, and 400 mg/kg males and females. Cholesterol concentrations were significantly decreased in 200 and 400 mg/kg males. Histopathological lesions of the skin at the site of application included epidermal hyperplasia, sebaceous gland hyperplasia, chronic active inflammation, parakeratosis, and ulcer. The incidences and severities of skin lesions generally increased with increasing dose in males and females. The incidences of renal tube regeneration in the 100, 200, and 400 mg/kg females were significantly greater than the vehicle control incidence, and the severities in the 200 and 400 mg/kg females were increased. Coconut oil acid diethanolamine condensate in ethanol was administered externally to 10 male and 10 female B6C3F1 mice at doses of the 50, 100, 200, 400, or 800 mg/kg body weight five times per week for 14 weeks. All mice survived until the end of the study. Final mean body weights and body weight gains of dosed males and females were similar to those of the vehicle controls. The only treatment-related clinical finding was irritation of the skin at the site of application in males and females administered the 800 mg/kg dose. Weights of the liver and kidney of 800 mg/kg males and females, the liver of the 400 mg/kg females, and the lung of the 800 mg/kg females were significantly increased compared to the controls. Epididymal spermatozoal concentration was significantly increased in the 800-mg/kg males. Histopathological lesions of the skin at the site of application included epidermal hyperplasia, sebaceous gland hyperplasia, chronic active inflammation, parakeratosis, and ulcer.

The incidences and severities of these skin lesions generally increased with increasing dose in males and females. Coconut oil acid diethanolamine condensate in ethanol was administered externally to 50 male and 50 female F344/N rats at doses of 50 or 100 mg/kg body weight five times a week for 104 weeks. The survival rates of treated male and female rats were similar to those of the vehicle controls. The mean body weights of dosed males and females were similar to those of the vehicle controls throughout most of the study. The only chemical-related clinical finding was irritation of the skin at the site of application in the 100-mg/kg females. There were marginal increases in the incidences of renal tubule adenoma or carcinoma (combined) in the 50-mg/kg females. The severity of nephropathy increased with increasing dose in the female rats. Nonneoplastic lesions of the skin at the site of application included epidermal hyperplasia, sebaceous gland hyperplasia, parakeratosis, and hyperkeratosis, and the incidences and severities of these lesions increased with increasing dose. The incidence of chronic active inflammation, epithelial hyperplasia, and epithelial ulcer of the forestomach increased with dose in female rats and the increases were significant in the 100-mg/kg group. Coconut oil acid diethanolamine condensate in ethanol was administered externally to 50 male and 50 female B6C3F1 mice at doses of 100 or 200 mg/kg body weight five times a week for 104 to 105 weeks. Survival of the mice was generally similar to that of the controls. Mean body weights of 100-mg/kg females from week 93 and 200-mg/kg females from week 77 were less than in the controls. The only clinical finding attributed to treatment was irritation of the skin at the site of application in males administered 200 mg/kg. The incidences of hepatic neoplasms (hepatocellular adenoma, hepatocellular carcinoma, and hepatoblastoma) were significantly increased in both sexes. Most of the incidences exceeded the historical control ranges. The incidences of eosinophilic foci in dosed groups of male mice were increased relative to that in controls. The incidences of renal tubule adenoma and renal tubule adenoma or carcinoma (combined) were significantly increased in the 200-mg/kg males. Several nonneoplastic lesions of the skin at the site of application were considered treatment related. Incidence of epidermal hyperplasia, sebaceous gland hyperplasia, and hyperkeratosis were greater in all dosed groups than in the controls. The incidence of ulcer in 200-mg/kg males and inflammation and parakeratosis in 200-mg/ kg females were greater than in the controls. The incidence of thyroid gland follicular cell hyperplasia in all of the dosed groups was significantly greater than those in the control groups.

Cardiovascular Effect

Coconut and coconut oil, administered to 32 coronary heart disease patients in 16 age- and sex-matched healthy controls with no difference in the fat, saturated fat, and cholesterol consumption, produced no effect. Hydrogenated oil, administered to young male Wistar rats, at a dose of 10% of diet for 10 weeks, produced an increase in the risk of ventricular arrhythmias under conditions of both ischemia and reperfusion. The incidence of ventricular fibrillation was 67% in the oil fed group. The time until the first occurrence of extrasystole, the incidence of ventricular tachycardia, and the incidence of reperfusion-induced ventricular fibrillation were influenced in a similar manner between groups. The fatty acid composition of myocardial tissue, the ratio of *n*-3 to *n*-6 fatty acids, and the double-bond index were significantly affected by the various diets. Oil was administered by gastric intubation to male Sprague–Dawley rats at a dose of 86:14 w/ w coconut oil/safflower oil for 10 days, containing either 0.8 mg Zn/ kg or 111 mg Zn/kg. Zinc-deficient rats that were fed coconut oil had higher concentrations of triglycerides and total fatty acids in the heart than the control rats. The concentrations of phospholipids and total cholesterol were not different between zinc-deficient and control rats. Concentrations of lauric acid, myristic acid, palmitic acid, palmitoleic acid, and oleic acid were 65 to 192% higher in the hearts of zinc-deficient rats fed coconut diet than in the control rats. The level of arachidonic acid in phospholipids, which may represent desaturation activity, was not different in the zinc-deficient rats and control rats. Oil was administered orally to rats fed a copper-deficient diet at doses of 10 g/100 g coconut oil or 10 g/100 g coconut oil with 1 g/100 g cholesterol. Rats fed the diet with coconut and cholesterol had left ventricular chamber volumes that were twofold larger than those of rats fed diet with coconut oil only.

Copper deficiency reduced left ventricular chamber volume only in rats fed coconut oil and cholesterol. The results indicated that preload and contractility in the hearts of coconut oil-fed rats were greater than cardiac response to cholesterol addition to the coconut oil diet. Hearts in copper-deficient rats fed coconut oil and cholesterol exhibited eccentric hypertrophy and ventricular dysfunction. Oil, administered orally to rats for 4 weeks, produced a significant decrease in 5'-nucleotidase, phosphodiesterase I and *p*-nitrophenylphosphatase activity of cardiac sarcolemma. Sarcolemma from coconut-fed rats contained a significantly lower concentration of total polyunsaturated fatty acids (PUFAs) and a higher concentration of total monounsaturated fatty acids than that from safflower-fed rats. The fatty acid composition of the phosphatidylcholine exhibited the largest alterations because of coconut oil feeding. No dietary effect was observed in the sarcolemma content of cholesterol and phospholipids.

Cholestatic Liver Disease

Medium-chain fat from the oil, administered to bile duct-ligated rats, resulted in a decrease of consumption of the fat source compared to control rats; however, carbohydrate and protein intakes were not affected. Body weight gain was significantly greater in coconut-fed rats than in rats fed long-chain fat (Crisco vegetable shortening). Mortality was 44% in bile duct-ligated animals fed the long-chain fat and 0% in those fed medium-chain fat.

Cholesterol Metabolism

Hydrogenated oil, administered orally to hamsters at a dose of 20% of diet for 4 weeks, induced hypercholesterolemia. Oil feeding had no effect on cholesterol synthesis but markedly inhibited cholesterol esterification in both the liver and the intestine. The diet-induced hypercholesterolemia was strongly correlated with an increase in acyl-CoA/cholesterol acyltransferase activity. The hypercholesterolemia increased aortic uptake of cholesterol and hence acyl-CoA/cholesterol acyltransferase activity. Coconut fat, administered orally to rabbits with partial ileal bypass, produced a significant increase of serum total cholesterol and phospholipids concentrations. The effect on serum lipids of the type of fat was similar in control and partial ileal bypass rabbits.

Coconut—a main source of energy for two Polynesian groups, Tokelauans and Pukapukans—was investigated. Tokelauans obtained a much higher percentage of energy from coconut than the Pukapukans, 63% compared with 34%, so their intake of saturated fat is higher. The serum cholesterol levels are 35–40 mg higher in Tokelauans than in Pukapukans. Analysis of a variety of food samples and human fat biopsies showed a high lauric (12:0) and myristic (14:0) acids content. Vascular disease is uncommon in both populations, and there is no evidence of the high saturated fat intake having a harmful effect in these populations. Oil, administered orally to growing male Sprague–Dawley rats fed a diet containing unoxidized or oxidized cholesterol (5 g/kg) with coconut or salmon oil (100 g/kg) for 5 weeks, produced an effect independent of the dietary fat. Hydrogenated coconut oil was administered to F(1)B Golden Syrian hamsters at a dose of 20 g/100 g for 10 weeks.

The hamsters were ranked according to their plasma VLDL and LDL cholesterol as a low or medium group. Low-coconut oil group had significantly higher aortic total and esterified cholesterol concentrations than the low-cholesterol-fed group. Hamsters in the low-cholesterolfed group had significantly higher aortic TNF-α concentrations than hamsters in the low-coconut oil group. Hamsters in the med-coconut oil group had significantly higher aortic IL-1-β concentrations than hamsters in the med-cholesterol group. Hamsters in both cholesterol fed groups had significantly lower plasma total cholesterol concentrations than hamsters in the low-coconut oil group.

Cholesterolemic Effect

Oil, administered to phospholipids transfer protein knockout (PLTP0)-deficient mice, produced an increase of phospholipids and free cholesterol in the VLDL-LDL region of PLTP0 mice. Accumulation of phospholipids and free cholesterol was dramatically increased in PLTP0/HL0 mice compared to PLTP0 mice. Turnover studies indicated that coconut oil was associated with delayed catabolism of phospholipids and phospholipids/free cholesterol-rich particles. Incubation of these particles with hepatocytes of coconut-fed mice produced a reduced removal of phospholipids and free cholesterol by scavenger receptor BI, even though scavenger receptor BI protein expression levels were unchanged.

Coagglutination Activity

Oil was administered to females at doses of a high-saturated fatty acids diet (HSAFA diet; 38.4% of energy from fat, polyunsaturated/saturated [P/S] fatty acid ratio 0.14) or low-saturated fatty acids diet (LSAFA; 19.7% of energy from fat, P/S ratio 0.17). The postprandial plasma concentration of tissue plasminogen activator (t-PA) antigen was decreased in the HSAFA-fed group. Plasma t-PA antigen was correlated with plasminogen activator inhibitor type 1 (PAI-1) activity when the participants consumed the HSAFA and LSAFA diet, although the diets did not affect the PAI-1 levels. There were no significant differences in postprandial variations in t-PA activity, factor VII coagulant activity, or fibrinogen levels as a result of the diets. Serum-fasting Lp(a) levels were lower in the HSAFA group and were lower in the LSAFA group. Serum Lp(a) concentrations did not differ when the women consumed the HSAFA and LSAFA diets.

Cytotoxic Activity

Oil, in cell culture at a concentration of 300 μg/mL, was active on Ca-colon-HT29. Coir fiber, administered intratracheally to guinea pigs, resolved the granulomas. Coir fiber and ash, in cell culture, produced a hemolytic activity and macrophage cytotoxicity more marked with ash compared with the fiber alone. Extract of the husk fiber, in cell culture at a concentration of 10 μg/mL, inhibited the growth of promastigote and amastigote developmental stages of *Leishmania amazonensis* after 60 minutes. Extract of the fiber husk (rich in catechins), in erythroleukemia cell line (K562) and normal human peripheral blood lymphocytes activated by phytohemagglutinin or phorbol ester, produced a dose-dependent effect. For phytohemagglutinin, this effect was irreversible, being already established on the first hours of culture. Oil, administered to C57B16 mice at a dose of 21% fat by weight, produced a significant decrease in macrophage-mediated death of P815 cells by nitric oxide and did not alter the death of L929 cell by macrophages. Liposaccharide-stimulated TNF-α production by macrophages decreased with increasing unsaturated fatty acid content of the diet: fish oil < safflower oil < olive oil < coconut oil < low-fat diet.

Desensitization Effect

Saline extract of the dried pollen, administered subcutaneously to 96 allergic adults at variable doses, produced a clinical improvement and decreased IgE levels.

Dietary Macronutrient Distribution Influence

Oil, administered to diet-induced overweight rats fed an energy-restricted diet with 60% of coconut oil, produced no difference between control and fat-fed group in weight loss and serum parameters. The coconut-fed group produced a greater reduction in the subcutaneous fat depot and of total body fat. Hepatic glycogen and glycogenic amino acid were altered in coconut-fed rats.

Diuretic Activity

Decoction of the dried fruit, administered nasogastrically to rats at a dose 1 g/kg, was active. Fruit juice, administered intravenously by infusion to dogs at a dose of 3 mL/minute for 100 minutes, produced weak activity. Ethanol extract of the leaf, administered intraperitoneally to saline-loaded male rats at a dose of 0.185 mg/kg, was active. Urine was collected for 4 hours after treatment.

Erythrocytic Effect

Hydrogenated coconut oil, administered orally to healthy rats for 10 weeks, produced a significant effect on five of the six classes of erythrocytes identified. The proportion of cells in each class was dependent on the diet. There was no significant effect of diet on erythrocyte filterability index and no statistical correlation between erythrocyte filterability index and morphology.

Exocrine Pancreatic Secretion

Oil, administered to piglets at a dose of 10 g/100 g of diet, did not affect the output of carboxylester hydrolase. Protein, chymotrypsin, carboxypeptidase A, elastase, and amylase outputs were not different among the dietary treatment groups. The outputs of trypsin and colipase were higher in the coconut-fed group. Oil, administered to three barrows at a dose of 15 g fat/100 g diet, produced a significant increase of chymotrypsin secretion.

Fatty Acid Composition Influence

Oil was administered orally to mice bearing the L1210 murine leukemia cells at a dose of 16% of diet. A microsome-rich fraction prepared from the L1210 cells produced more monoenoic fatty acids (37% vs 12%) compared to mice fed the sunflower oil. Oil, administered to chicken at concentrations of 10 and 20% of diet, produced changes in fatty acid composition of free fatty acid and triglyceride fractions of chick plasma parallel to that of the experimental diet. Plasma phospholipids incorporated low levels of 12:0 and 14:0 acids, whereas 18:0, the main saturated fatty acid of this fraction, increased after coconut oil feeding. The percentage of 18:2 acid significantly increased after coconut oil feeding. Oil, administered to rats at a dose of 15% of diet for 4 weeks, produced an increase of total cholesterol in the liver and the decrease of total liver phospholipids. Coconut/soy oil, administered by infusion into the duodenum of rats, produced the proportion of capric and lauric acids in the lymphatic triacylglycerol, reflecting the fatty acid composition in the diet. Results indicated that more than 50% of the capric and lauric acids could have been absorbed from the intestine as sn-monoacylglycerols. Hydrogenated coconut oil was administered to rats at a dose of 7% of the diet (essential fatty acids deficient in both *n*-6 and *n*-3) for 28 weeks (1 week gestation, 3 weeks lactation, and 24 weeks thereafter). The fatty acid compositional changes indicative of an essential fatty acid deficiency, such as the decreases in the levels of 18:2 *n*-6 along with an accumulation of 20:3 *n*-9 were observed in all the salivary glands. In the submandibular glands, the proportions of 16:1, 18:1 *n*-9 and 18:1 *n*-7 were higher in hydrogenated coconut oil-fed group than in the other groups. Some differences in the fatty acid composition of the three glands were found.

Fatty Acid Metabolism

Oil was administered to preruminant Holstein Friesian male calves fed a conventional milk diet containing coconut oil for 19 days. Fatty acid oxidation was determined by measuring the production of CO_2 (total oxidation) and acid-soluble products (partial oxidation). Production of CO_2 was 1.7–3.6-fold lower and acid soluble products tend to be lower in liver slices of coconut oil-fed than beef-tallow-fed calves. Fatty acid esterification as neutral lipids was 2.6- to 3.1-fold higher in liver slices of coconut oil-fed group. The increase in neutral lipid production did not stimulate VLDL secretion by the hepatocytes, leading to a triacylglycerol accumulation in the cytosol of the calves fed coconut oil. Oil, administered to preruminant calves fed a milk replacer containing coconut oil for 19 days, produced no significant difference in weight of the total body and tissue between the treated groups. Plasma glucose and insulin concentrations were lower in the coconut oil-fed group. Feeding on the coconut oil diet induced an 18-fold increase in the hepatic concentration of triacylglycerol. The perixomal oxidation rate of oleate was 1.5-fold higher in the hearts of calves fed coconut oil. The cytochrome C oxidase/citrate synthase activity ratio was lower in the liver of the coconut oil-fed animals. Oil, administered as a ratio 7:1 w/w coconut oil/safflower oil by gastric intubation to zinc-deficient rats, reduced δ-desaturase activity in liver microsomes of rats fed coconut oil. Zinc-deficient rats on the coconut oil diet had unchanged δ-6-desaturase activity with linoleic acid as substrate and lowered activity with α-linolenic acid as substrate.

Genotoxic Activity

Coconut oil acid diethanolamine condensate with or without S9 activation enzyme, in cell culture, produced no effect on *Salmonella typhimurium*. The treatment did not produce an increase in mutant L5178Y mouse lymphoma cell colonies, and no increased in the frequencies of sister chromatid exchanges or chromosomal aberrations in Chinese hamster ovary cells. In a peripheral blood micronucleus test in male and female mice from the 14-week test, positive results were obtained.

Genotype and Diet Effect

Coconut oil was administered to female Zucker rats throughout mating and lactation. Homozygous lean male and female rats, obese male, and lean heterozygous female rats were bred. Additional male rats were maintained on the same diet as their mothers until 11–12 days of age. Obese sucking rats had higher body weights than lean pups. Inguinal fat pad weights and pad-to-body weight ratios followed the pattern of obese greater than lean (FA/fa genes) pups that were greater than lean (FA/FA) pups. A similar relationship was found for adipose tissue lipogenic enzyme activities. At 11–12 weeks of age, measurements followed the general pattern of obese rats having greater value than lean rats (i.e., FA/fa = FA/FA). Coconut oil-fed fa/fa rats had lower hepatic lipogenic enzyme activities and lower fat cell numbers than safflower-fed fa/fa rats. High-fat diet did not result in a heterozygous effect in young adult lean male rats. Hydrogenated coconut oil, corn oil, or menhaden oil were administered to diabetes-prone BHE/ cdb and normal Sprague–Dawley rats. Both fat source and strain affected the temperature dependence of succinate-supported respiration. The transition temperature was greater in BHE/cdb rats than in the Sprague–Dawley rats. The efficiency of adenosine triphosphatase synthesis as reflected by the adenosine diphosphatase/O ratio was decreased in the BHE/cdb rats compared to Sprague–Dawley rats.

Glucose Metabolism

Hydrogenated oil, administered to prediabetic weanling BHE rats fed a 6% fat and 64% sucrose diet, produced an increase of the fractional irreversible glucose turnover rates, fractional glucose carbon recycling, hepatic fatty acid synthesis rates, adipose fatty acid synthesis rate, lower muscle glycogen, and lower rates of incorporation of glucose into muscle glycogen than corn oil-fed rats. The diet had no effect on glucose mass and space, hepatic glycogen, or blood glucose levels. Oil, administered to male Wistar rats at a dose of 25% by weight for 26 days, produced a significant increase in serum total cholesterol, LDLs, liver cholesterol, and liver weight. There was a decrease in serum HDLs, triacylglycerol levels, and abdominal fat weight; an increase in 3-hydroxy-3- methylglutaryl-CoA reductase activity in the liver; reduction in plasma lecithin-cholesterol acyltransferase activity; lower level of serum insulin; and liver glycogen in the coconut oil-fed rats. Glucose use was altered because lower glucose-6-phosphatase and increased glucokinase activities in the liver of coconut oil-fed rats were found. Coconut palm wine was administered to rats at a dose of 24.5 mL/kg body weight/d for 15 days before conception and throughout gestation. On the 19th day of gestation, hypoglycemia was observed in both wine- and ethanol-treated groups. Synthesis of glycogen was elevated on exposure to ethanol/wine, but its degradation was enhanced only in ethanol-exposed rats. Key enzymes of the citric acid cycle and gluconeogenesis were inhibited on administration in both groups. The activities of glycolytic enzymes were increased. Hydrogenated oil (6%), administered to BHE rats at a dose of 5% of diet, produced no influence on glucose. Hepatocytes isolated from rats fed coconut oil had a significantly lower affinity for insulin than menhaden oil-treated rats.

Glycemic Index

The glycemic index of different commonly consumed products supplemented with increasing levels of coconut flour was determined in 10 normal and 10 diabetic subjects. The test food with 200–250 g coconut flour/kg had significantly low, gastrointestinal (GI) (< 60); with 150 g coconut flour/kg had GI ranging from 61.3 to 71.4. A very strong negative correlation (r –0.85, $p < 0.005$) was observed between the GI and dietary fiber content of food. Oil, administered to lean Zucker rats at a dose of 5% of diet (87% saturated fatty acids) for 2 weeks, produced no differences in food intake, body weight, tissue level of glucagons-like peptide-1, plasma insulin, and glucagons levels in coconut oil- and olive oil-fed groups.

Hair Damage Prevention

Coconut oil application has a strong effect on hair as compared to sunflower and mineral oils. Among the three oils investigated, coconut oil was the only one that reduced the protein loss remarkably for both damaged and undamaged hair when used as a prewash and postwash grooming product.

Hemostatic Effect

Coconut water, in citrated plasma of eight healthy volunteers, was observed. Replacement of up to 50% of diluted plasma by water did not influence initiation of coagulation. Replacing 50% of citrated plasma by coconut water reduced maximum amplitude of thrombelastography recording dose by 39%.

Hepatic Activity

Hydrogenated oil, administered orally to male weanling rats at a dose of 10% of diet for 9 weeks, produced no significant alteration of hepatic 5-, 6-, and 9- desaturase activity. Addition of oil to the diet, simultaneously with lowering of the carbohydrate level, diminished the stimulatory effect of dietary sucrose vs glucose on 9-desaturase activity.

Hepatic Mitochondrial Effect

Oil, administered to chicken at a dose of 20% of the diet, produced a clear damage to the hepatic mitochondria accompanied by an accumulation of glycogen and lipid droplets in the hepatocyte cytoplasm. Pharmaceutical coconut oil induced a high percentage of cellular death when administered for 14 days. Fatty acid profiles in liver and hepatic mitochondria changed during 24 hours after pharmaceutical and cooking oils supplementation to the diet. The accumulation of shorter chain fatty acids (12:0) and (14:0) was higher after pharmaceutical than after cooking oil diet feeding. Mitochondrial ratios of saturated/unsaturated and saturated fatty acids (SUFAs)/PUFAs rapidly changed in parallel to these ratios in both diets. Most of the mitochondrial parameters measured recuperate to the control values when diets were supplied for 5–14 days. The maintenance of these ratios after 14 days pharmacological oil diet feeding was significantly higher than those in control.

Hyperalphalipoproteinemic Activity

Oil, administered orally to rabbits fed commercial chow or chow plus 14% w/w coconut oil, resulted in double increase the plasma levels of HDL cholesterol, phospholipids, and protein for up to 4 months without affecting HDL lipid and apoprotein composition. After 3 months also increased VLDL (107%) and LDL cholesterol (40%) levels, but the absolute increases in each of these lipoprotein fractions was less than half of that of HDL. Isotope kinetic studies of 125I-HDL protein indicated a double rate of production of HDL and no change in the efficiency of removal of HDL from plasma.

Hypercholesterolemic Activity

Seed oil, administered by gastric intubation to dogs, was active. Seed oil, administered orally to adults, was active. Oil, administered orally to rats at a dose of 79 g/day for 3 weeks, produced a significant increase of the final plasma cholesterol in groups that consumed yeast with coconut oil (91 mg/dL), in comparison to group consuming soybean protein with coconut oil (36 mg/dL). Coconut fat, administered to hyporesponsive and hyperresponsive inbred strains of rabbits with high or low response of plasma cholesterol to dietary SUFAs vs PUFAs, produced no influence on the efficiency of cholesterol absorption. Fat, administered orally to young normolipidemic males at a dose of 30% of the diet with a polyunsaturated/saturated fat ratio of 4 or 0.25 for 8 weeks, produced a significant increase of total plasma cholesterol level. Oil, administered to Mongolian gerbils, produced a hypercholesterolemia associated with elevations in VLDL, LDL, and HDL. The type of dietary fat did not influence lipoprotein composition and size. Fresh and thermally oxidized oil, administered to rats at a dose of 20% of diet, produced an increase of total cholesterol, LDL and VLDL cholesterol, and triacylglycerol and phospholipids levels and a decrease in HDL cholesterol. Oil, administered to 25 women at doses of 38.4% of energy from fat (HSAFA; PUFA/SUFA P/S ratio = 0.14) or 19.7% energy from fat (LSAFA; P/S ratio 0.17) for 3-week periods, produced no difference in serum total cholesterol, LDL cholesterol, and apo B concentrations between the diet periods. HDL cholesterol and apo A-I were 15% and 11%, respectively, higher during HSAFA diet period than during LSAFA diet period. The LDL/HDL-C and apo B/apo A-I ratios were higher during LSAFA diet period.

Hyperlipidemic Activity

Palm wine was administered orally to female albino rats at a dose of 24.5 mL/kg body weight/day for 15 days before conception and during pregnancy. On days 13 and 19 of gestation, liver function and hyperlipidemia were seen in the fetuses. Hyperlipidemia was caused by increased biosynthesis since the incorporation of ^{14}C acetate into lipids and activity of HMG-CoA reductase and lipogenic enzymes were elevated. Oil, administered to miniature pigs fed a pig chow supplemented with 17.1% of coconut oil for 30 d, produced a significant increase of cholesterol, triglyceride, HDL cholesterol and subfractions, LDL cholesterol and subfractions, and lipoprotein lipase activity in both genders. For cholesterol, triglyceride, HDL cholesterol (HDL-C and HDL[2]-C), LDL cholesterol (LDL-C, LDL[1 and 2]-C), and hepatic lipase, the female response to the diet was exaggerated compared to the male response.

Hyperthermic Effect

Coconut oil, administered orally to rats with human recombinant TNF-α at a dose of 190 g/kg for 12 weeks, produced an antihypothermic effect and changes in serum albumin and Cu content 8 hours after treatment and in muscle and liver protein after 24 hours. The results indicated that changes in ecosanoid metabolism may be involved in the modulatory effect of the coconut oil-enriched diet.

Hypertriglyceridemic Activity

Coconut oil, administered orally to rats at a dose of 14%:0.5% oil/cholesterol diet, produced an increase of lipids and apo B in the VLDL and intermediate-density lipoprotein fractions. The particle diameters of lipoproteins were similar in coconut oil- and olive oil-fed groups. The rates of triglyceride hydrolysis of both groups' VLDL by postheparin lipoprotein lipase in vitro were the same. The average fractional removal of apo B did not differ between diet groups. Oil, administered to rabbits at a dose of 14%: 0.5% coconut oil/cholesterol of diet, produced an increase of plasma triglycerides 15 times higher than basal level. Postprandial triglyceride responses after the first high-fat/ cholesterol meal were more prolonged in coconut-fed rabbits than in olive-fed group. Postprandial triglyceride responses after chronic coconut oil/cholesterol feeding were significantly greater compared to the olive oil-fed group. One coconut oil/cholesterol meal was associated with a 40% increase of postheparin plasma lipoprotein lipase activity and changed little in chronically fed coconut oil/cholesterol group.

Hypocholesterolemic Activity

Kernel protein, administered orally to rats on a coconut oil diet, lowered the levels of cholesterol, phospholipids, and triglycerides in the serum and most tissues when compared to casein-fed animals. The increase of hepatic degradation of cholesterol to bile acids and hepatic cholesterol biosynthesis and the decrease of esterification of free cholesterol were noted. In the intestine, cholesterogenesis was decreased. The kernel proteins also decreased lipogenesis in the liver and intestine.

Hypoglycemic Activity

Ethanol extract of the leaf, administered orally to rats at a dose of 250 mg/kg, produced less than a 30% drop in blood sugar level. Fruit juice, administered intravenously by infusion to dogs at a dose of 3 mL/min for 100 min, was active. Hot water extract of the dried shell, administered by gastric intubation to dogs at a dose of 200 mL/animal (20 g of air-dried plant material), produced weak activity. The neutral detergent fiber from kernel, administered orally to rats at doses of 5%, 15%, and 30% of diet, resulted in significant decrease in the level of blood glucose and serum insulin, with increasing in the intake of fiber. The increase of fecal excretion of Cu, Cr, Mn, Mg, Zn, and Ca was present. Neutral detergent fiber from coconut kernel, administered to rats at doses of 5, 15, and 30% of diet, produced an increase of fecal excretion of Cu, Cr, Mn, Mg, Zn, and Ca.

Hypolipidemic Activity

Protein, administered orally to hypercholesterolemic rats, reduced total, LDL, and VLDL cholesterol; triglycerides; and phospholipids levels in the serum and increased the level of serum HDL cholesterol. The concentration of total cholesterol, triglycerides, and phospholipids in the tissues was lower than in the control group. There was increased activity of superoxide desmutase and catalase. An increase of hepatic cholesterogenesis, conversion of cholesterol to bile acids and fecal excretion of bile acids, and excretion of urinary nitrate and an decrease of malonaldehyde level in the heart were observed. The neutral detergent fiber of kernel digested with cellulase and hemicellulase was administered orally to rats. Hemicellulose-rich fiber showed decreased concentration of total cholesterol and LDL and VLDL cholesterol and increased HDL cholesterol. Cellulose-rich fiber showed no significant alteration. There was increased HMG-CoA reductase activity and increased incorporation of labeled acetate into free cholesterol. Rats fed hemicellulose-rich fiber produced lower concentration of triglycerides and phospholipids and a lower release of lipoproteins into circulation. There were an increased concentration of hepatic bile acids and increased excretion of fecal sterols and bile acids.

Ileal Oleic Acid Uptake

Hydrogenated oil, administered to rats at a dose of 5 g/100 g of diet, produced saturable kinetics in ileal brush border membrane vesicles: $V_{max} = 0.23 \pm 03$ μmol/mg protein/5 minutes and $K_m = 196 \pm 50.3$ nmol for controls, and $V_{max} = 04 \pm 01$ μmol/mg protein/5 minutes and $K_m = 206 \pm 85.3$ nmol for coconut oil-fed group.

Immune Function

Coconut oil, administered to rats fed a fat-rich diet (corn oil) or a diet poor in linoleate (coconut oil) at high and low concentrations, completely abolished the responses to *Escherichia coli* endotoxin.

Intestinal Brush Border Membrane

Oil, administered orally to rats at a dose of 10% for 5 weeks, produced an increase in level of saturated fatty acids in the brush border membrane from coconut oil-fed animals. Membrane fluidity was as follows: coconut oil less than commercial pellet diet less than corn oil less than fish oil. The membrane hexose content was high in the coconut-fed rats. Hexamines were elevated in coconut- treated rat brush borders. The activities of alkaline phosphatase, sucrase, and lactase were increased.

Intestinal Esterase Activity

Oil was administered to rats at different doses with or without clofibrate for 15 days. The hypolipidemic action of clofibrate was not influenced by the amount of fat. Clofibrate did not affect lower cholesterol concentration in rats fed the low-fat diet, but it counteracted the rise in liver cholesterol seen in rats fed the high-fat diet. The high-fat diet produced slightly higher levels of butyryl cholinesterase in the small intestine but markedly raised intestinal esterase-1 activity.

Intestinal Neoplasia

Oil was administered to 8-week-old male Fischer rats divided into two groups of 60 each (sedentary and exposed to moderate exercise), at a dose of 21% of diet for 38 weeks. The exercising and sedentary rats fed coconut oil were significantly heavier than rats fed corn oil. In the rats fed coconut oil diet, nine carcinomas were recorded in the sedentary groups and five in the exercised rats, which developed significantly fewer neoplasms than corn oil-fed group.

Intestinal Transport

Coconut water in different stages of maturation, administered to rats, produced a jejunal water absorption (17 ± 0.45 μL/minutes/cm), sodium excretion (–1694 ± 296 μEq/minutes/cm), and glucose absorption (5212.70 ± 2098.47 μg%/minutes/cm) in all the stages studied.

Intravenous Hydration

The use of coconut water as a short-term intravenous hydration fluid for Solomon Island residents was investigated. Fresh young coconut water, administered to eight healthy male volunteers in three doses in separate trials representing 50, 40, and 30% of the 120% fluid loss at 30 and 60 minutes of the 2-hour rehydration period. The percent of body weight loss than was regained (used as index of percent rehydration) was 75 ± 5%. The rehydration index, which provided an indication of how much of what was actually ingested and used for body weight restoration, was 1.56 ± 0.14. There was no difference at any time in serum Na^+ and Cl^-, serum osmolality, and net fluid balance among the trials. Coconut water was significantly sweeter, caused less nausea and more fullness and no stomach upset, and was easier to consume in a larger amount compared to carbohydrate-electrolyte beverage and plain water. Water, administered to children with diarrhea, was inactive. The results indicated that coconut water composition, sodium and glucose concentrations, and osmolality values vary during maturation of the fruit. In no instance did the coconut water contain sodium and glucose concentrations of value as an oral rehydration solution.

Iron Bioavailability

Oil, administered to suckling rats dosed with ^{59}Fe-labeled diet, produced a higher percentage of ^{59}Fe in the blood than those fed other fat sources. Administration to weanling rats produced a significantly higher percentage of ^{59}Fe retention than rats fed a formula-blend fat diet.

Jejunal Oleic Acid Uptake

Hydrogenated oil, administered to rats at a dose of 5 g/100 g of diet, produced saturable kinetics in jejunal brush border membrane vesicles: V_{max} = 0.15 ± 01 μmol/mg protein/5 minutes, and K_m = 136 ± 29.1 nmol for controls, and V_{max} = 03 ± 01 μmol/mg protein/5 minutes and K_m = 124.5 ± 72.6 nmol for the coconut oil-fed group.

Lauric Acid Incorporation

Lauric acid (50%) from the oil, administered to rats for 6 weeks, produced no significant difference between the experimental distribution of triacylglycerol types and the random distribution, calculated from the total fatty acid composition.

Lipid metabolism

Oil, administered orally to female C57BL/6 mice weaned at 21 d of age at a dose of 15% w/w for 6 weeks, increased the total lipids, triglycerides, LDL and VLDL cholesterol, and thiobarbituric acid-reactive substances (TBARS) and reduced glutathione concentrations, without changes in phospholipids or total cholesterol concentrations compared to controls. The concentrations of total cholesterol, free and esterified cholesterol, triglycerides, and TBARS were increased in the macrophages of coconut-fed mice, whereas the content of total phospholipids did not change. The phospholipids composition showed an increase of phosphatidylcholine and a decrease of phosphatidylethanolamine. Incorporation of [^{3}H]-cholesterol into the macrophages and into the cholesterol ester fraction was increased. The coconut oil diet did not affect [^{3}H]-AA uptake, induced an increase in [^{3}H]-AA release, and enhanced AA mobilization induced by lipopolysaccharide. Oil, administered to 28 persons with moderately elevated cholesterol level, decreased total cholesterol and LDL cholesterol (6.4 ± 0.8 and 4.2 ± 0.7 mmol/L), respectively, compared to butter diet (6.8 ± 0.9 and 4.5 ± 0.8 mmol/L). Apos A-1 and B were significantly higher on coconut oil and on butter than on safflower oil. In the group as a whole, HDL did not differ significantly in the three diets, whereas levels in women fed coconut oil were significantly higher than in the safflower oil group. Triacylglycerol level was lower in coconut oil group, but results were significant statistically only in women.

Lipid Peroxide Formation Stimulation

Seed oil, administered to rats at a dose of 15% of diet for 6 weeks, was inactive on rat liver microsomes. Malondialdehyde concentration was unchanged among animals given different oils. Vitamin E level decreased among those fed soybean oil.

Lipogenetic Effect

Kernel protein, administered orally to rats on a coconut oil diet, decreased lipogenesis in the liver and intestine. The kernel proteins also lowered the levels of cholesterol, phospholipids, and triglycerides in the serum and most tissues when compared to casein-fed animals. There was an increase in hepatic degradation of cholesterol to bile acids and hepatic cholesterol biosynthesis and a decrease in esterification of free cholesterol. In the intestine, cholesterogenesis was decreased.

Lipoprotein Composition

Oil, administered to newborn chicken at a dose of 20% for 2 weeks, increased cholesterol concentration in all the lipoprotein fractions, whereas 10% coconut oil only increased cholesterol in LDL and HDL, an increase that was significant after 1 week of treatment. Similar results were obtained for triacylglycerol concentration after 2 weeks of treatment. Changes in phospholipids and total protein levels were less profound. Coconut oil decreased LDL and fluidity. Oil, administered with or without 0.5% cholesterol to 36 young male Syrian hamsters for 6 weeks, produced higher plasma total triglyceride and total cholesterol in coconut oil without cholesterol supplementation-fed group than in the fish oil-fed group. With cholesterol supplementation, there was no significant difference in plasma total triglyceride level among the three dietary groups. The hepatic cholesteryl ester content was higher, and there was lower liver microsomal acyl-CoA/cholesterol acyltransferase activity in the cholesterol-supplemented coconut oil group compared to other groups. There was no significant difference in the excretion of fecal neutral and acidic sterols among the three dietary groups.

Lipoprotein Lipase Activity

Oil, administered to preruminant calves, produced no effect on palmitate oxidation rate by whole homogenates and induced higher palmitate oxidation by intermyofibrillar mitochondria. Carnitine palmitoyltransferase I activity did not significantly differ between the groups. Heart and longissimus thoracis muscle of calves fed coconut oil had higher lipoprotein lipase activity but produced no differences in fatty acid-binding protein content or activity of oxidative enzymes.

Liver Function Effect

Palm wine was administered orally to female albino rats at a dose of 24.5 mL/kg body weight/day) for 15 days before conception and during pregnancy. On days 13 and 19 of gestation, liver function and hyperlipidemia were seen in the fetuses. Altered liver function was evidenced by the increased activity of alcohol dehydrogenase, aldehyde dehydrogenase, glutamic oxaloacetic transaminase (GOT) (aspartate amino transferase), and glutamic pyruvic transaminase (GPT) (alanine amino transferase).

Myocardial Infarction

Coconut oil, administered orally to rabbits with myocardial infarction induced by isoproterenol, produced a higher level of phospholipids in the heart and aorta. The concentrations of cholesterol and triglycerides were lower in the safflower oil fed group.

Nasal Absorption

Sucrose ester of coconut fatty acid in aqueous ethanol solution (sucrose cocoate SL-40) administered intranasally to anesthetized male Sprague–Dawley rats at a dose of 0.5% sucrose cocoate with insulin, produced a rapid and significant increase in plasma insulin level with a concomitant decrease in blood glucose levels. Administration of a dose of 0.5% sucrose cocoate with calcitonin produced a rapid increase in plasma calcitonin levels and a concomitant decrease in plasma calcium levels.

Nephrotoxic Activity

Fruit juice, administered by intravenous infusion to dogs at a dose of 3 mL/minute for 100 minutes, produced weak activity. Albuminuria was observed just before the end of infusion administration.

Neutrophil Functions

Oil, administered to rats 21 days old at a dose of 15% final fat content of the diet for 6 weeks, produced a reduction in spontaneous and phorbol myristate acetate-stimulated H_2O_2 generation in glycogen-elicited peritoneal neutrophils relative to neutrophils from rats fed the control diet. The activity of superoxide desmutase, glutathione peroxidase, and catalase did not change in animals fed the fat-rich diets. The initial rate of O_2 generation in both resting neutrophils and phorbol myristate acetate-stimulated cells was significantly reduced when animals were fed coconut oil.

Ophthalmic Absorption

Sucrose ester of coconut fatty acid in aqueous ethanol solution (sucrose cocoate SL-40), administered ophthalmically to anesthetized Sprague–Dawley male rats at a dose of 0.5% sucrose cocoate with insulin, produced an increase in plasma insulin level and a decrease in blood glucose levels.

Ornithine Decarboxylase Activity

Fixed oil (4.5%), *Clupeidae brevortia tyrannus* (4%), and *Zea mays* (1.5%); fixed oil (7.5%), *Clupeidae brevortia tyrannus* (1%), and *Zea mays* (1.5%); fixed oil (8.5%) and *Zea mays* (1.5%), administered orally to mice for 1 year, were active vs benzoyl peroxide-induced ornithine decarboxylase activity. Oil, administered to 30 ultraviolet (UV)-irradiated Sencar and SKH-1 mice at doses of 1/14% (A diet), 7.9/7.1% (B diet), and 15/0% (C diet) corn oil/coconut oil for 6 weeks, produced no increase in enzyme activity. The level of ornithine decarboxylase activity in the UV-irradiated mice fed diet A was significantly higher than in mice fed the B or C diet. In the SKH-1 mice, ornithine decarboxylase activity was increased by 3 weeks and was significantly higher in mice fed diet C than in mice fed diet A. There was no significant effect of dietary fat on UV-induced skin tumor incidence.

Oxidative DNA Damages

Oil, administered to Fischer F344 rats at a dose of 19.8% coconut oil and 2% corn oil for 12–15 weeks, produced an excretion of 8-oxo-7,8dihydro-2'-deoxyguanosine (8-oxodG) in male group equal to 954 ± 367 pmol/kg/ 24 hours in the coconut oil fed group compared to 403 ± 150 pmol/kg/24 hours in the control. Calculated per whole animal, the excretion was 328 ± 128 pmol/24 hours in the coconut oil-fed rats and 137 ± 51 pmol/24 hours in the control.

Phospholipidemic Effect

Oil, administered to phospholipids transfer protein knockout (PLTP0)-deficient mice, produced an increase of phospholipids and free cholesterol in the VLDL–LDL region of PLTP0 mice. Accumulation of phospholipids and free cholesterol was dramatically increased in PLTP0/HL0 mice compared to PLTP0 mice. Turnover studies indicated that coconut oil was associated with delayed catabolism of phospholipids and phospholipids/free cholesterol-rich particles. Incubation of these particles with hepatocytes of coconut-fed mice produced a reduced removal of phospholipids and free cholesterol by SRBI, even though SRBI protein expression levels were unchanged.

Plasma Fatty Acids

Oil, administered to chicken at doses of 10% and 20% of the diet, produced an increase in the percentages of lauric and myristic acids in free fatty acid and triacylglycerol fractions in chick plasma, whereas these changes were less pronounced in phospholipids and cholesterol esters. The percentage of arachidonic acid was higher in plasma phospholipids than in the other fractions and was drastically decreased by coconut oil feeding. Linoleic acid, the main fatty acid of cholesterol esters, was increased. Oil, administered to 37 children 1 year of age in the form of full vegetable-fat milk (3.5 g fat/dL, 100% vegetable fat from palm, coconut, and soybean oils), produced higher amounts of plasma linoleic acid and a plasma α-tocopherol concentrations than in the other milks tested. Oil, administered to male Wistar rats at a dose of 40% of diet for 2 months, increased apo A-I concentration in plasma and did not change apo A-I mRNA level.

Oil, administered to 38 healthy children in a form of full vegetable-fat milk (3.5 g fat/dL, 100% vegetable fat from palm, coconut, and soybean oils), produced a significantly lower percentage of SUFAs in plasma triglycerides than in children fed standard-fat milk. Plasma PUFA levels were significantly higher than in children fed standard-fat milk. Oil, administered to 41 healthy adults, produced a decrease of plasma lathosterol concentration, the ratio plasma lathosterol/cholesterol, LDL cholesterol, and apo B. Plasma total cholesterol, HDL cholesterol, and apo A levels were not significantly different between butter and coconut diets. Oil, administered to male golden Syrian hamsters at a dose of 15% w/w for 4 weeks, produced the highest triglyceride levels of the diets studied. Oil, administered to male golden Syrian hamsters at a dose of 4 g/kg of diet (12:0 and 14:0) for 7 weeks, produced the highest plasma cholesterol concentration compared to rapeseed and sunflower seed oil diets. Biliary lipids, lithogenic index, and bile acid profile of the gallbladder bile did not differ significantly among the six diets.

Platelets Aggregation Stimulation

Fruit juice, administered intravenously by infusion to dogs at a dose of 5 mL/min, was active. Total infusion was 300 mL. Oil, administered orally to six New Zealand white rabbits fed a commercial diet supplemented with 60 g/kg of coconut oil low in all PUFA for 60 days, produced a platelets aggregation induced by both thrombin and collagen significantly lower with either fish or linseed oil (*n*-3 PUFA), than with corn oil (*n*-6 PUFA) or the low PUFA coconut oil.

Prostaglandin Outflow

Oil was administered to weanling male rats fed *ad libitum* a semisynthetic diet supplemented with 10% by weight of primrose oil, replaced partly or completely (25, 50, 75, or 100%) by hydrogenated coconut oil for 8 weeks. The release of prostanoids from the mesenteric vasculature was significantly reduced in the animals on the diet with the oil replaced by coconut oil.

Protective Effect of Vitamin A

Vitamin A, 240,000 IU, predissolved in 11.7 g of coconut oil and bolused directly into the rumen of mature wethers along with 4 g of chromic oxide or predissolved in 11.7, 23.4, or 35 g of coconut oil, produced significantly higher recoveries of vitamin A when dissolved in coconut oil (55.6%) compared to safflower oil (35.5%). Recoveries in abomasal digesta increased linearly with the amount of carrier coconut oil.

Semen Cryopreservation

Coconut water extender, administered with glycerol to the semen of six adult dogs at concentrations of 4, 6, and 8%, produced satisfactory effect. There was no difference among groups in motility and vigor. A smaller percentage of total and secondary abnormalities were observed using 6% glycerol.

Sensitization (Skin)

Fruit juice, administered subcutaneously to guinea pigs at a dose of 02 mg/animal, was active on skin. Edema occurred at the site of injection and recovered within 200 minutes. Fruit juice, administered subcutaneously to adults at a dose of 02 mg/person, was active on skin. Inflammation occurred at the site of injection and recovered within 90 minutes. Aqueous extract of the husk fiber, administered externally to rabbits, produced no significant dermic or ocular irritation.

Sickness Behavior

Hydrogenated oil, administered to Swiss Webster mice at a dose of 17% w/w for 6 weeks, produced the bioactivity of plasma TNF-α equal to 32.6 ± 3.6 ng/mL in mice fed coconut oil diet compared to mice fed fish oil (98.2 ± 5.1 ng/mL).

Spasmogenic Activity

Ethanol (95%) extract of the fresh leaf and stem, administered to guinea pigs at a dose of 0.5 mL/L, was active on ileum. Water extract of the fresh leaf and stem, administered intraperitoneally to guinea pigs at a dose of 0.5 mL/L, was active on ileum.

Subcellular Membrane-bound Enzymes Activity

Oil, administered to male CFY weanling rats at a dose of 20% for 16 weeks, produced an increase of synaptosomal acetylcholinesterase activity in the coconut oil-fed group. The Mg^{2+}-adenosine triphosphate (ATPase) activity was similar among all groups in all the brain regions.

Toxicity Assessment

Ethanol extract of the leaf, administered intraperitoneally to mice, was active, LD_{50} 0.75 g/kg. Ethanol extract of the fresh leaf and stem, administered intraperitoneally to mice at the minimum toxic dose of 1 mL/animal, was active. Water extract of the fresh leaf and stem, administered intraperitoneally to mice at the minimum toxic dose of 1 mL/animal, was active. Aqueous extract of the husk fiber, administered orally to mice, was active, LD_{50} 2.30 g/kg.

Tricarboxylate Carrier Influence

Oil, administered to rats at a dose of 15% of the diet for 3 weeks, produced a differential mitochondrial fatty acid composition and no appreciable change in phospholipids composition and cholesterol level. Compared with coconut oil-fed rats, the mitochondrial tricarboxylate carrier activity was markedly decreased in liver mitochondria from fish oil-fed rats. No difference in the Arrhenius plot between the two groups was observed.

Tumor Prevention

The effect of kernel fiber on metabolic activity of intestinal and fecal β-glucuronidase activity during 1,2-dimethylhydrazine (DMH)-induced colon carcinogenesis was studied. Inclusion of fiber supported lower specific activity and less fecal output of β-glucuronidase than did the fiber-free diet. Kernel, administered to animals treated with DMH, resulted in higher average weight. A decrease of cholesterol and increase of phospholipids and cholesterol/phospholipids ratio in most of tissues was found. HMG-CoA reductase activity was decreased in most of the tissues of the kernel and DMH, kernel and chili, and kernel, chili, and DMH groups. Histopathological studies showed that kernel-fed animals had fewer papillae, less infiltration into the submucosa, and fewer changes in the cytoplasm with decreased mitotic figures. Oil, administered to rats for 4–8 weeks, produced a small but statistically insignificant reduction in TNF production. After 8 weeks, coconut oil suppressed production of the cytokine. Coconut oil produced no modulatory effect on the interleukin production. Oil, administered orally at a dose of 20% of diet to virgin female Balb/c mice treated with 7,12-dimethylbenz[a]anthracene (DMBA), produced no effect on body weight, feed intake, or survival to 44 weeks of age and 36 weeks after the six DMBA doses. Mammary tumor incidence was the same in the coconut oil or menhaden oil but significantly higher in the corn oil group. Oil was administered to female Wistar rats before mating and throughout pregnancy and gestation, and the male offspring were supplemented from weaning until 90 days of age. They were inoculated subcutaneously with Walker 256 tumor cells. Supplementation of the diet with coconut oil did not change cancer cachexia, except for a small decrease in serum triacylglycerol concentration.

Tumor-promoting Effect

Fixed oil (4.5%) with 4% *Clupeidae brevortia tyrannus* and 1.5% *Zea mays*; 7.5% fixed oil, 1% *Clupeidae brevortia tyrannus*, and 1.5% *Zea mays*; and 8.5% fixed and 1.5% *Zea mays*, administered to mice in the diet for 52 weeks, were active. Tumors were initiated with dimethylbenzanthracene and promoted with benzoyl peroxide for 52 weeks. Oil, administered orally to female Sencar mice at doses of 5, 10, 15, and 20%, with addition of 5% corn oil for 1 week after initiation with 7,12-dimethylbenzanthracene and 3 weeks before the start of promotion with 12-0-tetradecanoylphorbol- 13-acetate, produced no significant difference in latency or incidence of papillomas or carcinomas between the saturated fat diet groups. Oil, administered to 30 Sencar and SKH-1 mice at doses of 1:14% (A diet); 7.9%:7.1% (B diet) and 15:0% (C diet) corn oil/coconut oil for 3 weeks before UV irradiation, produced tumor incidence that reached a maximum of 60, 60, and 53% for diets A, B, and C, respectively, with an average one to two tumors per Sencar mouse. For the SKH-1 mice, the diet groups reached 100% incidence by 29 weeks, with approx 12 tumors per mouse. No significant effect of dietary fat was found for tumor latency, incidence, or yield in either strain. Oil (17%) with 3% of sunflower seed oil, administered to DMBA-treated female Sprague–Dawley rats in the diet, produced twice as many tumors as those fed 3% sunflower seed oil or 20% of either saturated fat alone. Tumor yields in the rats fed these mixed-fat diets were comparable to rats fed a 20% lard diet, which provided about the same amount of linoleic acid.

Uncoupling Protein Expression

Oil, administered to female Wistar rats fed *ad libitum* a high-fat diet with coconut oil for 7 weeks, promoted an increase in body fat content, body weight, and uncoupling protein levels. At the completion of experiment I, oil was administered to high-fat diet rats for 3 weeks. Adipose depots were strongly reduced in the rats fed the high fat diet enriched with coconut oil. Specific uncoupling protein was 3.4 times higher than in controls.

Vascular Permeability Increased

Fixed oil (4.5%), 4% *Clupeidae brevortia tyrannus*, and 1.5% *Zea mays*; 7.5% of fixed oil, 1% *Clupeidae brevortia tyrannus*, and 1.5% *Zea mays*; 8.5% fixed oil and 1.5% *Zea mays*, administered to mice in the diet for 52 weeks, was active vs vascular permeability induced by benzoyl peroxide.

7

FERULA ASSAFOETIDA

Ferula assafoetida is an herbaceous, monoecious, perennial plant of the Umbelliferae family. It grows to 2 m high with a circular mass of leaves. Flowering stems are 2.5–3 m high and 10 cm thick, with a number of schizogenous ducts in the cortex containing the resinous gum. Stem leaves have wide sheathing petioles. Compound large umbels arise from large sheaths. Flowers are pale greenish yellow. Fruits are oval, flat, thin, reddish brown and have a milky juice. Roots are thick, massive, and pulpy. It yields a resin similar to that of the stems. All parts of the plant have the distinctive fetid smell.

ORIGIN AND DISTRIBUTION

Asafoetida is native to central Asia, eastern Iran to Afghanistan, where it grows from 600 to 1200 m above the sea level. Although not native to India, it has been used in Indian medicine and cookery for ages. Today it is grown chiefly in Iran and Afghanistan, from where it is exported to the rest of the world.

TRADITIONAL USES

Afghanistan. Hot water extract of the dried gum is taken orally for hysteria and whooping cough and to treat ulcers.

Brazil. Hot water extract of the dried leaf and stem is taken orally by males as an aphrodisiac. Extract is taken orally as nerve and general tonics. Oleoresin powder, crushed with the fingertips, is used as a condiment.

China. Decoction of the plant is taken orally as a vermifuge.

Egypt. Dried gum is applied vaginally as a contraceptive before or after coitus. Fifty-two percent of the women interviewed practiced this method, and 48% of them depended on indigenous methods and/or prolonged lactation. Hot water extract of the dried root is taken orally as an antispasmodic, a diuretic, a vermifuge, and an analgesic.

Fiji. Paste made from the dried resin is applied to the chest for whooping cough. Fried *Ferula* is taken with *Allium sativum* and sugar to cleanse the new mother. Fried *Ferula*, *Piper nigrum*, and *Cinnamonum camphora* is taken orally for headache and toothache. Hot water extract of the dried resin is taken orally for upset stomach.

India. Extract of dried *Ferula assafoetida* with *Brassica alba* and rock salt is diluted with vinegar and taken orally as an abortifacient. Hot water extract of the dried gum is taken orally as a carminative, an antispasmodic, and an expectorant in chronic bronchitis. Mixed with cayenne pepper and sweet flag, it is used as a remedy for cholera. Exudate of the dried gum resin is eaten to prevent guinea worm disease. Gum resin with salt and the bark juice of *Moringa pterygosperma* is used externally for stomachaches. A dry *Lampyris noctiluca* without head is mixed with 200–300 mg of *Ferula* and taken mornings and evenings for gallstones and kidney stones. For old stones, potassium nitrate is added to the mixture. Hot water extract of the dried resin is taken orally as an emmenagogue.

Malaysia. Gum is chewed by females for amenorrhea.

Morocco. Gum is chewed as an antiepileptic.

Nepal. Water extract of the resin is taken orally as an anthelmintic.

Saudi Arabia. Dried gum is used medicinally for whooping cough, asthma, and bronchitis.

United States. Fluid extract of the resin is taken orally as an emmenagogue, a stimulating expectorant, an anthelmintic, an aphrodisiac, and a stimulant to the brain and nerves. It is claimed to be a powerful antispasmodic.

Medicinal Uses

Allergenic Activity

Oleoresin powder, administered externally to adults, was active. Reactions to patch test occurred most commonly in patients who were regularly exposed to the substance, or who already had dermatitis on the fingertips. Previously unexposed patients had few reactions (i.e., no irritant reactions).

Antibacterial Activity

Dried gum resin, on agar plate, was active on *Clostridium perfringens* and *Clostridium sporogenes*.

Anticarcinogenic Activity

Dried resin, administered orally to Sprague–Dawley rats at doses of 1.25 and 2.5% w/w of the diet, produced a significant reduction in the multiplicity and size of palpable *N*-methyl-*N*-nitrosourea-induced mammary tumors, and a delay in mean latency period of tumor appearance. Oral administration to mice increased the percentage of life span by 52.9%. Intraperitoneal administration did not produce any significant reduction in tumor growth. The extract also inhibited a two-stage chemical carcinogenesis induced by 7,12-dimethylbenzathracene and croton oil on mice skin with significant reduction in papilloma formation.

Antifertility Effect

A mixture of *Embella ribes* fruit, *Piper longum* fruit, borax, *Ferula* dried gum, *Piper betle*, *Polianthes tuberosa*, and *Abrus precatorius*, administered orally to female adults at a dose of 0.28 g/person starting from the second day of menstruation twice daily for 20 days, without sexual intercourse during the dosing period, produced the effect for 4 months. The biological activity reported has been patented[FA049]. Gum, administered by gastric intubation to male mice at a dose of 5 mg/kg for 32 days, was active. Methanol extract of the resin, administered orally to Sprague–Dawley rats at a dose of 400 mg/kg daily for 10 days, prevented pregnancy in 80% of the rats. When administered as a polyvinylpyrrolidone 1:2 complex, 100% pregnancy inhibition was observed at this dose. Lower doses of the extract produced a marked reduction in the mean number of implantations. Significant activity was observed in the hexane and chloroform eluents of sulfur-containing extract in an immature rat bioassay, the methanol extract was devoid of any estrogenic activity.

Antifungal Activity

Ethanol (95%) extract of the dried gum on agar plate was active. Essential oil of rhizome, on agar plate at a concentration of 400 ppm, was active on *Microsporum gypseum* and *Trichophyton rubrum*, and produced weak activity on *Trichophyton equinum*. Extract of asafetida, on agar plate at concentrations of 5–10 mg/mL, inhibited *Aspergillus parasiticus* aflatoxin production.

Antihepatotoxic Activity

A mixture of the methanol-insoluble fraction of the dried resin, fresh garlic, curcumin, ellagic acid, butylated hydroxytoluene, and butylated hydroxyanisole, administered by gastric intubation to ducklings at a dose of 10 mg/animal, was active vs aflatoxin B1-induced hepatotoxicity.

Antihypercholesterolemic Activity

Gum, administered to female rats at a concentration of 1% of diet, was inactive. A hot mixture of *Nigella sativa*, *Commiphora myrrha*, *Ferula assafoetida*, *Aloe vera*, and *Boswellia serrata*, administered by gastric intubation to rats at a dose of 0.5 g/kg for 7 days, was active vs streptozotocin-induced hyperglycemia.

Antihyperglycemic Activity

Hot water extract of the dried gum, *Nigella sativa*, *Myrrhis odorata*, and *Aloe* sp. in equal parts, administered by gastric intubation to rats at a dose of 10 mL/kg for 7 days, was active vs streptozotocin-induced hyperglycemia. Results were significant at $p < 0.05$ level. Hot water extract of the dried gum, administered by gastric intubation to rats at a dose of 10 mL/kg for 7 days, was inactive. A hot mixture of *Nigella sativa*, *Commiphora myrrha*, *Ferula assafoetida*, *Aloe vera*, and *Boswellia serrata*, administered by gastric intubation to rats at a dose of 0.5 g/kg for 7 days, was active vs streptozotocin-induced hyperglycemia.

Antimutagenic Activity

Water extract of the dried gum, on agar plate at a concentration of 2 mg/plate, was inactive on *Salmonella typhimurium* TA100 vs aflatoxin B1-induced mutagenesis and a concentration of 10 mg/plate, was inactive on *Salmonella typhimurium* TA98. Asafoetida, on agar plate at a dose of 0.5 μg/plate was active on *Salmonella typhimurium* TA98 and TA100 vs aflatoxin B1-induced mutagenesis. Asafoetida, on agar plate, was active on *Salmonella typhimurium* TA100 and TA1535 microsomal activation-dependent mutagenicity of 2-acetamidofluorene.

Antioxidant Activity

Asafetida, administered orally to Sprague–Dawley rats at doses of 1.25% and 2.5% w/w, significantly restored the level of antioxidant system, depleted by *N*-methyl-*N*-nitrosourea treatment. There was a significant inhibition in lipid peroxidation as measured by thiobarbituric acid-reactive substances in the liver of rats.

Antispasmodic Activity

Gum extract, administered to isolated guinea pig ileum at a dose of 3 mg/mL, produced a decrease of spontaneous contraction to 54 ± 7% of control. Exposure of precontracted ileum by acetylcholine, histamine, and KCl to *Ferula* gum extract produced a concentration-dependent relaxation. Preincubation with indomethacin, propanolol, atropine, and chlorpheniramine before exposure to the gum, did not produce any relaxation.

Antitumor Activity

Water extract of the dried oleoresin, administered by gastric intubation to mice at a dose of 50 mg/animal daily for 5 days, was active on CA-Ehrlich ascites, 53% increase in life span (ILS). Water extract administered intraperitoneally was inactive on Dalton's lymphoma, 4.8% ILS, and CA-Ehrlich ascites, 5.5% ILS.

Antiulcerogenic Activity

Colloidal solution, administered orally to rats at a dose of 50 mg/kg, 60 minutes before experiment, produced significant protection against gastric ulcers induced by 2 hours cold restraint stress, aspirin, and 4 hours pylorus ligation.

Chemomodulatory Influence

Asafetida, administered orally to Sprague–Dawley rats at doses of 1.25% and 2.5% w/w in diet, produced an increase in the development and differentiation of ducts/ductules and lobules and a decrease in terminal end buds as compared to both normal and *N*-methyl-*N*-nitrosourea-treated control animals. Asafetida treatment significantly reduced the levels of cytochrome P450 and b5. There was an enhancement in the activities of glutathione-*S*-transferase, deoxythymidine-diaphorase, superoxide desmutase, catalase, and reduced glutathione.

Cytotoxic Activity

Ethanol (90%) extract of the dried plant, in cell culture administered at a concentration 0.25 mg/mL, was active on human lymphocytes. The extract was active on Vero cells, effective dose (ED_{50} 0.15 mg/mL; Chinese hamster ovary (CHO) cells, ED_{50} 0.575 mg/mL; and Dalton's lymphoma, ED_{50} 0.6 mg/mL. Water extract of the dried gum, in cell culture at a concentration of 500 μg/mL, produced weak activity on CA-mammary-microalveolar cells.

Fibrinolytic Activity

Ether extracts of the dried gum and gum resin, administered orally to 10 healthy subjects fed 100 g of butter to produce alimentary hyperlipemia, were active.

Hypocholesterolemic Activity

A hot mixture of *Nigella sativa*, *Commiphora myrrha*, *Ferula assafoetida*, *Aloe vera*, and *Boswellia serrata*, administered by gastric intubation to rats at a dose of 0.5 g/kg for 7 days, was active vs streptozotocin-induced hyperglycemia.

Hypoglycemic Activity

Hot water extract of the dried gum, *Nigella sativa*, *Myrrhis odorata*, and *Aloe* sp. in equal parts, administered by gastric intubation to rats at a dose of 10 mL/kg for 7 days, was active. Results were significant at $p < 0.001$ level. Hot water extract of the dried gum, administered by gastric intubation to rats at a dose of 10 mL/kg for 7 days, was inactive. A hot mixture of *Nigella sativa*, *Commiphora myrrha*, *Ferula assafoetida*, *Aloe vera*, and *Boswellia serrata*, administered by gastric intubation to rats at a dose of 0.5 g/kg for 7 days, was active vs streptozotocin-induced hyperglycemia.

Hypolipemic Activity

A hot mixture of *Nigella sativa*, *Commiphora myrrha*, *Ferula assafoetida*, *Aloe vera* and *Boswellia serrata*, administered by gastric intubation to rats at a dose of 0.5 g/kg for 7 days, was active vs streptozotocin-induced hyperglycemia.

Hypotensive Activity

Tincture of the gland, administered intravenously to rabbits, was active. Water extract of the dried gum resin, administered intravenously to dogs at variable doses, was active. Gum extract, administered to anaesthetized rats at doses of 0.3–2.2 mg/100 g body weight, significantly reduced the mean arterial blood pressure.

Mutagenic Activity

Ethanol (95%) extract of the dried resin, on agar plate at a concentration of 15 mg/plate, produced weak activity on streptomycin-dependent strains of *Salmonella typhimurium* TA98. Metabolic activation has no effect on the result. Resin, on agar plate at a concentration of 200 µg/plate, was active on *Salmonella typhimurium* TA1537 and inactive on *Salmonella typhimurium* TA1538 and *Salmonella typhimurium* TA98.

Olfactory Status Influence

Asafoetida extract, administered to allergic (group I) and nonallergic rhinitis (group II) patients at a dose of 10% aqueous solution, produced an elevation of olfactory thresholds by 55.8% in group I and 66.8% for both groups.

Pancreatic Digestive Enzymes Effect

Asafetida, administered orally to albino rats at a dose of 250 mg% for 8 weeks, enhanced pancreatic lipase activity, stimulated pancreatic amylase and chymotrypsin. The stimulatory influence was not observed when their intake was restricted to a single oral dose.

Sister Chromatid Exchange Stimulation

Gum, administered by gastric intubation to mice at a dose of 1 g/kg, was active. The results were significant at p less than 0.01 level. A dose of 0.5 g/kg, produced weak activity. Asafoetida, administered orally to mice, produced weak activity in spermatogonia.

Toxic Effect

Gum, administered orally to adults, was active. A case of methemoglobinemia occurred in a 5-week-old male infant, after administration of asafetida preparation to alleviate colic. Treatment was with intravenous methylene blue and the infant recovered.

Tumor-promoting Activity

Water extract of the dried oleoresin, administered externally to mice at a dose of 200 μL/animal, was active vs 7,12-dimethylbenz[a]anthracene and croton oil treatment.

Uterine Stimulant Effect

Hot water extract of the plant, administered to female rats, was inactive on estrogen of uterus. Extract administered to pregnant rats, was inactive on uterus.

8

Coffea arabica

Coffee is a medium-size tree of Rubiaceae family. The plants can live up to 25 years and grows to a height of 6–15 m; commercially are kept to the height of 175–185 cm. The leaf is developed from the axil and arranged in pairs. The leaves on the main trunk develop in pairs and spirally, whereas leaves from the branch develop in a fan-like manner. The size of the mature leaf of *Liberica* coffee is approx 15-30 cm × 5-15 cm, with 7–10 veins. The dorsal surface is smooth and shiny. The mature leaf of the *Robusta* coffee is about the same size, except that it has 8–13 veins, whereas the dorsal surface is shiny and wavy. The tree starts flowering at the age of 18–36 months. The flowers develop from the axil of the leaves in the form of several in a bunch. Coffee berries are green when immature and turn yellow and red at maturity and ripening. Usually, each berry will contain two cotyledon or beans. In the case of a single cotyledon, it is called peaberry. The time of maturity is approx 8–13 months for *Liberica* and 9–10 months for *Robusta*. Fruits and beans are round, 0.8–1.5 cm (*Robusta*) and 2–2.5 cm (*Liberica*), bean size 0.7–0.9 cm (*Robusta*) and 1.3–1.5 cm (*Liberica*).

Origin and Distribution

Coffee originated from the tropical region of the African continent. In the first centuries, it was cultivated in Arabic countries: Aden and Yemen, later in Iran and India. At the end of the 17th century, the Dutch started to cultivate coffee on Jawa, Ceylon, and Surinam. In the 18th century, it was cultivated in Latin America and Brazil, then in Kenya, Tanzania, Malawa, and Uganda. Main producers currently are Brazil and Columbia.

TRADITIONAL USES

Brazil. Decoction of the seed is taken orally for influenza.

Cuba. Hot water extract of the seed is taken orally by males as an anaphrodisiac.

Haiti. Decoction of the grilled fruit and leaf is taken orally for anemia, edema, asthenia, and rage. The fruit is taken orally for hepatitis and liver troubles. The soaked fruit is used externally for nervous shock. For headache, the leaf decoction is taken orally or the leaf is applied to the head.

Mexico. The leaves are made into a poultice and used to treat fever. Hot water extract of the roasted seed is taken orally by nursing mothers to increase milk production.

Nicaragua. Leaves are used externally for headache, and the hot water extract is taken orally for stomach pain. Decoction of the seed is taken orally for fever and used externally for cuts and hemorrhage.

Peru. Hot water extract of the dried fruit is taken orally as a stimulant for sleepiness and drunkenness. Infusion of the leaf is taken orally to induce labor, and the hot water extract is taken orally as an antitussive in flu and lung ailments.

Thailand. Hot water extract of the dried seed is taken orally as a cardiotonic and neurotonic.

West Indies. Hot water extract of the seed is taken orally for asthma. Root juice is taken orally for scorpion sting.

MEDICINAL USES

Abortifacient Effect

In a population-based, case-controlled study of early spontaneous abortion in Sweden, 562 women who had spontaneous abortion at 6–12 weeks of gestation and 953 women who did not have abortion, indicated that the ingestion of caffeine may increase the risk of an early spontaneous abortion among nonsmoking women carrying fetus with normal karyotypes. Information on the ingestion of caffeine was obtained from in-person interviews. Plasma cotinine was measured as an indicator of cigarette smoking, and fetal karyotypes were determined from tissue samples. Multivariate analysis was used to estimate the relative risks associated with caffeine ingestion after adjustment for smoking and symptoms of pregnancy, such as nausea, vomiting, and tiredness. Among the nonsmokers, more spontaneous abortions occurred in women who ingested at least 100 mg of caffeine per day than in women who ingested less than 100 mg/day, with the increase in risk related to the

Fig. 8.1. A–A branch of coffee with clusters of berries in the leaf axils; B–a cluster of jasmine-like flowers; C–an individual flower.

amount ingested; 100–299 mg/day: odds ratio, 1.3; 300–499 mg/day: odds ratio 1.4; 500 mg or more per day: odds ratio 2.2. Among smokers, caffeine ingestion was not associated with an excess risk of spontaneous abortion. When the analysis was stratified according to the results of karyotyping, the ingestion of moderate or high levels of caffeine was associated with an excess risk of spontaneous abortion when the fetus had a normal or unknown karyotype but not when the fetal karyotype was abnormal.

Allergenic Activity

Ground coffee contains polyphenol haptens that activate the factor XII (Hageman factor)-dependent pathways of coagulation, fibrinolysis, and kinin generation in normal human plasma. Extract of the dried aerial part, administered by inhalation to female adult who developed rhinitis and conjunctivitus on exposure to coffee plant, was active. A skin prick test and rhinoconjunctival provocation test to coffee leaf allergen extract were positive. Seeds, administered by inhalation to adults at variable doses, were active. A 37-year-old worker in a coffee-roasting facility developed rhinoconjunctivis as a result of exposure to the dust of green, unroasted coffee bean. Extract of the dried seed, administered by inhalation to female adults, was active. Extract of the fresh seed, administered by inhalation to females with rhinitis and conjunctivitus, produced a positive skin prick test and rhinoconjunctival provocation test to coffee leaf allergen extract.

α-Amylase inhibition

Powder of the dried seed, administered intragastrically to mice at a dose of 100 mg/kg, was active vs *N*-methyl-*N*'-nitro-*N*-nitroso-guanidine-induced mutagenesis.

Ambulatory Blood Pressure

The effect of regular coffee drinking on 24-hour ambulatory blood pressure in 22 men and women who were normotensive and 26 men and women who were hypertensive with a mean age of 72.1 years (range 54–89 years) was investigated. After 2 weeks of drinking caffeine-containing drinks or instant coffee (five cups/day, equivalent to 300 mg caffeine per day), changes in systolic blood pressure (SBP) and diastolic blood pressure (DBP) in the hypertensive group rise in mean SBP was greater by 4.8 (Standard error of the mean, 1.3) mmHg ($p = 0.031$) and increase in mean DBP was higher by 3 (1) mmHg ($p = 0.010$) in coffee drinkers than in abstainers. There were no significant differences between coffee drinkers and abstainers in the normotensive group. In the group of 52 participants, the effect of coffee on blood pressure was estimated with the use of a random-effects model. In 11 trials, median duration was 56 days (range 14–79 days) and median dose of coffee was five cups per day. SBP and DBP increased by 2.4 (range 1–3.7) mmHg and 1.2 (range 0.4–2.1) mmHg, respectively, with coffee treatment compared to controls. Multiple linear regression analysis identified an independent, positive relationship between coffee consumption and changes in SBP. The effect on SBP and DBP was greater in trials with younger participants.

Anti-adhesive Effect

Green and roasted coffee, used in a treatment mixture and as a pretreatment on beads, inhibited the *Streptococcus mutans*' sucrose-independent adsorption to saliva-coated hydroxyapatite beads. The inhibition of *Salmonella mutans* adsorption indicated that coffee-active molecules may adsorb to a host surface, preventing the tooth receptor from interacting with any bacterial adhesions. Among the known tested coffee components, trigonelline and nicotinic and chlorogenic acids are very active. Dialysis separation of roasted coffee components also showed that a coffee component fraction commonly considered as low-molecular weight coffee melanoidins may sensibly contribute to the roasted coffee's antiadhesive properties.

Antibacterial Activity

Extract of the dried seed, on agar plate at a concentration of 0.1 mL/plate, was inactive on *Pseudomonas aeruginosa*, *Salmonella typhi*, *Salmonella typhimurium*, *Shigella dysenteriae*, *Shigella flexneri* 2A, *Vibrio mimicus*, *Yersinia enterolitica*, and *Escherichia coli*. Enteroinvasive, enterohemorrhagic, enteropathogenic and enterotoxic *Escherichia coli* was used. Extract of the dried seed, on agar plate at a concentration of 0.1 mL/plate, produced weak activity on *Enterobacter cloacae*, *Aeromonas sobria*, *Clavibacter michiganense* ssp. *nebraskense*, *Staphylococcus aureus*, *Staphylococcus epidermidis*, *Vibro cholera* 0-1 V86 EL TOR, *Vibro cholerae* 0-1 569B classical strain, *Vibro cholera* non 0-1 strain, *Vibrio fluvialis*, and *Vibrio parahaemolyticus*. Extract of the dried seed, on agar plate at a concentration of 0.1 mL/plate, was active on *Plesiomonas shigelloides*. Ethanol (95%) extract of the dried seed, on agar plate at a concentration of 1 mg/disc, was active on *Bacillus subtilis*. EtOAc extract of the roasted seed, on agar plate at a concentration of 0.76 mg/mL, was active on *Streptococcus mutans*, and a concentration of 0.84 mg/mL was active on *Staphylococcus aureus*. Water extract of the roasted seed, on agar plate at a concentration of 12.7 mg/mL, was active on *Streptococcus mutans*, and a concentration of 13.5 mg/mL was active on *Staphylococcus aureus*. Infusion of the roasted seed, on agar plate at a concentration of 1.56 mg/mL, was active on *Streptococcus mutans* and *Staphylococcus aureus*. Decoction of the dark-roasted seed, on agar plate at concentrations of 3 and 6 mg/mL, was active on *Staphylococcus aureus*. Decoction of the medium-roasted seed, on agar plate at concentrations of 4, 6, 10, 11, and 34 mg/mL, were active on *Staphylococcus aureus*. Decoction of the light-roasted seed, on agar plate at concentrations of 6, 11, 12, 15, and 17 mg/mL, was active on *Staphylococcus aureus*.

Anticarcinogenic Activity

The effects of coffee consumption on thyroid carcinomas and adenomas were investigated using a standard questionnaire is a case–control study in southwestern Germany, a know iodine-deficient area. The protective role of coffee drinking and the consumption of cruciferous vegetables, such as broccoli, were confirmed for both genders. Treatment for goiter and decaffeinated coffee consumption were associated with an increased risk for malignant tumors, but less so for adenomas.

Antifertility Effect

Decoction of the dried seed, administered orally to female adults at a dose of 0.96 L/day, was active. Coffee intake delayed the time to conception and increased relative risk of failure to conceive. Decoction of the dried seed, administered orally to female adults at variable doses, was active. There was a correlation between heavy coffee drinking and difficulty in becoming pregnant in women in the United States. Hot water extract of the roasted seed, administered in the drinking water of male rats at variable doses daily for 30 weeks, was inactive.

Antigen Modification

Pig-to-rhesus monkey vein transplants were studied to identify the efficiency of green bean α-galactosidase in delaying hyperacute rejection. Biopsies were taken after occluding the grafts for light microscopy (hematoxylin and eosin), scanning electron microscopy, and immunostaining with *Griffonia simplicifolia* IB4 lectin, and for immunoglobulin (Ig) M, IgG, and IgC_3. Galactosidase is effective in removing the terminal α-galactosidase and delays the onset of hyperacute rejection; however, its effect is temporary and it prolongs the survival of pig organs transplanted into primates.

Antihemolytic Activity

Water extract of the dried seed, administered to rabbits' red blood cells at variable concentrations, was inactive vs *Staphylococcus aureus* α-toxin-induced hemolysis and produced weak activity vs *Vibrio parahaemolyticus*-induced hemolysis.

Antimutagenic Activity

Extract of the seed, on agar plate at a concentration of 2.5 μg/mL, was active on *Salmonella typhimurium* TA100 and TA102 vs *T*-butyl peroxide-induced mutagenesis. Hot water extract of the seed, on agar plate at a concentration of 3 mg/mL, was active on *Salmonella typhimurium* TA1535. Infusion of the seed, on agar plate at a concentration of 100 μL/disc, was inactive on *Salmonella typhimurium* TA98 vs 2-amino-anthracene induced mutagenicity. Metabolic activation was required for activity. Lyophilized extract of the seed, on agar plate at a concentration of 6.8 mg/mL, was active on *Salmonella typhimurium* TA1535 vs aflatoxin-2; 4-NQO-, MNNQ-, and ultraviolet light-induced mutagenicity. Lyophilized extract of the seed, on agar plate at a concentration of 15 mg/mL, was active on *Salmonella typhimurium* TA100. Addition of catalase decreased the activity. Decoction of the dried seed, at a concentration of 2% was active on *Drosophila melanogaster* vs cyclophosphamide-induced genotoxicity. A concentration of 5% was active vs urethane-induced genotoxicity. Methylene chloride/2-propanol (1:1) extract, at a concentration of 2%, was active on *Drosophila melanogaster* vs mitomycin C-induced genotoxicity CA185. Hot water extract of the roasted coffee, on agar plate at a concentration of 1%, was active on *Salmonella typhimurium* TA 100 vs benzopyrene, AF-2, and 4NQO mutagenicity. Hot water extract of the roasted coffee, on agar plate at a concentration of 1%, was active on *Salmonella typhimurium* TA98 vs TRP-P-2, Glu-P-1,2-acetylaminofluorene, and IQ mutagenicity. Hot water extract of the roasted coffee, on agar plate at a concentration of 1%, was inactive on *Salmonella typhimurium* TA100 vs β-propiolactone and glycidol mutagenicity. Hot water extract of the roasted coffee, on agar plate at a concentration of 1%, was inactive on *Salmonella typhimurium* TA100 vs acrolein mutagenicity.

Antioxidant Activity

Green and roasted coffee beans were evaluated in relation to degree of roasting and species (*Coffea arabica* and *Coffea robusta*). The properties were evaluated by determining the reducing substances of coffee and its antioxidant activity in vitro and in vivo as protective activity against rat liver cell microsome lipid peroxidation measured as thiobarbituric acid-reacting substances. Reducing substances of *Robusta* samples were significantly higher when compared to those of *Arabica* samples ($p < 0.001$). Antioxidant activity for green coffee samples was slightly higher than for the corresponding roasted samples ($p < 0.001$). Extraction with three different organic solvents (ethyl acetate, ethyl ether, and dichloromethane) showed that the most protective compounds are extracted from acidified dark-roasted coffee solutions with ethyl acetate. Analysis of acidic extract by gel filtration chromatography produced five fractions. Higher molecular mass fractions showed protective activity. The small amounts of these acidic low-molecular-mass protective fractions isolated indicated that they contain strong protective agents. Coffee and the sum of coffee and red wine on healthy subjects showed detectable capacity to scavenge radical cations in the colonic lumen, suggesting that antioxidant activity occurs in the colonic lumen. Fourteen subjects recorded their food intake three times for a period of 2–4 days, each time collecting all of the feces passed during the next 24 hours. Total antioxidant activity (6-hydroxy-2,5,7,8-tetramethulchroman-2-carboxylic acid) of fecal suspension was measured using the 2,2'-azinobis-(3-ethylbenzothiazoline)-6-sulfonic acid radical cation decolorization assay. The average total antioxidant activity of feces was 26.6 mmol/kg wet feces. The total amount of antioxidant equivalents excreted over 24 hours, derived by multiplying the total antioxidant activity by the amount of feces passed during 24 hours, was 3.24 mmol, and this was significantly correlated with the average 24- hour intake of coffee and red wine, particularly to the sum of coffee and red wine. Hot water extract of the seed, produced an inhibition of Fenton-catalyzed oxidation of 2'-deoxyguanosine.

Anti-tumor Activity

Water extract of the dried seed, administered intraperitoneally to mice, was active on CA-755 cells. Hot water extract of the dried seed, administered in the drinking water of mice at a concentration of 0.5%, was active on spontaneous mammary tumors.

Anti-yeast Activity

Ethanol (100%) extract of the seed, on agar plate at a concentration of 18.7 mg/mL, was active on *Candida albicans*. Water extract of the seed, on agar plate was inactive on *Candida albicans*.

Birth-weight Effect

Caffeinated coffee alone had an adjusted odds ratio of 1.3 (95% confidence limits [CL] = 1.0, 1.7) for preterm delivery; mothers who consumed both caffeinated and decaffeinated coffee had an adjusted odds of 2.3 (95% CL = 1.3, 4), whereas those who consumed only decaffeinated coffee showed no increased odds of small-for-gestational age birth, low-birth-weight or preterm delivery. A reduction in mean birth-weight of –3 g per cup per week (95% CL = –5.9, –0.6) for caffeinated coffee and an increase of +0.4 g per cup per week (95% CL = 3.7, 4.5) for decaffeinated coffee was found.

Bone Mineral Density

The association of caffeine consumption and bone mineral density has been investigated in 177 healthy women, age 19–26 years. Average caffeine intake was calculated from self-reports of the consumption of coffee, tea, colas, chocolate products, and selected medications during the previous 12 months. Mean caffeine intake was 99.9 mg/day. Bone mineral density at the femoral neck and the lumbar spine was measured by dual-energy X-ray absorptiometry. After adjusting for potential confounders, including height, body mass index, age and menarche, calcium intake, protein consumption, alcohol consumption, and tobacco use, caffeine consumption was not a significant predictor of bone mineral density. For every 100 mg of caffeine consumed, femoral neck bone mineral density decreased 6.9 mg/cm^2 and lumbar spine bone mineral density decreased 11.9 mg/cm^2. No single source of caffeine was significantly associated with a decrease in bone mineral density. Furthermore, the association between caffeine consumption and bone mineral density at either site did not differ significantly between those who consumed low levels of calcium (≤836 mg/day) and those who consumed high levels of calcium (>836 mg/ day).

Bone Mineral Effect

Coffee, taken by 258 healthy occupationally active men aged 40–63 years, significantly reduced the trabecular bone mineral content. The extent of alcohol intake did not differentiate bone mineral content values at the distal radius, whereas the significant detrimental effects of both smoking and coffee drinking on trabecular (but not cortical and total) bone mineral content were revealed. Simultaneously, smokers and ex-smokers, when compared to lifelong nonsmokers, had lower trabecular bone mineral content.

Brain Metabolic Response

Changes in brain lactate resulting from the combined effects of caffeine's stimulation of glycolysis and reduction of cerebral blood flow were determined by a rapid proton echoplanar spectroscopic imaging technique in a group of nine heavy caffeine users and nine caffeine-intolerant persons. They were studied at baseline and 1 hour after ingestion of caffeine citrate (10 mg/kg). Five of the caffeine users were restudied after a 1- to 2-month caffeine holiday. Significant increases in global and regionally specific brain lactate and psychological and physiological distress in response to caffeine ingestion were observed only among the caffeine-intolerant persons. Reexposure of the regular coffee drinkers to caffeine after a caffeine holiday resulted in little or no adverse clinical reaction but did result in significant rises in brain lactate, which were of a magnitude similar to that observed for the caffeine-intolerant group.

Caffeine Intake, Tolerance, and Withdrawal

Caffeine in the form of brewed and instant coffee, tea, and caffeinated drinks was taken by 1934 individual twins from female–female pairs, including 486 monozygotic and 335 dizygotic pairs. The resemblance in twin pairs for total caffeine consumption, heavy caffeine use, caffeine intoxication, tolerance, and withdrawal was substantially greater in monozygotic than in dizygotic twin pairs and could be ascribed solely to genetic factors, with estimated broad heritabilities of between 35 and 7 7%.

Carcinogenesis Inhibition

Water-soluble fraction of the dried fruit, administered to female mice at a dose of 0.25% of diet, inhibited mammary tumor development in SHN virgin mice. Water-soluble fraction of the dried fruit, administered to female mice at a dose of 0.25% of diet, inhibited mammary glands in SHN virgin mice. Decoction of the dried seed, administered in drinking water to rats at a concentration of 57.0 g/L, produced no effect on dimethylnitrosamine-induced glutathione *S*-transferase positive foci after subtotal hepatectomy. Decoction of the dried seed, administered intragastrically to pregnant rhesuses at a dose of 10 mL/kg for 90 minutes before dosing with cyclophosphamide, *N*-nitrosodiethylamine, *N*-nitroso-*N*-ethylurea, or mitomycin, produced micronuclei and polychromatophilic nucleated erythrocytes in fetal liver, marrow, and blood. Seeds, administered in ration of high mammary tumor strain of SHN/MEI virgin female mice, were active. Hot water extract of the dried seed, administered in drinking water to rats at a dose of 6000 ppm, was active. Hot water extract of the dried seed, administered orally to adults at variable doses, produced no effect on pancreatic cancer. Hot water extract of the dried seed, administered orally to adults at variable doses, was inactive. Patients with newly diagnosed breast cancer (n = 818) were compared to surgical and neighborhood controls in a dietary case–control study of the relationship of dietary intake of coffee and total methylxanthine from coffee, tea, chocolate, and cocoa drinks. A nonsignificant negative association was found between methylxanthine consumption and breast cancer. This pattern was stronger in patients with high-fat diets after controlling several confounding hormonal factors. A diminished risk was found when consumption of methylxanthine of patients with breast cancer is compared to that of patients with benign disease. Decoction of the dried seed, administered to male rats at a dose of 5% of diet, was active vs dimethylnitrosamine-induced carcinogenesis. Lyophilized extract of the dried seed, administered intragastrically to mice at a dose of 50.0 g/kg of diet, was active. Animals were exposed to coffee *in utero*, as mother's diet was 1% instant coffee. After weaning, animals were given an instant coffee for 2 years. Incidence of neoplasms decreased from 70.6 to 34.8% in males and from 56.8 to 36.2% in females. The incidence of benign tumor was 2.72 vs 0% for controls. Seed oil, administered to hamster at a concentration of 2.25% of diet, was active vs 7,12-dimethylbenz[a]anthracene (DMBA) - induced oral tumors. Seed, administered to hamsters at a concentration of 15% of diet, was active vs DMBA-induced oral tumors. Seed, administered to rats at a concentration of 20% of diet, was active vs DMBA-induced carcinogenesis. Decoction of the roasted coffee, administered orally to adults, was active on risk of colon or rectal cancer. Risk of colon cancer was reduced in drinkers of four or more cups of coffee per day. There was no effect on rectal cancer. Methylene chloride/2-propanol (1:1) extract of the roasted coffee, administered in drinking water of male rats at a concentration of 10%, was inactive on urinary bladder. Seed oil, administered to male rats at a dose of 0.10%, was active on the colon.

Carcinogenic Activity

Decoction of the seed, administered orally to adults, was inactive. There was no association between colorectal adenomas and consumption of extract. Roasted seed, administered to male rats at a dose of 6% of diet for 2 years, was inactive. Water extract of the roasted seed, administered to female rats at a dose of 6% of diet, was inactive. Regular and decaffeinated instant coffees were studied. Coffees with highest caffeine content showed lower tumor incidence. Hot water extract of the roasted seed, administered orally to 18 rats at a dose of 2% for 120 days of dosing with cycasin orally (150 mg/kg) on day 121, produced five tumors. Hot water extract of the roasted seed, administered orally to rats at a dose of 2%, was inactive.

Carcinogenic Risk Analysis

Pooled data of 564 cases and 2929 hospitals or population controls who had never smoked were enrolled in epidemiological studies to examine the association of coffee with an excess bladder cancer risk. The data were evaluated from 10 studies conducted in Denmark, Germany, Greece, France, Italy, and Spain. Information on coffee consumption and occupation was recoded following standard criteria. Unconditional logistic regression was applied adjusting for age, study center, occupation, and gender. Seventy nine percent of the study population reported having consumed coffee, and 2.4% were heavy drinkers, reporting having ingested on average 10 or more cups per day. There was no excess risk in coffee drinkers compared to nondrinkers. The risk did not increase monotonically with dose, but a statistically significant risk was seen for subjects having ingested 10 or more cups per day. This excess was seen in both males and females. There was no evidence of an association of the risk with duration or type of coffee consumption. Nonsmokers who are heavy coffee drinkers may have a small excess risk of bladder cancer. Although these results cannot be attributed to confounding by smoking, the possibility of bias in control selection cannot be discarded. On the basis of the data, only a small proportion of cancers of the bladder among nonsmokers could be attributed to coffee drinking.

Cardiac Mechanoenergetics

Caffeine in a concentration higher than 0.05 m*M*, corresponding to the maximum blood concentration after a healthy human subject consumed a cup of coffee, depresses left ventricular systolic and diastolic functions and decreases a measure of total mechanical energy per beat in terms of SBP-volume area more severely in failing hearts at concentrations lower than those in normal hearts.

Cardioexcitatory Activity

Hot water extract of the dried seed, administered orally to adults, produced weak activity. There was no change in electrocardiogram pattern, but some subjects showed sinus arrhythmia, sinus tachycardia, and incomplete right bundle branch block, premature ventricular contraction, and premature atrial contraction.

Cardiovascular Effects

Caffeinated coffee was taken by 72 males and 72 females with a mean age of 21 years. Ingestion of caffeine had no effect on initial mood or working memory, but it improved encoding of new information, counteracted the fatigue, and increased blood pressure and pulse rate.

Cerebral Blood Flow

The possibility of caffeine-mediated changes in blood flow velocity in the middle cerebral artery induced by tests of cerebrovascular responsiveness was examined by transcranial doppler sonography. Velocity in the middle cerebral artery measures were obtained as healthy college students hypoventilated, hyperventilated, and performed cognitive activities (short-term remembering, generating an autobiographical image, and solving problems), each in 31-second tests. The measures were obtained from the same persons, in separate testing sessions, when they were noncaffeinated and under two levels of caffeine (45 mg/12 oz and 117 mg/8 oz). Compared with the no-caffeine control condition, a smaller amount of caffeine had no significant effects on global velocity in the middle cerebral artery but a larger amount suppressed the velocity by 5.8%. Time course analyses indicated that the velocity followed a triphasic pattern to increase over baselines during hypoventilation, regardless of caffeine condition; slowed below baselines during hyperventilation; and increased over baselines during all cognitive activities (ranges 3.8–6.9%).

Chemopreventive Effect

Chlorogenic acid had a regressive effect on induced aberrant crypt foci, as well as on development of aberrant crypt foci in azoxymethane-induced colorectal carcinogenesis in rats. Rice germs and γ-aminobutyric acid-enriched defatted rice germ inhibited azoxy-methane-induced aberrant crypt foci formation and colorectal carcinogenesis in rats. Ferulic acid, also known to be contained in coffee beans and rice, prevented azoxymethane aberrant crypt foci formation and intestinal carcinogenesis in rats.

Cholesteryl Ester Transfer Protein Activity

French press or filtered coffee, consumed by 46 healthy normolipidemic subjects for 24 weeks, produced a long-term increase in cholesteryl ester transfer protein, as well as phospholipid transfer protein activity; the increase in cholesteryl ester transfer protein activity may contribute to the rise in low-density lipoprotein (LDL) cholesterol. Relative to the baseline values, French-press coffee significantly increased average cholesteryl ester transfer protein activity by 12% after 2 weeks, by 18% after 12 weeks, and by 9% after 24 weeks. Phospholipid transfer protein activity was significantly increased by 6% after 2 weeks and by 10% after 12 weeks. Lecithin/cholesterol acyltransferase activity was significantly decreased by 6% after 12 weeks and by 7% after 24 weeks. The increase in cholesteryl ester transfer protein clearly preceded the increase in LDL cholesterol, but not the increase in total triglycerides (TGs). However, consumption of French-press coffee produced a persistent rise in cholesteryl ester transfer protein activity, whereas the rise in serum TGs was transient.

Water extract of the green seed, administered intravenously to male rats at a dose of 70 mg/kg, was inactive. Water extract of the roasted seed, administered intravenously to male rats at a dose of 0.84 mg/kg, was active.

Chromosome Aberration Induced

Lyophilized extract of the roasted seed, in cell culture at a concentration of 3.9 mg/mL, was active on human lymphocytes. Caffeinated and decaffeinated coffees without S9 mix was tested. The extract produced weak activity with S9 mix. Extract of the roasted seed, in cell culture at variable concentrations, was active on human lymphocytes. Metabolic activation reduced the effect.

Cognitive and Psychomotor Performance

Coffee and tea, consumed four times during the day by 30 healthy volunteers, maintained aspects of cognitive and psychomotor performance throughout the day and evening when caffeinated beverages were administered repeatedly. Tea, coffee, or water was administered in a randomized five-way crossover design. A psychometric battery consisting of critical flicker fusion, choice reaction time, and subjective sedation tests was administered predose and at frequent time points postdose. The Leeds sleep evaluation questionnaire was completed each morning, and a wrist Actigraph was worn for the duration of the study. Caffeinated beverages maintained critical flicker fusion threshold throughout the whole day, independent of caffeine dose or beverage type. During the acute phase of the beverage ingestion, caffeine significantly sustained performance compared with water after the first beverage of critical flicker fusion and subjective sedation and after the second beverage for the recognition component of the choice reaction time task.

There were significant differences between tea and coffee at 75 mg caffeine dose after the first drink. Compared to coffee, tea produced a significant increase in critical flicker fusion threshold between 30 and 90 minutes postconsumption. After the second beverage, caffeinated coffee at 75 mg dose significantly improved reaction time, compared with tea at the same dose, for the recognition component of the choice reaction time task. Caffeinated beverages had a dose-dependent negative effect on sleep onset, time, and quality. Day-long tea consumption produced similar alert effects as coffee, despite lower caffeine levels, but it is less likely to disrupt sleep.

Colonic Cancer Risk

French-press coffee, consumed by men and women with mean age of 43 ± 11 years, did not influence the colorectal mucosal proliferation rate but may increase the detoxification capacity and antimutagenic properties in the colorectal mucosa through an increase in glutathione concentration.

Comutagenic Activity

Hot water extract of the roasted seed with methylglyoxal, DL-glyceraldehyde, dihydroxyacetone, and autoxidized linoleic acid, on agar plate at a concentration of 1%, were active on *Salmonella typhimurium* TA100.

Coronary Heart Disease

In a study of 20 randomly selected groups of 179 Finnish men and women aged 30–59 years, it was determined that coffee drinking did not increase the risk of coronary heart disease or death. In men, the effects of smoking and a high serum cholesterol level largely explain slightly increased mortality from coronary heart disease and all causes in heavy coffee drinkers. Habitual coffee drinking, health behavior, major known coronary heart disease risk factors, and medical history were assessed at the baseline examination. Each subject was followed up 10 years after the survey using the national hospital discharge and death registers.

Multivariate analyses were performed using the Cox proportional hazards model. In men, the risk of nonfatal myocardial infarction was not associated with coffee drinking. The highest coronary heart disease mortality was found among those who did not drink coffee at all. Also, in women, all-cause mortality decreased by increasing coffee drinking. The prevalence of smoking and the mean level of serum cholesterol increased with increasing coffee drinking. Non-coffee drinkers more often reported a history of various diseases and symptoms, and they were also more frequently users of several drugs compared with coffee drinkers. A risk of coronary events (death, nonfatal infarction, or coronary artery surgery) was estimated in a group of more than 11,000 men and women aged 40–59 years by approx 7.7 years of study. Coffee and tea consumption showed a strong inverse relation. Coffee showed a weak but beneficial gradient with increasing consumption, associated with beneficial effects for mortality and coronary morbidity, although there was a residual benefit of coffee consumption in avoiding heart disease among men. Decoction of the dried seed, administered to adults of both sexes at variable doses, produced equivocal effect.

In a 12-year cohort study on the influence of coffee intake on coronary heart disease in 38,500 subjects it was indicated that during the first 6 years a strong correlation between high coffee intake and coronary death was found. After the first 6 years, the correlation was significantly decreased.

Cytotoxic Activity

Ethanol (50%) extract of the aerial parts, in cell culture, was inactive on CA-9KB, effective dose$_{50}$ greater than 20.0 μg/mL.

Dermatitis-producing Effect

Hot water extract and powder of the dried seed, administered externally to adults, were active.

Down Syndrome Effect

Data from a case–control study of 997 live-born infants or fetuses with Down syndrome and 1007 live-born controls with a birth defect indicated that among nonsmoking mothers, high coffee consumption is more likely to reduce the viability of a Down syndrome conceptus than that of a normal conceptus.

Embryotoxic Effect

Hot water extract of the Folger's instant coffee, administered by gastric intubation to pregnant mice at a dose of 1.28 mg/animal, was inactive. Hot water extract of the roasted seed, administered in drinking water of pregnant rats at variable doses daily for 30 weeks, was inactive.

Estrogenic Effect

Unsaponifiable fraction of the seed oil, administered subcutaneously to immature female rats at a dose of 117 mg/ animal, was inactive. Subcutaneous administration to ovariectomized female guinea pigs was active.

Fatalities

Extract of the roasted seed, administered rectally to a 37-year-old woman with breast cancer after radical mastectomy and chemotherapy at a dose of 0.95 L/person four times daily, was active. Death was attributed to fluid and electrolyte imbalance. Sodium and chloride could not be detected. Extract of the roasted seed, administered rectally to a 46-year-old woman at a dose of 10–12 coffee enemas, three to four an hour, produced convulsive seizures and eventually death. Decoction of the dark-roasted seed, on agar plate, was active on *Staphylococcus aureus*, with lethal dose$_{50}$ of 16 mg/mL. Concentrations of 23, 35, and 40 mg/mL, were active on *Escherichia coli*. Decoction of the medium-roasted seed at concentrations of 29, 41, 50, and 52 mg/mL, were active on *Escherichia coli*. Decoction of the light-roasted seed at concentrations of 40, 46, 50, and 57 mg/mL, were active on *Escherichia coli*. Decoction of the roasted seed, on agar plate at concentrations of

28 and 41 mg/mL, was active on *Escherichia coli*. Decoction of the medium-roasted seed, on agar plate at a concentration of 4 mg/mL, was active on *Sarcina lutea*.

Fertilization Inhibition

Hot water extract of the roasted seed, administered in the drinking water of female rats at variable doses daily for 30 weeks, was inactive.

Fibrinogen level increase

Hot water extract of the dried seed, administered to adults at a dose of five cups/day, produced weak activity.

Fungal Activity

Coffee leaves, fruits, and soil were cultured and inoculated into mice. A fungus isolated from the liver of a mouse inoculated with soil showed temperature- dependent dimorphism and in vitro mycelium and yeast phases characteristic of *Paracoccidioides brasiliensis*. Yeast cells of the fungus produced disseminated infection after intraperitoneal inoculation in Wistar rats from which the fungus was reisolated. An antigen reacting with sera from patients with paracoccidioidomycosis was obtained from this *Paracoccidioides brasiliensis* strain; antigen identity with strain 339 and with four other *Paracoccidioides brasiliensis* strains was detected by gel immunodiffusion. However, when the exoantigen was submitted to sodium dodecyl sulfate-polyacrylamide gel electrophoresis, a low gp43 expression in the new strain, which was called Ibia, was observed.

Gallbladder Diseases

The relation of ultrasound-documented gallbladder disease with coffee drinking in 13,938 adult participants was examined between 1988 and 1994. The prevalence of total gallbladder disease was unrelated to coffee consumption in either men or women. However, among women, a decreased prevalence of previously diagnosed gallbladder disease was found with increased coffee drinking. These findings do not support a protective effect of coffee consumption on total gallbladder disease, although coffee may decrease the risk of symptomatic gallstones in women.

γ-Glutamyltransferase Effect

In a cross-sectional study involving 1353 males aged 35–59 years, it was concluded that coffee consumption is inversely related to serum γ-glutamyltransferase and that coffee may inhibit the inducing effects of aging and possibly of smoking on serum γ-glutamyltransferase in the liver.

Gastroesophageal Reflux Effect

Coffee, a known lower esophageal sphincter relaxant, was tested in 185 and 258 cases of esophageal adenocarcinoma and gastric adenocarcinoma, respectively, and 815 controls. There was no association between lower esophageal sphincter-relaxing foods and symptoms of chronic reflux. There was no association between dietary factors known to cause lower esophageal relaxation and the risk of adenocarcinoma of the esophagus or gastric cardia. The results indicated that dietary factors associated with lower esophageal sphincter relaxation and transient gastroesophageal reflux are not associated with any important risk of esophageal malignancy.

Gastrointestinal Effect

It was demonstrated that coffee promotes gastroesophageal reflux. It stimulated gastrin release and gastric acid secretion, but studies on the effect on lower esophageal sphincter pressure yielded conflicting results. Coffee also prolonged the adaptive relaxation of the proximal stomach, suggesting that it might slow gastric emptying. However, other studies indicated that coffee does not affect gastric emptying or small bowel transit. It induced cholecystokinin release and gallbladder contraction, which may explain why patients with symptomatic gallstones often avoid drinking coffee. Coffee increased rectosigmoid motor activity within 4 minutes after ingestion in some people. Their effects on the colon were comparable to those of a 1000-kcal meal. Because coffee contains no calories and its effects on the gastrointestinal tract cannot be ascribed to its volume load, acidity, or osmolality; it must have pharmacological effects. Caffeine alone could not account for these gastrointestinal effects.

Genotoxicity Inhibition

Hot water extract of the fruit, administered intragastrically to mice at a dose of 500 mg/kg, was active vs adriamycin-, cyclophosphamine-, procarbazine-, and mitomycin-induced genotoxicity. Genotoxicity was measured by the presence of micronucleated polychromatic erythrocytes in bone marrow.

Homocysteine Effects

Elevated homocysteine concentration is considered an independent risk factor for cardiovascular diseases and has been associated with neural tube defects. In a study of 290 young women aged 25–30 years and in 288 older women aged 60–65 years total homocysteine concentrations were measured. All of the participants completed questionnaires about factors, including lifestyle, health, and use of vitamin supplements. Smoking status, coffee consumption, SBP, and body mass index were positively associated, and estrogen replacement therapy and tea consumption were inversely associated with total homocysteine in some of the models. According to the criteria used, between 1 and 36% of the women had suboptimal folate intake. Folic acid is a strong predictor of total homocysteine concentration; however, several dietary and other lifestyle factors are important as well. Coffee, consumed by 26 volunteers (18–53 years of age) at a dose of 1 L/day for 4 weeks, raised plasma concentrations of total homocysteine in healthy individuals. Coffee increased homocysteine concentrations in 24 of 26 individuals. Circulating concentrations of vitamin B_6, vitamin B_{12}, and folate were unaffected. Infusion of the seed oil, administered orally to more than 15,000 adults of both sexes at variable doses, was active on plasma.

Hypercholesterolemic Effect

Triacylglycerols has been determined to be the major lipid constituents of the coffee oil, along with sterol esters, sterols/triterpene alcohol, hydrocarbons, and the hydrolyzed products of triacylglycerols as the minor components. Fatty acid composition of total oil, neutral lipids, polar lipids, and pure triacylglycerols showed the presence of fatty acids of C14, C16, C18, and C20 carbon chains. Palmitic and linoleic acids were the major fatty acids and comprise approx 38.7% and 35.9%, respectively. Pancreatic lipase hydrolysis revealed that the linoleoyl and palmityl moieties are preferentially esterified at the Sn-2 and Sn-1,3 positions of triacylglycerols, respectively. The presence of high amounts of palmitic acid at Sn-1,3 position in coffee oil may be partly responsible for its hypercholesterolemic effects. Coffee oil was administered orally to 11 healthy normolipemic volunteers at a dose of 2 g/day for 3 weeks. After a 2-week washout period, the reverse treatments were applied for another 3 weeks. Six subjects received oil supplying 72 mg/day of cafestol and 53 mg/day of kahweol, and five received oil that provided 40 mg of cafestol, 19 mg of 16-*O*-methyl-cafestol, and 2 mg of kahweol/day.

The average cholesterol level increased by 0.65 mmol/L (13%) on coffee oil. The TG level increased by 0.49 mmol/L (61%). No effects on serum lipids or lipoprotein cholesterol levels were significantly different between variety *Arabica* or *Robusta* oils. The treatments elevated serum lipid levels; therefore, cafestol must be involved and kahweol cannot be the sole cholesterol-raising diterpene. Coffee total lipids, coffee nonsaponifiable matter, and coffee diterpene alcohols have been examined in adult Syrian hamsters. The animals were fed either a commercial laboratory chow diet containing 5% fat and low in saturated fat (1.46 g/100 g diet) and cholesterol (0.03 g/ 100 g diet) or a semisynthetic diet set in gelatin, containing 10% fat and high in saturated fat (4 g/100 g diet) and cholesterol (90.5 g/100 g diet). The coffee lipid extracts were dissolved in olive oil (concentration either 5 mg of total lipid, 0.5 mg nonsaponifiable matter or 0.5 mg diterpene alcohols for 250 μL olive oil) in study 1 and in coconut oil (concentrations either 20 mg total lipid, 2 mg nonsaponifiable matter, or 2 mg diterpene alcohols per 250 μL) in study 2. A dose of 250 μL of these solutions was administered daily by gavage. Control animals received 250 μL vehicle only. For serum lipid analysis, blood samples were obtained on days 0, 7, and 14 in study 1 and on days 0, 7, 14, and 21 in study 2. The results indicated a tendency of serum total cholesterol (TC) and high-density lipoprotein (HDL) cholesterol to increase with administration of coffee total lipid, nonsaponifiable, and diterpene alcohols.

In contrast, in study 2 there were no significant differences in serum lipids between control and coffee lipid-treated groups across time. The results support the concept that coffee lipids may be hypercholesterolemic and indicate that diterpene could be the lipid component responsible for such an effect. However, it appears that this hypercholesterolemic effect is apparent only when the background diet is low in saturated fat and cholesterol. A high-saturated fat/high-cholesterol diet may mask the hypercholesterolemic effect of coffee lipid. Hot water extract of the dried kernel, administered intragastrically to male hamsters, was active vs feeding high-fat diet. Hot water extract of the boiled seed, administered in drinking water of hamster and rats at a concentration of 0.5 g/mL, was inactive. Decoction of the dried seed, administered orally to adults, produced equivocal results. Decoction of the roasted coffee, administered orally to 20 healthy volunteers at a dose of 600 mL/day for 4 weeks, produced a significant increase of LDL and TG levels, and LDL–HDL ratio. Decoction of the boiled coffee passed through a conventional paper filter, administered orally to 20 healthy volunteers at a dose of 600 mL/day for 4 weeks, produced no change in LDL and TG levels and LDL– HDL ratio. Filtering removed more than 80% of the lipid-soluble substances present in boiled coffee. Heartwood, administered orally to adults for 24 hours, produced an increase of cholesterol level.

Hot water extract of the seed, administered orally to 1629 middle-aged adults, produced an increase of serum cholesterol level and intake of fat. Hot water extract of the seed, administered orally to 1625 middle-aged adults, produced equivocal effect. Consumers of filtered coffee had no significant change in serum cholesterol level. Powder of the seed, administered orally to adults at a dose of 8 g/day dosed daily in unfiltered coffee, was active. Hexane-diethyl ether extract of the roasted seed, administered intragastrically to hamsters at a dose of 2 mg/animal, was active. Lipid fraction of the roasted seed, administered to hamsters at a dose of 20 mg/animal, was active. Nonsaponifiable fraction of the roasted seed, administered to hamsters at a dose of 2 mg/animal, was active. The effect was found only in diet low in saturated fat and cholesterol. Decoction of the dried seed, administered orally to adults of both sexes at variable doses, decreased the level of cholesterol. Decoction of the dried seed, administered orally to adults at a dose of five cups per day, produced weak activity, and when administered to new coffee drinkers of both sexes at variable doses, decreased cholesterol level.

Immunostimulant Activity

Hot water extract of fruit, in the drinking water of mice at a concentration of 0.5%, increased the percentage of thymocytes expressing mature CD4 or CD8 markers and increased the proportion of peripheral lymphocytes expressing CD25, a marker of activation. Hot water extract of the fruit, administered orally to adults at variable doses, was active on lymphocytes vs suppressor T-cells and natural-killer cells and inactive vs helper T-cells. Methanol extract of the dried pericarp, administered in drinking water of mice at a concentration of 0.5%, was active on lymphocytes B. The extract enhanced lipopolysaccharide-induced activation. Extract of the dried seed, administered intramuscularly to adult calves at a concentration of 10 mL/animal, was active.

K-*ras* gene mutagenesis

The relationship between consumption of coffee and mutations in the K-*ras* gene in exocrine pancreatic cancer was investigated in 185 patients, 121 for whom tissue was available. Mutations in codon 12 of K-*ras* were detected by the artificial restriction fragment-length polymorphism technique. Mutations were found in tumors from 94 of 121 patients (77.7%) and were more common among regular coffee drinkers than among non-regular coffee drinkers (82% vs 55.6%, $p = 0.018$, $n = 107$). The weekly intake of coffee was significantly higher among patients with a mutated tumor (mean of 14.5 cup/week vs 8.8 among patients with a wild-type tumor, $p < 0.05$). Regarding non-regular drinkers, the odds ratio of a mutated tumor adjusted by age, sex, smoking, and alcohol drinking was 3.26 for drinkers of 2–7 cups/ week, 5.77 for drinkers of 8–14 cups/week, and 9.99 for drinkers of more than 15 cups/week ($p = 0.01$).

Lipoprotein Modification

Decoction of the seed, administered orally to 22 adults at a dose of five to six strong cups for 1 day, was active. Consumption of cafestol and kahweol resulted in decreased lipoprotein A levels. Filtering coffee removed the diterpenes. Decoction of the dried stem bark, administered orally to 150 healthy adults of both sexes who consumed five or more cups of boiled coffee and 159 filter coffee consumers at a dose of 1.2 L/day, was active on human serum. Median level of serum lipoprotein was higher in the boiled coffee drinkers.

Liver Dysfunction

In a 4-year study in 1221 liver dysfunction-free (serum aspartate aminotransferase [AST] and alanine aminotransferase [ALT] <39 IU/L and no medical care for or no past history of liver disease) males aged 35–56 years, was investigated for the association of coffee consumption with the development of increased serum AST and/or ALT activities. From the analysis using the Kaplan-Meier method, the estimated incidence of serum AST and/ or ALT ≥ 40 IU/L, ≥ 50 IU/L, and ≥ 60 IU/L decreased with an increase in coffee consumption. From the Cox proportional hazards model, coffee drinking was independently inversely associated with the development of serum AST and/or ALT ≥ 40 IU/L, ≥ 50 IU/L, and ≥ 60 IU/L, controlling for age, body mass index, alcohol intake, and cigarette smoking.

Maternal Risks

Three-hundred six mothers who gave birth to babies with cleft lip, or palate, or both were matched with 306 mothers who gave birth to healthy babies in the same area during the same period. Significantly more babies in the cleft palate group had a family history of clefts (48/306 compared with 7/306) in the cases studied; combined cleft lip and palate was significantly more common among boys (82/157 compared with 57/149) and cleft palate alone among girls (48/149 compared with 22/157). There was no difference between the groups regarding dietary preferences, but during pregnancy the mothers who gave birth to babies with defects tended to drink less alcohol and less coffee.

Metabolism

Decoction of the seed, administered orally to adults at variable doses, was active. Volunteers consumed food containing hydroquinone or glycopyranoside derivative (arbutin). Blood and urine levels of the compounds and conjugates were assayed.

Miscellaneous Effects

Decoction of the dried seed, administered orally to 171 healthy nonsmoking adults of both sexes over the age of 50 years, indicated that coffee may decrease postprandial falls in SBP and can increase DBP in untreated hypertensives.

Mitogenic Activity

Hot water extract of the fruit, administered orally to adults at variable doses, was inactive on lymphocytes vs T-lymphocyte proliferation. Hot water extract of the fruit, administered orally to adults at variable doses, was inactive on lymphocytes B vs B-cell proliferation.

Mood Effects

In a full crossover design study, the effect of coffee and tea on acute physiological responses and mood indicated that caffeinated beverages acutely stimulate the autonomic nervous system and increase alertness. In the study, caffeine levels in tea were 37.5 and 75 mg and in coffee 75 and 150 mg in one group. In another group caffeine, level was manipulated. SBP, DBP, heart rate, skin temperature, skin conductance, and mood were monitored over each 3-hour study session. Tea and coffee produced mild autonomic stimulation and an elevation on mood. There were no effects of tea vs coffee or caffeine dose, despite a four-fold variation in the latter. In one study, increasing beverage strength was associated with greater increases in DBP and energetic arousal. In the other, caffeinated beverages increased DBP, SBP, and skin conductance and lower heart rate and skin temperature compared to water. Significant dose–response relationships to caffeine were seen only for SBP, heart rate, and skin temperature. There were significant effects of caffeine on energetic arousal but no consistent dose–response effects.

Mutagenic Activity

Freeze-dried roasted and instant coffee, at a dose of 20 mg/plate, induced between 6 and 10 times the revertants found in negative controls of *Salmonella typhimurium*. Green coffee beans had no mutagenic activity. Mutagenicity increased with roasting time to 4 minutes, the time normally used roast coffee. The genotoxic compounds were quickly formed at temperature of 220°C. Mutagenic activity was independent of the roasting procedure. Water extract of the dried fruit, in cell culture at a concentration of 2 mg/mL, was active on hamster lung cells without microsomal activation. Hot water extract of the seed, on agar plate at concentration of 40 mg/plate, was active on *Salmonella typhimurium* TA102 and inactive on *Salmonella typhimurium* TA100. Hot water extract of the seed, on agar plate at concentration of 50 mg/plate, was active on *Salmonella typhimurium* TA100 and inactive on *Salmonella typhimurium* TA1535, TA1537, TA1538, and TA98. Hot water extract of the seed, administered intragastrically to mice at a dose of 6 g/animal, was inactive on *Escherichia coli* K12 and *Salmonella typhimurium* TA1530. The effect was assayed on bacteria injected intravenously coincidentally with extract administration and harvested 1.5 hours later. Ethanol (95%) and hot water extracts of the solid residue of brewed coffee, on agar plate at a concentration of 12.5 mg/plate, were inactive on *Salmonella typhimurium* TA100 and TA98. Hot water extract of the distillates of brewed coffee overheated to 150°C, on agar plate at a concentration of 5 mg/plate, was inactive on *Salmonella typhimurium* TA100. Metabolic activation had no effect on the results. $MeCl_2$ extract of distillates of brewed coffee overheated to 300°C, on agar plate at a concentration of 100 μg/plate, was inactive on *Salmonella typhimurium* TA100 and TA98. Extract was toxic at higher doses. $MeCl_2$ extract of distillates of brewed coffee overheated to 300°C, on agar plate at a concentration of 300 μg/ plate, was active on *Salmonella typhimurium* TA98. Metabolic activation was required for activity. $MeCl_2$ extracts of distillates and solid residue of brewed coffee, on agar plate at a concentration of 5 mg/plate, were inactive on *Salmonella typhimurium* TA100 and TA98. $MeCl_2$ extract of distillates of brewed coffee overheated to 150°C, on agar plate at a concentration of 750 μg/plate, was active on *Salmonella typhimurium* TA98. Metabolic activation was required for activity. Hot water extract of the seed, on agar plate at a concentration of 10 g/L, was active on *Salmonella typhimurium* TA98. Hot water extract of the roasted seed, on agar plate at a concentration of 14 mg, was active on *Salmonella typhimurium* TA100. The activity shown was the result of caffeine.

Myocardial Infarction

A group of 340 of age-, sex-, and community-matched individuals drinking caffeinated and decaffeinated coffee was investigated. The odds ratio for drinking four or more cups per day of caffeinated coffee was 0.84 (95 % confidence interval [CI], 0.49–1.42) compared with drinking one cup or less per week, after adjustment for coronary risk factors. The odds ratio for drinking more than one cup per day of decaffeinated coffee vs nondrinkers was 1.25 (95% CI, 0.76–2.04).

Nuclear Aberration Reduction

Decoction of the seed, administered intragastrically with methylurea and sodium nitrite to mice at a dose of 1 g/animal, was active. Decoction of the seed, administered intragastrically with methylurea and sodium nitrite to mice at a dose of 600 mg/animal, was active on colon.

Occupational Respiratory Allergy

There was a significant correlation between sensitization to green coffee bean and work-related symptoms (asthma and/or rhinitis) ($p < 0.01$), common allergic symptoms ($p < 0.05$), and atopy by prick test ($p < 0.01$).

Ovarian Cancer Risk

In a study of 549 women with newly diagnosed epithelial ovarian cancer and 516 control women, it was concluded that coffee and caffeine consumption may increase the risk of ovarian cancer among premenopausal women. Coffee and alcohol consumption was assessed through a semiquantitative food-frequency questionnaire, and information on tobacco smoking was collected through personal interview. There was no risk for ovarian cancer overall associated with tobacco or alcohol use in either premenopausal or post-menopausal women. Association of border-line significance for tobacco and invasive serous cancers and alcohol and mucinous cancers were observed but reduced after adjustment for coffee consumption.

Pancreatic Cancer Risk

In a study of 583 individuals with histologically confirmed pancreatic cancer and 4813 controls, it was determined that consumption of total alcohol, wine, liquor, beer, and coffee was not associated with pancreatic cancer.

Parkinson's Disease

The association of smoking, alcohol, and coffee consumption with Parkinson's disease was investigated in 196 subjects who developed Parkinson's disease from 1976 to 1995. Each incident case was matched by age (± 1 year) and sex to a general population control subject. The findings suggest an inverse association between coffee drinking and

Parkinson's disease; however, this association did not imply that coffee has a direct protective effect against Parkinson's disease. In a study of 8004 Japanese-American men aged 45–60 years, it was indicated that higher coffee intake is associated with a significantly lower incidence of Parkinson's disease. Data were analyzed from 30 years of follow-up. During the follow-up, 102 men were identified as having Parkinson's disease. Age-adjusted incidence of Parkinson's disease declined consistently with increased amounts of coffee intake from 10.4 per 10,000 person-years in men who drank coffee to 1.9 per 10,000 person-years in men who drank at least 450 g/day. Similar relationships were observed with total caffeine intake and for caffeine from noncoffee sources. Consumption of increasing amounts of coffee was also associated with lower risk of Parkinson's disease in men who were never, past, and current smokers at baseline. Other nutrients in coffee, including niacin, were unrelated to Parkinson's disease incidence. The relationship between caffeine and Parkinson's disease was unaltered by intake of milk and sugar.

Pharmacokinetic Interactions

The most serious coffee (caffeine)-related central nervous system (CNS) effects include seizures and delirium. Other symptoms affecting the cardiovascular system range from moderate increases in heart rate to more severe cardiac arrhythmia. Although tolerance develops to many of the pharmacological effects of caffeine, tolerance may be overwhelmed by the nonlinear accumulation of caffeine when its metabolism becomes saturated. This might occur with high levels of consumption or as the result of pharmacokinetic interaction between caffeine and medications. The polycyclic aromatic hydrocarbon-inducible cytochrome P450 IA2 participated in the metabolism of caffeine, as well as of several clinically important drugs. A number of drugs, including certain selective serotonin reuptake inhibitors (particularly fluvoxamine), antiarrhythmics (mexiletine), antipsychotics (clozapine), psoralens, idrocilamine and phenylpropanolamine, bronchodilators (furafylline and theophylline), and quinolones (enoxacin), have been reported to be potent inhibitors of this isoenzyme. Thus, pharmacokinetic interactions at the cytochrome P450 IA2 enzyme level may cause toxic effects during concomitant administration of caffeine and certain drugs used for cardiovascular, CNS, gastrointestinal, infectious, and respiratory and skin disorders. Decoction of the dried seed was administered orally to nine healthy patients of both sexes with ileostomies at a dose of 720 mL (one, two, or three cups of French-press coffee with a standardized breakfast) for three separate days in random order. Ileostomy effluent was collected for 14 hours and urine for 24 hours. Stability of cafestol and kahweol was assessed under simulated gastrointestinal tract conditions. Corrected mean absorption of diterpenes expressed as percentages of the amount consumed and the amount entering the duodenum were 67% and 88%, respectively, for cafestol and 72% and 93%, respectively, for kahveol. There was a loss of diterpenes during incubation in vitro with gastric juice (cafestol 24% and kahweol 32%), during storage with ileostomy effluent (cafestol 18% and kahweol 12%), and during freeze-drying (cafestol 26% and kahweol 32%). Mean excretion of glucuronidated plus sulphated conjugates in urine was 1.2% of the ingested amount for cafestol and 0.4% of the ingested amount for kahweol. Approximately 70% of the ingested cafestol and kahweol was absorbed in ileostomy volunteers. Only a small part of the diterpenes was excreted as a conjugate of glucuronic acid or sulphate in urine.

Prophylactic Effect

The therapeutical effect of a 30% extract from coffee beans has been investigated in newborn calves in herds that endemically showed a high proportion of infections within the gastroenteric and/or respiratory system in calves. Fifty newborn calves were given a subcutaneous injection of 10 mL of the extract on the first and third days of life. Another 50 calves received physiological saline as control. On the first 2 days of life, the group treated with the extract had fewer animals with body temperature below physiological values; during the first period of diarrhea (between days 4 and 6), there was a significantly lower tendency of diarrhea, and after the second period of diarrhea (day 9), a better and quicker recovery and a lower tendency of exsiccation as the control calves. The average duration of illness was shorter (4.7 instead of 7 days), and the average number of therapeutical interventions were less (3.1 instead of 4.5) than in control calves. Within the four herds endemically showing a high number of cases of diarrhea in newborn calves, the morbidity could be dropped by 35% by the administration of one to three subcutaneous injections of 10 mL of the extract.

Prostaglandin Inhibition

Water extract of the dried seed, administered in drinking water of rats at a dose of 5%, was active. Prostaglandin I-2 synthesis in the rat thoracic aorta was assayed.

Psychomotor Performance

Coffee, taken by 17 introverts and 19 extroverts at doses of 2 and 4 mg caffeine/kg during the morning and evening, did not support the hypothesis that caffeine differentially affects extroverts and introverts. In this randomized, double-blind, crossover study, the subjects drank coffee during the mornings and evenings. At 30-minute intervals for 180 minutes after drinking coffee, the participants completed the Profile of Mood States, a battery of self-report visual analog scales, and the Digital Symbol Substitution Test. Caffeine affects on mood and task performance did not significantly interact with extroversion, except for nonsignificant trends for caffeine to increase happiness and vigor more among extroverts than introverts. No three-way interactions of group, time, and dose were found on any scales or on the Digital Symbol Substitution Test.

Renin–Angiotensin–Aldosterone System Activity

Plasma renin activity and enzyme-converting angiotensin I to angiotensin II activity, serum concentration of aldosterone, and catecholamines in patients with hypertension and patients with low or normal plasma renin activity, drinking one cup of coffee were measured by radioimmunoassay, and blood pressure was measured by ambulatory monitoring. Drinking one cup of coffee produced, after 1–2 hours, elevated SBP and after 1 hour DBP in patients with low renin–angiotensin–aldosterone system activity who habitually drink coffee. Patients with normal renin–angiotensin–aldosterone system activity had elevations only in DBP from 1 to 2 hours after drinking. Plasma catecholamines, aldosterone concentrations, blood renin activity, and enzyme-converting angiotensin I to angiotensin II activities were not elevated.

Respiratory Effect

Respiratory consequences of work in coffee processing were studied in 764 female workers exposed to dusts associated with the processing of green and roasted coffee. A group of 387 females not exposed to respiratory irritants served as controls for the prevalence of acute and chronic respiratory symptoms. A greater prevalence of all acute and chronic respiratory symptoms was consistently found among exposed workers than among control workers. The highest prevalence of chronic respiratory symptoms was recorded for chronic cough (40%), followed by acute symptoms of dry cough (58.7%). Mean acute reductions of lung function throughout the work shift were recorded in all of the studied groups; the mean across-shift decrease as a percentage of preshift values was particularly marked in forced expiratory flow at 75% ([FEF25]; –26.7%), forced expiratory flow at 50% ([FEF50]; –20.6%), followed by forced expiratory volume in 1 second (–9.9) and forced vital capacity (–3.7%). The preshift (baseline) values of ventilatory capacity were decreased in comparison to the predicted ones, and were lowest for FEF50 and FEF25. Disodium cromoglycate significantly diminished across- shift reductions for FEF50 and FEF25 in a subgroup of the examined workers.

Rheumatoid Arthritis Risk

Coffee, consumed by 6809 subjects with no clinical arthritis, indicated that the amount of coffee consumed was directly proportional to the prevalence of rheumatoid factor positivity. The consumption of coffee was first studied for its association with rheumatoid factor (sensitized sheep cell agglutination titer) in a cross-sectional survey and second for its prediction of rheumatoid in a cohort of 18,981 men and women who had neither arthritis nor a history of it at the baseline examination. In the cross-sectional survey, the amount of coffee consumes was directly proportional to the prevalence of rheumatoid factor. Adjusted for age and sex, this association was significant, but after further adjustment for smoking, the linear trend declined below significance. In the cohort study, there was an association between coffee consumption and the risk of rheumatoid factor-positive rheumatoid arthritis that did not result from age, sex, level of education, smoking, alcohol intake, body mass index, or serum cholesterol. After adjusting for these potential confounders, the users of four or more cups a day still have a relative risk of 2.2 (95% CI at 1.13–4.27) for developing rheumatoid factor-positive rheumatoid arthritis compared with those drinking less. Coffee consumption did not predict the development of rheumatoid factor-negative rheumatoid arthritis.

RNA (messenger) Polymerase Inhibition

Petroleum ether extract of the green seed, administered in ration of female mice at variable doses for 12 days, produced strong activity. The seed, administered in ration of female mice at variable concentrations for 12 days, was active.

Serum Homocysteine Concentration

The influence of nutritional factors associated with total homocysteine in 260 school teachers, 151 women, and 109 men with a median age of 64 years was performed by observational analyses of baseline and 2–4 months follow-up tests that designed to test the feasibility of conducting a large-scale clinical trial of vitamin supplements. In multivariable linear regression and generalized linear models, there was a positive, significant dose–response relationship between coffee consumption and total homocysteine ($p = 0.01$).

Serum Lipids and Lipoproteins

Serum concentrations of TC, TGs, and HDL cholesterol were measured in a group of 4587 males aged 48–56 years, drinking instant and brewed coffee. LDL cholesterol levels were calculated from the values of TC, TG, and HDL cholesterol. The consumption of brewed coffee was unrelated to any parameter, and instant coffee consumption showed a highly significant positive association with serum LDL cholesterol levels and an inverse association with serum TG levels. For each cup of instant coffee a day, LDL cholesterol levels were 0.82 mg/dL (95% CI, 0.29–1.35) higher, and TG levels in a natural log-scale were 0.014 mg/dL (95% CI, 0.006–0.022) lower. A tendency for positive association between instant coffee intake and serum TC levels ($p = 0.09$) was observed. HDL cholesterol levels were unrelated to instant coffee consumption.

Sex Hormone-Binding Globulin Level Increase

Water extract of the seed, administered to 50 premenopausal women at variable doses daily, was active. Blood samples were obtained from each woman on days 11 and 22 of her menstrual cycle. High intakes of caffeinated coffee, green tea, and total caffeine were commonly correlated with increasing sex hormone-binding globulin on days 11 and 22 of the cycle after controlling for potential confounders. Green tea but not caffeinated coffee intake was inversely correlated with estradiol on day 11 of the cycle. Although the effect of caffeine cannot be distinguished from the effects of coffee and green tea, consumption of caffeine-containing beverages favorably altered hormone levels associated with the risk of developing breast cancer.

Sister Chromatid Exchange Stimulation

Lyophilized extract of the essential oil, administered by gastric intubation to mice at a dose of 50 mL/kg in two doses, was inactive. Lyophilized extract of the dried seed, administered by gastric intubation to hamsters at a dose of 2.5 g/kg, was inactive.

Skin Depigmentation Effect

Extract of the dried seed, administered externally to adults at a dose of 5%, was active. Skin-lightening cosmetics contained extract of *Coffea arabica* seeds (containing chlorogenic acid) as melanin-formation inhibitors. The extract has been incorporated into cosmetics for skin-aging prevention or into hair preparations for hair protection. Biological activity reported has been patented.

Smooth Muscle Relaxant Activity

The aqueous extracts of green and roasted coffees were assayed on isolated guinea pig tracheal spirals. Contractile and relaxant activities were compared with histamine and theophylline, respectively. Green coffee extracts induced concentration dependent contraction, but the maximal tension never exceeded 76.3% ± 5.2 of a maximal histamine contraction (0.69 ± 0.07 g/mm^2 vs 0.52 ± 0.05 g/mm^2; $p = 0.01$). One gram of green coffee dust had a biological activity equivalent to 1.23 ± 0.1 mg of histamine. The pD2 value of histamine was –5.17 ± 0.05.

The potency of green coffee was unaffected by mepyramine maleate (1 μg/mL, final bath concentration), whereas that of histamine was reduced 500-fold. Tissues contracted with histamine were not significantly relaxed by green coffee extracts. By contrast, roasted coffee extracts induced concentration-dependent relaxation of uncontracted and histamine contracted tissues. Tissues contracted with green coffee extracts were also completely relaxed by roasted coffee extracts. The pD2 value of theophylline was –4.10 ± 0.03. The relaxant activity of 1 g of roasted coffee was equivalent to 1.95 ± 0.16 mg of theophylline. The potency of these extracts was significantly reduced after propranolol (1 μg/L; dose ratio 1.56).

Stress Relief

A survey of 261 house staff, nurses, and medical oncologists in a cancer research hospital and oncologists in outside clinical practices was carried out to measure burnout, physiological distress, and physical symptoms. Each participant completed a questionnaire that quantified life stressors, personality attributes, burnout, psychological distress, physical symptoms, coping strategies, and social support. The results indicated that house staff experienced the greatest burnout.

They also reported greater emotional exhaustion, a feeling of emotional distance from patients, and a poorer sense of personal accomplishment. Nurses reported more physical symptoms than house staff and oncologists. However, they were less emotionally distant from patients. Women reported a lower sense of accomplishment and greater distress. The four most frequent methods of relaxing were talking to friends, using humor, drinking coffee or eating, and watching television.

Suicidal Risk

Data from 36,689 adult men and women (25–64 years of age) who participated in a population survey between 1972 and 1992 indicated that clustering of the heavy use of alcohol, cigarettes, and coffee could serve as a new marker for increased risk of suicide. The mortality of the cohort was monitored for a mean of 14.4 years, which yielded 169 suicides. Criteria for heavy use of each psychoactive substance were defined as: alcohol more than 120 g/ week, cigarettes more than 21/ day, and coffee more than 7 cups/day. Approximately 50% of the men and 80% of the women did not use any of the psychoactive substances heavily. Every third man and every fifth woman used one substance heavily. The prevalence for those who exceeded criteria for joint heavy use of two substances was 9% for men and 1% for women. Joint-heavy use of all three substances was rare. The adjusted risk of suicide increased linearly with increasing level of joint heavy use of alcohol, cigarette, and coffee. Among subjects with heavy use of one substance, the risk was 1.55, with joint-heavy use of two substances 2.22, and with joint-heavy use of all three substances 3.99 compared with no heavy use.

Symptomatic Gallstone Disease

In a 10- year study, 46,008 men aged 40–75 years without history of gall stone disease were investigated for the association of coffee and caffeinated drink consumption with symptomatic gallstone disease or cholecystectomy, diagnosed by ultrasonography or X-ray. After adjusting for other known or suspected risk factors, compared with men who did not consume regular coffee, the adjusted relative risk (RR) for those who consistently drank two to three cups of regular coffee per day was 0.60 (95% CI, 0.42–0.86); four or more cups per day the RR was 0.55 (95% CI, 0.33–0.92). The risk of gall stone disease declined with increasing caffeine intake. RR for men in the highest category of caffeine intake (>800 mg/day) compared with men in the lowest category (≤25 mg/day) was 0.55 (95% CI, 0.35–0.87). Decaffeinated coffee was not associated with a decreased risk. The effect of coffee drinking in relation to alcohol drinking, smoking, and obesity was investigated in the population of 7637 males, aged 48–59 years; 1360 men with a possible pathologic condition influencing liver enzyme levels, and 182 former alcohol drinkers; the effect of coffee on serum γ-glutamyltransferase (GGT) was examined by a multiple linear regression model and analysis of variance adjusting for alcohol drinking, smoking and body mass index. The adjusted percentage of difference in serum GGT was –4.3 (95% CI, –5 to –3.5) per cup. The inverse coffee–GGT relationship was most prominent among men drinking 30 mL or more of ethanol and smoking 15 or more cigarettes/day, and positive associations of alcohol and smoking with GGT were attenuated by coffee drinking, more clearly among men with body mass index of 25 kg/m^2 or greater. Adjusted percentages of difference in serum GGT were –2.6% ($p = 0.0003$) per cup of brewed coffee and –5.1% ($p = 0.0001$) per cup of instant coffee.

Toxicity

Atractyloside, a diterpenoid glycoside that occurs naturally in plants, may be present at levels as high as 600 mg/kg of dried plant material. Consumption of plants containing atractyloside or carboxyatractyloside has caused fatal renal proximal tubule necrosis and/or centrilobular hepatic necrosis in man and farm animals. Although pure atractyloside and crude plant extracts disrupt carbohydrate homeostasis and induce similar pathophysiological lesions in the kidney and liver, it is also possible that the toxicity of atractyloside may be confounded by the presence of other natural constituents in plants. Atractyloside competitively inhibits the adenine nucleoside carrier in isolated mitochondria and thus blocks oxidative phosphorylation. This has been assumed to explain changes in carbohydrate metabolism and the toxic effects in liver and kidney. In vitro proximal tubular cells are selectively sensitive to atractyloside, whereas other renal cell types are quite resistant. There are also differences in the response of liver and renal tissue to atractyloside. Thus, not all of the clinical, biochemical, and morphological changes caused by atractyloside can simply be explained on the basis of mitochondrial phosphorylation. The relevance to a wider human risk is shown by the presence of atractyloside analogues in dried roasted coffee beans (17.5–32 mg/kg). Ethanol (50%) extract of the aerial parts, administered intraperitoneally to mice, was active, lethal dose$_{50}$ of 1 g/kg. Water extract of the roasted seed, administered to female rats at a dose of 6% of diet for 2 years, was inactive. Both regular and decaffeinated instant coffees were tested.

Urinary Diterpenes Excretion

Absorption and excretion of the cholesterol-raising coffee diterpenes cafestol and kahweol were observed in nine healthy patients with ileostomies. Ileostomy effluent was collected for 14 hours, and urine was collected for 24 hours. Approximately 70% of the ingested cafestol and kahweol was absorbed. Only small part of the diterpene was excreted as a conjugate of glucuronic acid or sulphate in urine, mean excretion was 1.2% of the ingested amount for cafesterol and 0.4% for kahweol.

Urinary Hydrogen Peroxide

Instant coffee, taken by healthy human volunteers, indicated that coffee drinking is rapidly and reproducibly followed by increased levels of hydrogen peroxide detectable in the urine for up to 2 hours after drinking coffee. The levels of hydrogen peroxide indicated that exposure of human tissues to hydrogen peroxide might be greater than is commonly supposed. It is possible that hydrogen peroxide in urine could act as an antibacterial agent and that hydrogen peroxide is involved in the regulation of glomerular function.

White Blood Cell-Macrophage Stimulant

Water extract of the freeze-dried fruit, at a concentration of 2 mg/mL, was inactive on macrophages. Nitrate formation was used as an index of the macrophage stimulating activity to screen effective foods.

Weight-Gain Inhibition

Lyophilized extract of the dried seed, administered intragastrically to mice at a dose of 50 g/kg of diet, was active. Animals were exposed to coffee *in utero*, as mother's diet was 1% instant coffee. After weaning, animals were given instant coffee in diet for 2 years. Increase in energy expenditure was shown by increase in caloric intake and depressed growth.

9

DAUCUS CAROTA

Carrot is an erect (30–120 cm high) annual or biennial herb of the Umbelliferae family with branched stem arising from a large, succulent, thick, fleshy 5–30 cm long tap root. The color of the root in the cultivated varieties ranges from white, yellow, orange, light purple, or deep red to deep violet. The shape varies from short stumps to tapering cones. Leaves are finely dissected, twice or thrice-pinnate, segments are linear to lanceolate, 0.5–3 cm long. Upper leaves are reduced, with a sheathing petiole. Stem is striate or ridged, glabrous to hispid, up to 1 m tall. Flowers are borne in compound, more or less globose, to 7-cm-in-diameter umbels. Rays are numerous, bracts 1–2 pinnated, lobes linear, 7–10 bracteoles similar to bracts. Flowers are white or yellowish; the outer are usually the largest. Sepals are minute or absent, there are five petals and stamens, ovary inferior with two cells and one ovule per cell, two styles. Fruits are oblong, with bristly hairs along ribs, 2–4 mm long.

ORIGIN AND DISTRIBUTION

Cultivated carrot originated in Afghanistan then spread to China in the 13–14th century and reached England in the 15th century. It was introduced to North America by settlers and is grown widely in temperate and tropical regions of the world.

TRADITIONAL USES

Algeria. Hot water extract of the seed, mixed with *Euphorbia* species and a beetle, is taken orally to facilitate childbirth.

Arabic countries. The dried seeds are used as an abortifacient in the form of a pessary in Unani medicine.

Belgium. Dried root is taken orally for diabetes.

Brazil. Water extract of the dried root is taken orally as a nerve tonic and stimulant.

Canary Islands. Infusion of the dried aerial parts is taken orally for cystitis.

China. Decoction of the seed is taken orally as an emmenagogue. Root juice is taken orally for cancer of the stomach, bowel, and uterus, and for ulcers.

Egypt. Hot water extract of the fruit is taken orally to facilitate pregnancy and as an emmenagogue, aphrodisiac, diuretic, and antispasmodic. Hot water extract of the dried fruit is taken orally as a diuretic and for urinary colic.

England. Hot water extract of the root and seed are taken orally to induce the menstrual cycle.

Europe. Decoction of the dried leaf is taken orally for diabetes mellitus. Hot water extract of the root is taken orally as an emmenagogue and anthelmintic. Hot water extract of the seed is taken orally to induce menstruation.

Fiji. Fresh leaf juice is used as a nose drop for headache. Fresh root is taken orally for heart diseases.

France. Hot water extract of the fruit is taken orally as an emmenagogue.

Greece. Infusion of the dried flowers is taken orally as a tonic and to relieve sluggishness.

India. Decoction of the fresh root is taken orally for jaundice and inflammation, as an anthelmintic, and externally for leprosy. Dried seeds are mixed with crude sugar and eaten to terminate early pregnancy. Hot water extract of the dried root is taken orally as a tonic, expectorant, diuretic, stomachic, and liver cleanser. Hot water extract of the leaf is taken orally as a uterine stimulant during parturition. Hot water extract of the seed is taken orally as an abortifacient, emmenagogue, and aphrodisiac. The dried seeds are used as a powerful abortifacient. The root is taken orally as a hypotensive medication.

Iran. Water extract of the fruit is taken orally as an emmenagogue.

Italy. Decoction of the root is used as a gargle for loss of speech. Root juice is taken orally as an anthelmintic and cicatrizing agent, for leukorrhea, and to improve sight. The fresh root is used externally for dermatitis and burns. The fresh root juice is taken orally for loss of voice and persistent coughs, and the decoction is taken orally for diuresis. The root is taken orally as a diuretic and a digestive and to treat uricemia and constipation.

Kuwait. The seeds are taken orally as an emmenagogue.

Madeira. Infusion of the entire plant is taken orally for jaundice.

Mexico. Hot water extract of the fresh root is taken orally as a cardiotonic. The flowers or root, boiled together with *Cassia fistula* and "Rosa de Castilla," are taken orally before breakfast to induce abortion. To correct delayed menstruation, the liquid is taken daily for 40 days.

New Caledonia. Infusion of the fruit is taken orally as an emmenagogue.

Pakistan. Hot water extracts of the leaf and seed are taken orally as stimulants of the uterus during parturition.

Peru. Hot water extracts of the dried root and dried aerial parts are taken orally as a carminative, emmenagogue, and vermifuge.

Philippines. Hot water extract of the leaf is taken orally as a stimulant of the uterus during parturition.

Rodrigues Islands. Decoction of the entire plant is taken orally for gout, jaundice, and mouth ulcers.

South Korea. Hot water extract of the dried fruit is taken orally as an abortifacient and emmenagogue.

Tunisia. Dried leaf is used externally for chilblains.

Turkey. The seed, ground with the seeds of *Brassica rapa* and *Raphanus sativus*, is taken orally as a tonic.

United States. Hot water extract of the fruit is taken orally to stimulate menstruation. Hot water extract of the seed is taken orally as an emmenagogue. Seeds are taken orally as an emmenagogue, diuretic, and abortifacient. The fresh root is taken orally for general nervousness, and the hot water extract is taken orally as a diuretic in dropsy and as a tonic. Hot water extract of the dried root and seed is taken orally as a carminative, diuretic, and stimulant.

Medicinal Uses

Abortifacient Effect

Ethanol (95%) extract of the dried seed, administered by gastric intubation to pregnant mice at doses of 30 mg/animal and 40 mg/kg on days 4–6, was inactive. Petroleum ether extract of the dried seed, administered subcutaneously to pregnant rats beginning on day 7 of pregnancy, was active. Petroleum ether extract of the dried seed, administered subcutaneously to pregnant rats at a dose of 2 mL/kg, was active. The effect was blocked by progesterone given on days 7–19 of pregnancy. Seed essential oil, administered to pregnant mice at a dose of 5 mg/kg, was active. Acetone extract of the fresh root, at a concentration of 1 μg/mL, produced weak activity and the propanol extract was inactive on *Salmonella typhimurium* TA98 vs 2-amino-3-methylimidazo (4,5-F) quinoline-induced mutagenicity. Fresh fruit juice, administered by gastric intubation to male mice at a dose of 0.5 mL/animal, was active on *Schizosaccharomyces pombe*. The animals were treated with the juice and nitrosation precursors, then yeast cells were injected into the venous plexus of the orbit. Four hours later, the animals were sacrificed and the livers removed, plated with yeast and examined. Results were significant at $p < 0.00\ 1$ level. Infusion of the stem, on agar plate at a concentration of 100 μL/disc, produced strong activity on *Salmonella typhimurium* TA98 vs 2-amino-anthracene-induced mutagenicity. Metabolic activation was not required for activity. Weak activity was produced on *Salmonella typhimurium* TA100 vs ethyl methanesulfonate-induced mutagenicity. Metabolic activation was not required for activity. Methanol extract of the dried root, on agar plate at a concentration of 50 μL/disc, was inactive on *Bacillus subtilis* NIG-1125 His Met and *Escherichia coli* B/RWP2-TRP. Root juice, on agar plate at a concentration of 500.0 μL/plate produced weak activity on *Salmonella typhimurium* TA98 vs 2-nitrofluorine- and 1-nitropyrene-induced mutagenesis. Water extract of the fresh root, on agar plate plus S9 mix at a dose of 0.4 mL/plate, was active on *Salmonella typhimurium* TA100 vs TRP-P-2 mutagenicity. Water extract of the fresh root, on agar plate at a concentration of 500 μg/plate, produced weak activity on *Salmonella typhimurium* TA100 vs *N*-nitrosoamine-induced mutagenicity.

AIDS Therapeutic Effects

Water extract of the dried rhizome taken orally by adults was active. A pharmaceutical solution containing fruit bodies of *Tremella fuciformis*, *Daucus carota* rhizome, *Astragalus mongholicus* root, and *Zizyphus jujuba* fruits, honey, vitamin A palmitate, zinc sulfate, and vitamin C was useful for controlling acquired immunodifficiency syndrome (AIDS), cancer, and infections.

Anti-allergenic Activity

Water extract of the fresh root, in cell culture at a concentration of 100 μL/mL, was inactive on Leuk-RBL 2H3 vs biotinylated antideoxyriboneucleoprotein immunoglobulin E/avidin-induced β-hexosaminidase release.

Anti-amoebic Activity

Essential oil, in broth culture at a concentration of 0.5 μL/ mL, was active on *Entamoeba histolytica*.

Antibacterial Activity

Essential oil, on agar plate at a concentration of 0.43 mg/mL, produced weak activity on *Streptococcus-β-hemolytic* and *Staphylococcus aureus*, equivocal on *Escherichia coli*, minimal inhibitory concentration (MIC) 1.74 mg/mL and inactive on *Proteus mirabilis*, MIC 17.4 mg/ mL. Fruit essential oil on agar plate was active on *Staphylococcus aureus*, MIC 0.12 mg/ mL, and *Streptococcus-β hemolytic*, MIC 0.23 mg/mL, inactive on *Proteus mirabilis*, MIC 18.5 mg/mL, and equivocal on *Escherichia coli*, MIC 9.25 mg/mL. Ethanol (70%) extract of the fruit, on agar plate, was active on *Bacillus megaterium*, *Staphylococcus albus*, *Staphylococcus aureus*, and *Bacillus cereus*. Ethanol (95%) and water extracts of the entire plant, on agar plate, were inactive on *Escherichia coli* and *Staphylococcus aureus*. Fresh root, macerated, in pieces and shredded, was active on *Listeria monocytogenes*. Fresh shredded root dipped in chlorine and packaged under an atmosphere containing 3% oxygen and 97% nitrogen, was active on *Listeria monocytogenes*. Bacterial growth was inhibited on shredded carrots more than on whole carrots. There was no inhibition on cooked carrots. The root, on agar plate, was active on *Streptococcus mutans*.

Anticytotoxic Activity

Ethanol (95%) extract of the fresh root, at a concentration of 80 mg/mL in cell culture, was active on Vero cells vs *N*-nitrosopiperidine, *N*-nitrosodibutylamine, and *N*-nitrosodimethylamine cytotoxicity. A dose of 20 mg/ mL was inactive vs *N*-nitrosopiperidine, nitrosodimethylamine, *N*-nitrosopyrrolidine, and *N*-nitrosodibutylamine cytotoxicity.

Anti-edema Activity

Methanol extract of the root, applied externally to mice at a dose of 2 mg/ear, produced inhibition ratio of 37.

Anti-estrogenic Effect

Ethanol (95%) extract of the dried seed, administered by gastric intubation of ovariectomized mice at a dose of 40 mg/kg daily for 3 days, produced weak activity. Petroleum ether extract of the dried seed, at a dose of 10 mg/kg, was active.

Antifungal Activity

Acetone, water, and ethanol (95%) extracts of the dried fruit, on agar plate at a concentration of 50%, were inactive on *Neurospora crassa*. The essential oil, at a concentration of 1000 ppm on agar plate, produced weak activity on *Aspergillus flavus*. Ethanol (50%) extract of the dried root, on agar plate at a concentration of 500 mg/mL, was active on *Botrytis cinerea* and inactive on *Aspergillus fumigatus*, *Aspergillus niger*, *Fusarium oxysporum*, *Penicillium digitatum*, *Rhizopus nigrans*, and *Trichophyton mentagrophytes*. Seed essential oil, in broth culture at variable concentrations, was active on *Cladosporium werneckii*. The root, on agar plate, was active on *Porphyromonas gingivalis*.

Antihepatotoxic Activity

Water extract of the fresh root, administered intragastrically to male rats at a dose of 20 mL/kg, was active vs lindane-induced hepatotoxicty. Supernatant of the fresh root, administered intragastrically to male mice at a dose of 50 mL/kg, decreased serum bilirubin, urea, lactic dehydrogenase, serum glutamic pyruvic transaminase, and serum glutamic oxaloacetic transaminase levels vs carbon tetrachloride (CCl_4) -induced hepatoxicity.

Antihyperglycemic Activity

Decoction, ethanol (80%), and water extracts of the dried root, administered intragastrically to mice at a dose of 25 g/kg, were active vs glucose-induced hyperglycemia. Dried leaf, administered to male mice at a concentration of 6.25% of the diet for 28 days, was inactive vs streptozotocin-induced hyperglycemia. Fresh root, taken orally by 15 adults of both sexes with type II diabetes at a dose of 280 g/person, was active.

Anti-implantation Effect

Chloroform–methanol (9:1) fraction of ethanol (95%) extract, a chloroform-soluble fraction and an ethyl acetate-soluble fraction of a water extract, methanol-soluble fraction of a petroleum ether extract and chloroform soluble fraction of petroleum ether extract of the seed, administered orally to female rats at a dose of 50 mg/kg, were active vs foot shock. Water and petroleum ether extracts of the seed, administered orally to female rats at doses of 100 and 20 mg/kg, respectively, were active. Petroleum ether extract of the seed, administered orally to female rats at a dose of 500 mg/kg, was inactive. Ethanol (50%) extract of the dried seed, administered orally to female rats at a dose of 500 mg/kg, was inactive. Ethanol (95%) extract of the dried fruit, administered orally to pregnant rats at a dose of 500 mg/kg, produced 60% inhibition of implantation. Petroleum ether extract of the dried seed, administered subcutaneously to pregnant rats at a dose of 6 mL/kg, was inactive. Seed essential oil, administered to pregnant mice at a dose of 5 mg/kg, was active.

Antimycobacterial activity. Ethanol (95%) and water extracts of the entire plant, on agar plate, were inactive on *Mycobacterium tuberculosis*. Leaf juice, on agar plate, produced weak activity on *Mycobacterium tuberculosis*, MIC less than 1:20.

Anti-nematodal Activity

Methanol extract of the fruit, at a concentration of 1 mg/ mL, was active, and the water extract, at a concentration of 10 mg/mL, produced weak activity on *Toxacara canis*.

Antioxidant Activity

Plant juice, at a dose of 100 μL/kg, produced weak activity vs Fenton's reagent-induced lipid peroxidation. The root, at a concentration of 1%, produced weak activity at 120°F. Water extract of the fresh root, at a concentration of 10 μM trolox equivalent per gram, produced weak activity vs oxygen radical absorption capacity assay with hydroxyl and peroxyl radical generators, and Cu^{2+} (reactive species) activity. Fresh root homogenate produced 31.8% inhibition of lipid peroxidation.

Antispasmodic Activity

Petroleum ether fraction chromatographed and fraction eluted with chloroform, at a concentration of 0.50 mg/mL, was active on the guinea pigileum vs histamine-induced contractions. Tertiary alkaloid fraction of the dried seeds produced weak activity on the dog trachea vs acetylcholine (ACh)- and KCl-induced contractions, and active on the guinea pig ileum. A concentration of 25 μg/mL was active on the rat uterus vs ACh and oxytocin-induced contractions, results significant at $p < 0.02$ level. Methanol extract of the dried seed, at a concentration of 0.1 mg/mL, was active on the guinea pig ileum vs histamine-induced contractions.

Anti-tumor Activity

Petroleum ether extract of the dried seed, administered intraperitoneally to male mice at a concentration of 3 mg/kg, was active on Chinese hamster cells-V79. The root, administered in the ration of female rats for 1 month before 7, 12-dimethybenz[a]anthracene treatment, reduced tumor size. Water extract of the aerial parts, administered intraperitoneally to mice at a dose of 400 mg/kg, was inactive on Leuk (Friend Virus-Solid) and Leuk-L1210. A dose of 500 mg/kg was inactive on Sarcoma 180 (ASC). Water extract of the dried root taken orally by adults was active. A pharmaceutical solution containing fruit bodies of *Tremella fuciformis*, *Daucus carota* root, *Astragalus mongholicus* root, and *Zizyphus jujuba* fruits, honey, vitamin A palmitate, zinc sulfate, and vitamin C was useful for controlling AIDS, cancer, and infections. Hot water extracts of the fresh leaf and fresh root, in cell culture, produced strong activity on Raji cells vs phorbol myristate acetate-promoting expression of Epstein-Barr virus (EBV) early antigen. Methanol extract of the fresh root, at a concentration of 200 mg/mL, was inactive on Raji cells vs EBV activation induced by 12-*O*-hexadecanoylphorbol (40 ng/mL).

Catalase Inhibition

Water extract of the fresh root, administered intragastrically to infant mice at a dose of 50 mL/kg, was active. The treatment was administered for seven successive days, followed by a single dose of 20% v/v CCl_4 in olive oil subcutaneously at 1 mL/kg on the last day 1 hour after the administration of the carrot extract. Chloroform–methanol (9:1) fraction of the ethanol (95%) extract, ethyl acetate fraction of the water extract, and chloroform-soluble fraction of the water extract of the seed were active on the nonpregnant rat uterus.

Central Nervous System (CNS) Depressant Activity

Ethanol (95%) extract of the seed, administered orally to mice and rats at a dose of 50 mg/kg, was inactive.

CNS Stimulant Activity

Ethanol (95%) extract of the seed, administered orally to mice and rats at a dose of 50 mg/kg, was inactive.

Conditioned Taste Aversion

Frozen leaf and stem, administered intragastrically to rats at a dose of 562 mg/kg, was inactive. The test substance was temporarily paired with the introduction of sodium saccharin solution. Consumption of the saccharin solution 2 days after the test was used to estimate aversiveness of the test substance.

Cytotoxic Activity

Ethanol (50%) extract of the root, in cell culture, was inactive on CA-9KB, effective dose $(ED)_{50}$ greater than 20 μg/mL. Methanol extract of the fresh root, in cell culture at a concentration of 200 μg/mL, was inactive on macrophage cell line RAW 264.7. Water extract of the aerial parts, in cell culture, was inactive on CA-9KB, ED_{50} greater than 0.1 mg/mL.

Dermatitis-producing Effect

Ether extract of the fresh entire plant, applied by patch at a concentration of 1%, was active. Fresh root, in a mixture containing *Apium graveolens*, *Aromatica rusticana*, *Solanum tuberosum*, and *Petroselinum crispum*, was active.

Desmutagenic Activity

Fresh plant juice, on agar plate at a concentration of 0.5 mL/disc, was inactive on *Salmonella typhimurium* TA98. Homogenate of the fresh root, at a concentration of 100 μL/disc on agar plate, was active on *Salmonella typhimurium* TA98 and TA100 vs 1,4-dinitro-2-methyl pyrrole mutagenesis.

Diuretic Activity

Ethanol (70%) extract of the dried fruit, administered intravenously to dogs at a dose of 150 mg/kg, increased diuresis 1.7-fold. Ethanol (95%) extract of the seed, administered orally to rats at a dose of 100 mg/kg, was inactive. Seed essential oil, administered intravenously to dogs at a concentration of 4 μL/kg, produced 2.4-fold increase in urine flow and an increase in K^+, Na^+, and Cl^- excretion. Ethanol (70%) extract of the seed essential oil, administered intravenously to dogs at a dose of 20 mg/kg, produced 1.6-fold increase in urine flow.

Embryotoxic Effect

Ethanol (95%) extract of the dried seed, administered by gastric intubation to pregnant rats at a dose of 40 mg/animal on days 4–6, was inactive. A dose of 0.10 g/kg, administered by gastric intubation to pregnant rats on days 1–10, was inactive, and a dose of 0.25 mg/kg on days 1–10 was equivocal. Six of 10 animals were pregnant vs 10 of 10 in the control group. Ethanol (50%) extract of the dried seed, administered by gastric intubation to pregnant rats at a dose 200 mg/kg, was equivocal. The water extract, at a dose of 0.5 g/kg on days 1–10, was inactive. Petroleum ether extract, administered subcutaneously to pregnant rats at a dose of 1 mL/kg, was inactive and a dose of 2 mL/kg was active. Results were significant at $p < 0.05$ level. Powdered dried seed, administered by gastric intubation to pregnant rats at doses of 2–4.5 g/kg on days 1–10, was inactive. Petroleum ether extract of the seed chromatographed and fraction eluted with chloroform, administered orally to female rats at a dose of 20 mg/kg, was active.

Ethanol (95%) extract of the seed, administered orally to female rats at a dose of 50 mg/kg, was inactive. Chloroform-soluble and ethyl acetate-soluble fractions of a water extract of the seed, administered orally to female rats at a dose of 50 mg/kg, were active vs foot shock. Chloroform- and methanol-soluble fractions of the petroleum ether extract of the seed, administered orally to female rats at a dose of 50 mg/kg, was inactive. Petroleum ether extract of the seed, administered subcutaneously to pregnant rats at a dose of 0.6 mL/animal, was active. Seed essential oil, administered subcutaneously to pregnant rats on days 1, 2, 7, and 8, was active.

Estrogenic Effect

Ethanol (95%) extract of the dried seed, administered by gastric intubation to infant female mice at doses of 20 and 40 mg/animal daily for 3 days, produced weak activity. Doses of 50, 100, and 500 mg/kg, administered subcutaneously to ovariectomized rats, produced weak activity. Petroleum ether extract of the dried seed, administered subcutaneously to ovariectomized rats at doses of 2 and 6 mL/kg, was active, results significant at $p < 0.01$ level. Ethanol (95%) extract of the root, administered subcutaneously to female infant mice, was active. Ethanol (95%) extract of the root, administered subcutaneously to female infant mice, was inactive. The seed essential oil was inactive. Root, in the ration of female infant mice, produced activity equivalent to 2 Pg stilbestrol in 100 g carrot (dry weight).

Glutathione Formation Induction

Water extract of the fresh root, administered intragastrically to infant mice at a dose of 25 mL/kg, was active. The treatment was administered for seven successive days followed by a single dose of 20% v/v CCl_4 in olive subcutaneously at 1 mL/kg on the last day, 1 hour after the administration of the carrot extract.

Glutathione Peroxidase Inhibition

Water extract of the fresh root, administered intragastrically to infant mice at a dose of 50 mL/kg, was active. The treatment was administered for seven successive days followed by a single dose of 20% v/v CCl_4 in olive subcutaneously at 1 mL/kg on the last day, 1 hour after the administration of the carrot extract.

Glutathione Reductase Stimulation

Water extract of the fresh root, administered intragastrically to infant mice at a dose of 25 mL/kg, was active. The treatment was administered for seven successive days followed by a single dose of

20% v/v CCl_4 in olive subcutaneously at 1 mL/kg on the last day, 1 hour after the administration of the carrot extract.

Glutathione S-transferase Inhibition

Water extract of the fresh root, administered intragastrically to infant mice at a dose of 50 mL/kg, was active. The treatment was administered for 7 successive days followed by a single dose of 20% v/v CCl_4 in olive subcutaneously at 1 mL/kg on the last day, 1 hour after the administration of the carrot extract.

Hypocholestrolemic Activity

Fresh root, taken orally by human adults at a dose of 200 g/person, was active. Daily ingestion at breakfast for 3 weeks decreased cholesterol in serum by 11%, increased fecal bile acid and fat excretion by 50%, and increased stool weight by 25%.

Hypoglycemic Activity

Dried leaf, in the ration of male mice at a concentration of 6.25% of the diet for 28 days, was inactive vs streptozotocin-induced hyperglycemia. Ether extract of the fresh root, administered subcutaneously to dogs, rabbits, and human adults, was active.

Hypotensive Activity

Essential oil, administered intravenously to dogs at a dose of 3 μL/kg, was active. The ethanol (70%) extract, administered intravenously to dogs at a dose of 75 mg/kg, was active. There was a dip followed by rise in blood pressure. Ethanol (80%) extract of the aerial parts, at a dose of 10 mg/kg, was not blocked by atropine. The extract did not inhibit pressor response of norepinephrine either. Ethanol (95%) extract of the seed, administered intravenously to dogs at a dose of 10 mg/kg, produced a transient effect that was blocked by atropine. Petroleum ether fraction chromatographed and fraction eluted with chloroform, administered intravenously to rabbits at a dose of 0.80 mg/kg, was inactive. Methanol extract, administered intravenously to dogs and rabbits at a dose of 2–5 mg/kg, was inactive. Seed essential oil, administered intravenously to dogs at a concentration of 4 μL/kg, was active.

Immunostimulant Activity

Fresh plant juice, in the ration of female mice, was active in ovalbumin-sensitized mice. Water extract of the dried root, taken orally by human adults, was active. A pharmaceutical solution containing fruit bodies of *Tremella fuciformis*, *Daucus carota* root, *Astragalus mongholicus* root and *Zizyphus jujuba* fruits, honey, vitamin A palmitate, zinc sulfate, and vitamin C is claimed useful as an immunostimulant for controlling AIDS, cancer, and infections.

Lipid Peroxidase Inhibition

Water extract of the fresh root, administered intragastrically to infant mice at a dose of 50 mL/kg, was active. The treatment was administered for 7 successive days followed by a single dose of 20% v/v CCl_4 in olive subcutaneously at 1 mL/kg on the last day, 1 hour after the administration of the carrot extract.

Lipid Peroxide Formation Inhibition

Fresh fruit juice, taken orally by human adults at a dose of 16 mg of β carotene/person, was active. Hot water extract of the fresh leaf produced strong activity vs *t*-butyl hydroperoxide/heme-induced luminolenhanced chemiluminescence. Hot water extract of the fresh root produced weak activity vs *t*-butylhydroperoxide/ heme-induced luminol-enhanced chemiluminescence.

Liver Regeneration Stimulation

Seed essential oil, administered subcutaneously to partially hepatectomized male rats at a dose of 100 mg/animal daily for 7 days, was inactive.

Mutagenic Activity

Methanol extract of the fresh root, on agar plate, was inactive on *Salmonella typhimurium* TA98 and TA100. Root juice, on agar plate at a dose of 200 μL/mL, produced weak activity on *Salmonella typhimurium* TA100 and a dose of 400 μL/mL was inactive on *Salmonella typhimurium* TA98.

Nematocidal Activity

Water extract of the dried fruit, in cell culture at a concentration of 10 mg/mL, and methanol extract at a concentration of 1 mg/mL, were active on *Toxacara canis*.

Quinone Reductase Induction

Acetonitrile extract of the dried root, in cell culture at a concentration of 7.9 mg/g, was active on Hematoma-Mouse-ICIC7. The sample was assayed for the induction of detoxifying enzymes that may have anticarcinogenic activity.

Sister Chromatid Exchange Inhibition

Fresh root juice, in the drinking water of rats at a dose of 300 mL animal, was active on Chinese hamster ovary cells. The plasma from the rats treated with carrot juice and cyclophoshamide reduced sister chromatid exchange in DNA-repair-deficient and in normal human lymphocytes in the Chinese hamster ovary cells when compared to cyclophoshamide treated only.

Skeletal Muscle Relaxant Activity

Petroleum ether fraction chromatographed and fraction eluted with chloroform, at a concentration of 0.50 mg/mL, was inactive on the frog rectus abdominus muscle vs ACh-induced contractions.

Skeletal Muscle Stimulant Activity

Ethanol (95%) extract of the seed, at a concentration of 10 mg/mL, was inactive on frog rectus abdominus muscle.

Smooth Muscle Relaxant Activity

Alkaloid fraction (tertiary) of the dried seed, at a concentration of 25 μg/mL was active on the rabbit and rat ileum.

Spasmogenic Activity

Ethanol (95%) extract of the seed, at a concentration of 1.5 mg/mL, was active on the dog tracheal chain. The extract, at a concentration of 1 mg/mL, was active on the guinea pig, rabbit, and rat ileum. The activity was equal to approx 0.1 μg/mL of acetylcholine. The action was blocked by atropine.

Spasmolytic Activity

Ethanol (80%) extract of the aerial parts, at a concentration of 3 mg/mL, was active on the rabbit aorta vs K^+-induced contractions.

Tumor Promotion Inhibition

Methanol extract of the dried leaf, in cell culture at a concentration of 200 μg/disc, was inactive on EBV vs 12-*O*-hexadecanoylphorbol-13-acetate-induced EBV activation.

Uterine Relaxation Effect

Alkaloid fraction of the dried seed, at a concentration of 25 μg/mL, was active on the rat uterus. Methanol extract of the dried seed, at a concentration of 0.5 mg/mL, was equivocal. Petroleum ether fraction of the seed, chromatographed and fraction eluted with chloroform, at a concentration of 0.50 mg/mL, was active on the nonpregnant rat uterus vs oxytocin-induced contractions.

Uterine Stimulant Effect

Petroleum ether fraction of the seed, chromatographed and fraction eluted with chloroform, at a concentration of 0.5 mg/mL, was active on the nonpregnant rat uterus vs oxytocin-induced contractions. Ethanol

(95%) extract of the seed, at a concentration of 1 mg/mL, was inactive on the rat uterus. The amplitude was diminished but not the tone.

Vasodilator Activity

Alkaloid fraction (tertiary) of the dried seed, administered to frog by perfusion at a dose of 2.5 mg/kg, was active vs barium chloride-induced vaso-constriction.

White Blood Cell Stimulant

Root juice, administered intraperitoneally to mice at a dose of 0.2 mL/animal, increased neutrophil accumulation by 71%.

White Blood Cell-macrophage Stimulant

Water extract of the freeze-dried root, at a concentration of 2 μg/mL, was inactive. Nitrite formation was used as an index of the macrophage stimulating activity.

INDEX